MORTUARY LANDSCAPES OF THE CLASSIC MAYA

THE LINDA SCHELE SERIES IN MAYA AND PRE-COLUMBIAN STUDIES

This series was made possible through the generosity of William C. Nowlin Jr. and Bettye H. Nowlin, the National Endowment for the Humanities, and various individual donors.

MORTUARY LANDSCAPES OF THE CLASSIC MAYA

RITUALS OF BODY AND SOUL | *Andrew K. Scherer*

UNIVERSITY OF TEXAS PRESS AUSTIN

Printed in the United States of America
First edition, 2015

Requests for permission to reproduce material
from this work should be sent to:
Permissions
University of Texas Press
P.O. Box 7819
Austin, TX 78713-7819
http://utpress.utexas.edu/index.php/rp-form

♾ The paper used in this book meets the minimum requirements
of ANSI/NISO Z39.48-1992 (R1997) (Permanence of Paper).

Library of Congress Cataloging-in-Publication Data
Scherer, Andrew K., author.
Mortuary landscapes of the Classic Maya: rituals of
body and soul / Andrew K. Scherer. — First edition.
pages cm — (Linda Schele series in
Maya and Pre-Columbian studies)
Includes bibliographical references and index.
ISBN 978-1-4773-0051-0 (cloth : alk. paper) —
ISBN 978-1-4773-0052-7 (library e-book) —
ISBN 978-1-4773-0053-4 (non-library e-book)
1. Mayas—Funeral customs and rites. 2. Mayas—Antiquities.
3. Human remains (Archaeology)—Mexico. 4. Human
body—Symbolic aspects—Mexico. I. Title. II. Series: Linda
Schele series in Maya and pre-Columbian studies.
F1435.3.M6S34 2015
972'.601—dc23 2014046940

doi:10.7560/300510

For those who endure, Alcuin, Vera, and especially Fotini

CONTENTS

FIGURES

ACKNOWLEDGMENTS

The seeds of this book were planted many moons ago during my graduate school years. Lori Wright gave me my introduction to Maya bones and burials firsthand over many weeks at Tikal and the Museo Nacional de Arqueología e Etnología de Guatemala. For that experience I am ever grateful. Stephen Houston and Héctor Escobedo invited me to work as the project osteologist at Piedras Negras in the late 1990s. In the final season of the project I had the invaluable opportunity to oversee the excavation of over thirty burials. The work that I continue to conduct along the Usumacinta River owes much to their pioneering efforts. In those early years I also benefited from conversations in the field and laboratory regarding the burials of Piedras Negras with fellow project members, especially James Fitzsimmons, Alejandro Gillot, Charles Golden, Zachary Hruby, A. René Muñoz, Zachary Nelson, and David Webster.

Since those early years in graduate school, I have continued my work on Maya mortuary remains, primarily in the Usumacinta River region. This research has been conducted through the Proyecto Regional Arqueológico Sierra del Lacandón (PRASL, 2003–2010) and the Proyecto Arqueológico Busilja-Chocolja (PABC, 2011–2013, Mexico), led by Charles Golden and me. In Guatemala this work was co-directed by the very capable Ana Lucía Arroyave, Luz Midilia Marroquín, and Rosaura Vásquez. We have had the opportunity to work with many fantastic colleagues and students over the years in Mexico and Guatemala. I especially want to acknowledge the collaborative efforts of Jeffrey Dobereiner, Melanie Kingsley, Betsy Marzahn-Ramos, Alan Méndez, Griselda Pérez, Socorro Jiménez, Varinia Matute, Juan Carlos Meléndez, Fabiola Quiroa, and Whittaker Schroder. I have also been fortunate to work with a number of other projects working in the Central Petén, many of which are cited throughout this volume. Steve Houston, Edwin Román, and Thomas Garrison have graciously allowed me to collaborate with their project at El Zotz, and Sarah Newman has valuably contributed to work on burials at the site. I also had the wonderful opportunity to work with Laura Gámez at Yaxha in recent years. These efforts have been a constant reminder of just how diverse Maya mortuary practices are.

Although all of these collaborations have in some way influenced my thinking on Maya burials, conversations with Charles Golden and Steve Houston have had the most impact. Burials have been central to the work that Charles and I have done on the sociopolitical organization of the kingdoms of Piedras Negras and Yaxchilan. Despite professing a deep dislike of burials, Charles has spent many, many hours laboring over some of the burials discussed in this book. Over the years Steve has been a most gracious colleague and took the time to read and comment on an earlier draft of this book. I have done my best to credit Steve's most valuable insights, though, of course, all of the bad ideas are mine.

Initial support for the preparation of this manuscript came in the form of a Dumbarton Oaks Research Fellowship. Final support was provided by a Brown University Henry Merritt Wriston Fellowship. Funding for PRASL and PABC was provided by Foundation for the Advancement of Mesoamerican Studies, Inc. (Grants #02020, 05027, and 07043), the National Geographic Society (Grants #7575-04 and 7636-04), the National Science Foundation (Grant #0406472 and 0715463), Dumbarton Oaks Research Library and Collection, H. John Heinz III Charitable Trust Grants for Latin American Archeology, the Kaplan Fund through the World Monuments Fund, Baylor University, Brown University, and the Norman Fund and the Jane's Fund at Brandeis University. None of this work could have happened without the gracious oversight of the Instituto de Antropología e Historia of Guatemala and the Instituto Nacional de Antropología e Historia de México.

Photographs, maps, and illustrations are original data and reflect the labors of the scholars who produced them. All images are credited as such throughout the book. I am especially indebted to Justin Kerr's collection of rollout pottery images, maintained at www.mayavase.com by Justin and his wife, Barbara (who will be greatly missed). Each vase is given a unique identifier, the Kerr number, which is indicated by a "K" followed by a number. Kerr numbers appear throughout this book. Other illustrations were contributed by Joseph Ball and Jennifer Taschek, Ed Barnhart, Claude Baudez, James Doyle, James Fitzsimmons, Laura Gámez, Charles Golden, Ian Graham, Nikolai Grube, Yeni Gutiérrez, Lucia Henderson, Stephen Houston and the Piedras Negras and El Zotz Archaeological Projects, Heather Hurst, Zachary Hruby, Takeshi Inomata, Simon Martin, Peter Mathews, Patricia McAnany on behalf of the K'axob Archaeological Project, Mary Miller on behalf of the Bonampak Documentation Project, the Foundation for the Advancement of Mesoamerican Studies, Inc. (FAMSI) on behalf of John Montgomery, A. René Muñoz, Jorge Peréz de Lara, David Schele and FAMSI on behalf of Linda Schele, Holly and Jeremiah Scherer, David Stuart, Karl Taube, the University Museum of the University of Pennsylvania on behalf of William Coe and the Tikal Project, Mark Van Stone, Lori Wright, and the Yale University Art Gallery.

I am grateful to the University of Texas Press for supporting the publication of this book. This manuscript greatly benefited from the capable editorial guidance of Casey Kitrell and Theresa May. Kathy Lewis carefully pored over the manuscript, correcting missing accent marks and other frequent mistakes. Two anonymous reviewers offered excellent advice that greatly improved this manuscript. But again, any errors of interpretation are mine.

LANGUAGE

Classic period and modern Maya languages are vital for illuminating Maya concepts of the body and soul. For the transcription of Classic Maya glyphs, I follow current convention and use boldface. Logograms are written in full capitals and syllabograms are lowercase. Transliterations of those glyphs are presented in italics. Not being a linguist, I defer to current orthography and spellings employed by epigraphers—specifically, Andrea Stone and Marc Zender, *Reading Maya Art: A Hieroglyphic Guide to Ancient Maya Painting and Sculpture* (New York: Thames and Hudson, 2011); and Simon Martin and Nikolai Grube, *Chronicle of the Maya Kings and Queens*, 2nd ed. (New York: Thames and Hudson, 2008). For contemporary Maya languages, I use the spellings and orthography as presented in the ethnographic publication that I cite, which may differ from source to source.

MORTUARY LANDSCAPES OF THE CLASSIC MAYA

INTRODUCTION

I began preparations for this book with a simple thesis. Although the Classic Maya held certain universal beliefs about death and burial, localized traditions developed over the course of the Classic period in regard to the proper treatment of the dead in funerary (and nonfunerary) contexts. I first began to appreciate the diversity of Classic Maya deathways in my first year of graduate school (longer ago than I care to admit) when I helped Lori Wright build her database of Tikal burials. After two semesters spent poring through the Penn Tikal reports, I felt that I had a very good idea of what constituted typical Classic Maya mortuary practice. That summer I joined Stephen Houston and Héctor Escobedo's project at Piedras Negras. Working with the skeletons and burials from that site, I came to realize that Classic Maya mortuary practices were far more diverse than is generally acknowledged. As a simple example, most burials at Tikal include pottery, and its royal tombs are overflowing with bowls, vases, and other ceramic wares. At Piedras Negras, however, ceramics were rarely placed within the burials and even the Late Classic period kings were buried with only a single dish.

Since those early graduate school years, I have spent the past decade excavating burials along the Usumacinta River of Mexico and Guatemala, studying graves and skeletons from Classic period communities that were subject to the competing royal courts of Piedras Negras and Yaxchilan. Over the years my colleague Charles Golden and I began to see an interesting pattern in the graves. Burials from different communities within the same kingdom looked surprisingly like one another and were distinct from those in other kingdoms. For example, burials at Piedras Negras are most often oriented 30° east of north (E of N). At Yaxchilan they are mostly 120° E of N. As further evidence of difference, burials within the Yaxchilan kingdom usually contain ceramics, most commonly a perforated dish located over the head of the deceased. In over 130 burials excavated in the kingdom of Piedras Negras, such a practice has never been observed.

In discussing the graves of Maya elites and nonelites, scholars have generally emphasized their differences, drawing a divide between the royals and the people they governed. In her landmark book *Living with the Ancestors*, Patricia McAnany suggests that royal mortuary practices were a co-option of more traditional, lineage-based rites. She sees "the realms of kinship and kingship as an arena of conflict"; relative to villages and towns, "the city is seen as the aberration."[1] Similarly, in his masterful *Death and the Classic Maya Kings*, James Fitzsimmons "primarily examines royal rituals of death, the process by which a ruler is turned into an ancestor."[2] Yet the respective commonalities in the funerary practices among the royal and commoner Maya of Yaxchilan and Piedras Negras make it clear that something is lost when we focus only on the differences. Archaeological preoccupations with divisions of power, wealth, and prosperity have led us to overlook the very processes that bound Maya kingdoms together. The burial evidence suggests that to some degree king and commoner, city and village dweller, participated in a greater community of shared ritual practice.

As I pored over the mortuary data—burial orientations, placement of pots, and the like—a nagging question emerged. What does this all mean? Why did the people of Piedras Negras place their dead at 30° E of N while bodies 40 km upriver at Yaxchilan were placed on a perpendicular axis, 120° E of N? Are these differences simply reflective of local idiosyncrasies or was something more complex (and interesting) at work? Were the Maya concerned with articulating their political allegiances when they buried their dead? I find this simple explanation unlikely. Rather, something more complex was afoot. An interdisciplinary approach is necessary to grapple with these questions and gain a better understanding of Classic Maya perceptions of life, death, and the afterlife. My strengths are in archaeology and bioarchaeology, so these approaches are quite prominent in this book. However, I also draw extensively on pertinent research in epigraphy, iconography, ethnohistory, and ethnography to illuminate my own field and laboratory work.[3] The result is a book that explores how the Classic Maya understood bodies and their relationship to souls in life and death.

CONTEXT

For millennia the Maya have occupied a remarkably diverse landscape that includes the flat scrub forests of the Yucatán peninsula, the jungles of the Petén and Belize, and the volcanic highlands of Chiapas and Guatemala (fig. 0.1). Within this domain the Maya established one of the world's most enduring cultural traditions. Today the Maya are synonymous with lost cities and ruined pyramids, inscriptions and prophecy, tombs and human sacrifice, reflecting a popular fascination with an era of history that archaeologists call the Classic period (AD 250-900). This particular bracket of time presumably meant nothing to the Maya. It was conceived by scholars in the early twentieth century, during the earliest days of scientific archaeology in Mexico, Guatemala, Belize, and Honduras. For these early explorers and archaeologists, the Classic period was a time of artistic and architectural fluorescence, reminiscent of the aesthetic of the "Classic civilizations" of the ancient Old World.[4] Today the idea of a Maya Classic period is entrenched and is used to reference the era when the Maya carved hieroglyphic inscriptions on stone monuments, built monumental pyramids, were ruled by sacred lords (*k'uhul ajaw*, glossed as "kings" throughout this book), and used the long count—a calendric system that places an event into linear time by counting forward from the Maya date of creation in 3114 BC. Archaeologists further subdivide the Classic period into the Early (AD 250–600), Late (AD 600-800), and Terminal Classic periods (AD 800-900/1100).

The Maya long count fell out of use by the early tenth century. Many of the southern cities in modern-day Chiapas, Tabasco, Petén, and parts of Belize were abandoned as the royal dynasties came to an abrupt end. The abandonment of these cities, the great political

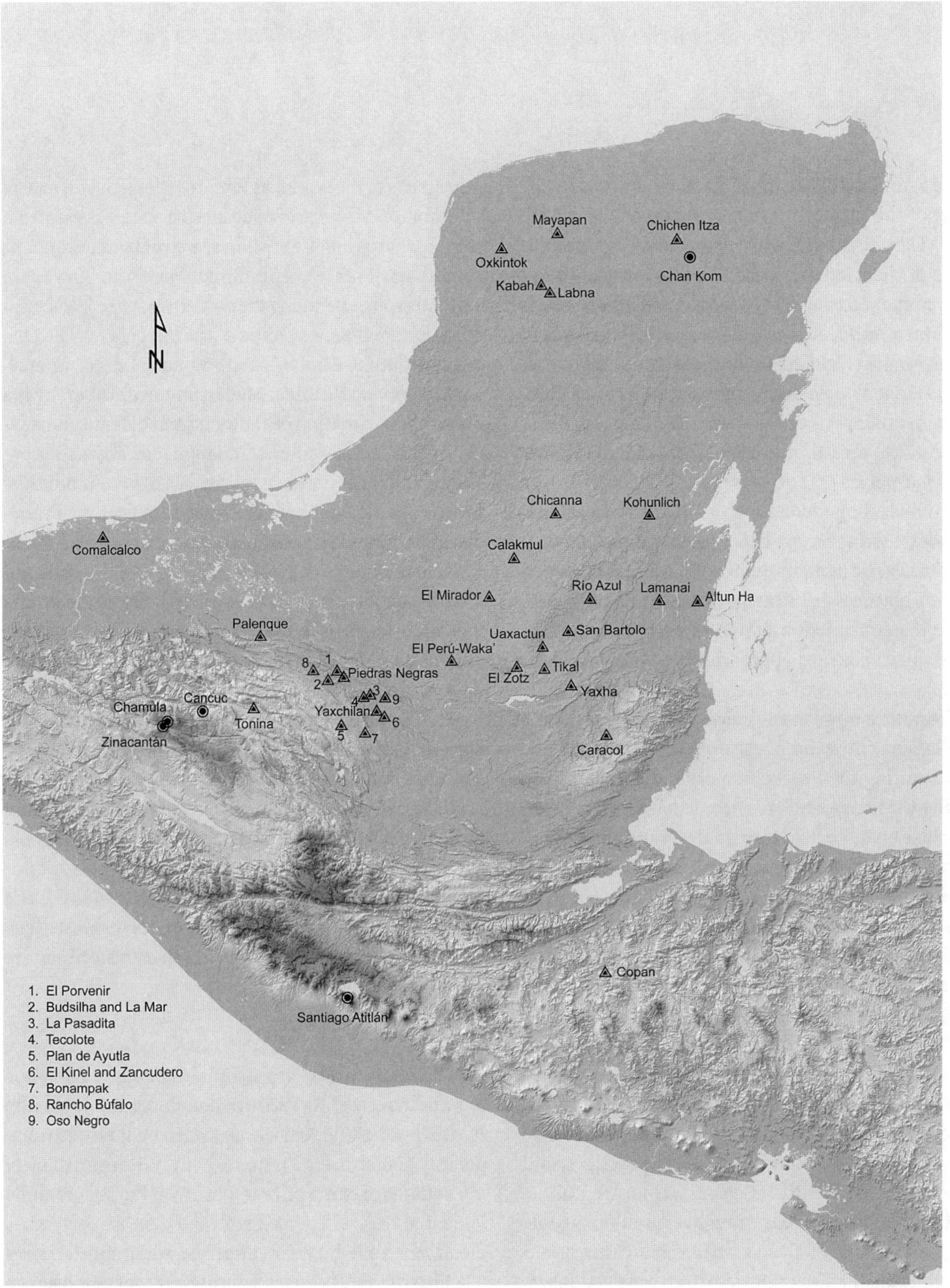

FIGURE 0.1. The Maya area showing the location of archaeological sites (triangles) and modern communities (circles) mentioned in this book (map by Andrew Scherer and Charles Golden).

rupture caused by the death of the royal courts, and the neglect of the long count define the great Maya "collapse."[5] Despite the political collapse and demographic decline of the southern lowlands, vibrant occupation and monumental construction continued in other parts of the Maya world, notably the Yucatán peninsula and the Guatemalan highlands. The material culture of this later period looks decidedly different and shows influence from other parts of Mesoamerica. This era, which ends with the arrival of the Spanish, is known as the Postclassic period (AD 900–1521).

The Preclassic period (2000 BC–AD 250) is traditionally understood as the era before the kingship and high art of the Classic period, when the Maya experimented with early practices that would later become entrenched, such as the construction of pyramids. In recent years scholars have come to appreciate the Late Preclassic period (350 BC–AD 250) as more socially and politically complicated than previously imagined.[6] Arguably, some of the largest cities and finest works of Maya art were created during the Preclassic period, and the divide from the Classic period has become blurred.

Because of the popular trope of the "vanishing Maya," many unfortunately fail to recognize that Maya cultural tradition endures to this day. Elements of modern Maya culture—particularly core religious beliefs—can be traced back at least as far as the Late Preclassic period.[7] Even the Spanish conquest is now understood to have been less destructive to the fabric of Maya society than scholars once perceived.[8] Yet it would also be a grave injustice to treat the Maya as a fossilized people, frozen in time as the modern world moves on. As with all people, they embrace tradition that helps make sense of the world, while adopting new ideas, practices, and technologies that are advantageous or meaningful.

Today the "Maya" are the roughly 11 million people that speak a group of closely related but distinct languages throughout Guatemala, eastern Mexico, Belize, and parts of Honduras and El Salvador.[9] An unknown number of Maya are also living in the United States and other parts of the world, as émigrés who fled violence in their homelands in recent decades, as adoptees of foreign parents, and as opportunists in search of better economic conditions. Over twenty different Mayan languages are currently spoken. Among the modern Maya referenced in this book are the speakers of K'iche', Kaqchikel, Q'eqchi, Mam, Tz'utujil, Chol, Cho'rti', Mopan, Tzeltal, Tzotzil, Lacandon, and Yucatec. Mayan linguistic diversity is a rough approximation of their cultural diversity; speakers of closely related Mayan languages are also more likely to share common experiences of geography, history, economy, and belief relative to speakers of more distantly related Mayan languages.

Constrained by the limits of space and the biases of my own expertise, this book focuses primarily on the mortuary practices of the Classic period. Greatest treatment is given to the western Maya lowlands, especially the Usumacinta River kingdoms of Piedras Negras and Yaxchilan, where I have worked, as well as other neighboring polities, including Palenque and Tonina. I have also conducted research with human remains from El Zotz, Yaxha, and Tikal, however, so the Central Petén of Guatemala is also an important presence in this book. By focusing on the familiar, I am able to present primary data that have not been published previously. The diminished presence of the northern lowlands, Belize, and the highlands does not mean that these areas are not equally important.

SOULS OF THE CLASSIC MAYA

Classic Maya "souls" are an important yet problematic topic of inquiry.[10] By Western definition, the soul is the spiritual side of the self. It is an incorporeal essence temporarily bound inside a physical body. We do not think of our souls as existing prior to our birth, but we certainly hope that they will persist after our deaths. Our soul is the essence of who we are; our body is simply the shell in which the soul is housed during our mortal existence on earth. The contemporary Maya, however, see things very differently. For the Tzeltal, the body is the self, there are multiple souls, those souls existed before birth, and, most confusing of all for non-Maya, the soul is in some sense the other.[11] The first chapter examines these concepts and explores how contemporary Maya belief can help us understand Classic Maya bodies and souls.

Because Western and Maya concepts of the spiritual aspect of the self are not really the same, ethnographers have long recognized that to call such essences "souls" is problematic. But to replace the term with indigenous vocabulary would be equally flawed. Various contemporary groups use different terms to refer to these entities. For example, the Tzeltal speak of *ch'ulel* and *lab*, whereas the Yucatec, Mopan, and Lacandon call such beings *pixan*. *Ch'ulel* may be a loose cognate of the Classic period *k'uhul*. Nevertheless, I follow the lead of the ethnographers and grudgingly employ the word "souls" (and similar terms, such as spirits, essences, and co-essences) to refer to the supernatural beings that were bound to the Classic Maya physical human body and gave it life. As Evon Vogt, the eminent ethnographer of the Tzotzil, explained, "I use the term 'soul' advisedly in quotes to indicate that the familiar European concepts of 'souls' and 'spirits' are inadequate for precise ethnographic description of these concepts."[12] For ease of readability, I have abandoned the quotation marks around the term. Yet I stress that the term "souls" as applied to the Maya does not accurately capture their understanding of the body and self.

For ethnographers, the Maya soul is elusive. As Pedro Pitarch explains, "there is, of course, no canonical body of Tzeltal knowledge that defines the soul."[13] Nevertheless, the rich body of ethnographic literature reveals that souls are fundamental for understanding Maya society; they are at the core of indigenous philosophies of life, death, illness, and interpersonal relations. Quite simply, by ignoring the Classic Maya soul, we handicap our capacity to understand how the Classic Maya perceived themselves and the world in which they lived. Just as the ethnographers struggle to define contemporary concepts of the soul, identifying the Classic Maya soul is fraught with difficulty. Nevertheless, if we consider how important souls were in the Classic period worldview, we can reasonably hope to find traces of them in ancient Maya thought and practice. To do so requires marshaling many disparate lines of evidence.

We must also bear in mind that the contemporary Maya do not share a universal perspective as to whether they are or are not the modern descendants of the people that archaeologists define as the "Classic Maya." On one hand, current pan-Maya movements have sought to appropriate the Classic period as part of their identity. Some archaeological sites, such as the Great Plaza of Tikal, are places of pilgrimage for contemporary ritual performances.[14] Yet none of the various Q'eqchi, Kaqchikel, Chol, and Tzeltal men who have worked on my archaeological projects over the years saw themselves as particularly connected to the Classic period Maya that we excavated—or if they did they never expressed it to me. Whenever I inquired what they thought about the ruins we explored or the burials we excavated, they expressed fascination with some of the parallels with their modern life (especially the foods that they ate) yet still perceived these people as distant, remote, and, in many respects, a supernatural other.[15] The lack of personal connection is much like my own feelings about my Germanic ancestors. When I look to medieval Catholicism I see the obvious roots of my family's faith—especially my mother's fascination with patron saints and the power of medallions. Yet I do not feel any particular closeness to the Franks, any more than my indigenous colleagues see themselves in the Classic period peoples of the Usumacinta River area. We should not assume that indigenous people are more or less connected to their past than are modern peoples. Our relationship to the past is what we make of it.

BURIALS AND BODIES

The primary data for this book are the burials of the Classic Maya and the bodies they contain. Most Maya burials can be adequately described by using the typology that Ledyard Smith developed in his description of the burials of Uaxactun.[16] Smith's typology has been widely adopted by Mayanists, and I use a modified version of it in this volume. *Simple burials* are graves with no formal burial architecture. They are typically bodies placed in the ground or within the rubble fill of architecture. *Cists* are inhumations with only minor elaboration to delimit the mortuary space from the surrounding architecture or soil, such as rocks placed around the body. *Simple crypts* are formal masonry-walled containers, complete with a lid that was typically made of large, flat limestone slabs (*lajas*). The internal height of crypts is less than 0.5 m. *Elaborate crypts* are larger, more complex versions and have internal heights of 0.5 m to 1.0 m. *Tombs* are corbel vaulted burial chambers with an interior height greater than 1.0 m. Obviously this is an etic typology, though arguably one with some emic value; all of the epigraphically confirmed graves of Classic period *k'uhul ajaw* are tombs.

To excavate a Classic Maya site is to uncover burials. In part this is because they are everywhere: in plazas and pyramids, in houses and middens. Burials have also been targeted for their value as sealed contexts, however, as products of intentional human behavior that can be studied in a comparative perspective when a sizable number are uncovered. In the early days of Mayanist archaeology, priority was given to the objects within the burials, especially the whole ceramic vessels that are essential for constructing chronologies. Such burial objects are the focus of chapter 3.

In recent decades there has been a growing interest in the human skeleton, the physical remains of the Maya body. As a subject of anthropological inquiry, the human body is at the junction of the biological and sociocultural arms of the discipline. Biological anthropologists have traditionally been concerned with the body as a topic of anatomy, as a manifestation of our evolutionary history, and as evidence of adaptation to different environments.[17] Sociocultural anthropologists, in contrast, have been concerned with the body as a canvas for cultural behavior and as a component of broader studies of embodiment, self, and personhood.[18] The study of the ancient body has a rich history, focusing on ancient skeletons, mummies, and other archaeologically recovered human remains, conducted primarily by specialists trained in the domain of biological anthropology.[19] The study of ancient text and image has also been an important window into culturally specific ideas about the body—particularly the body ideal—in past societies.[20]

The term "body" most readily calls to mind our corporeal mass: a thorax with four limbs and a head. In perceiving the body, we focus on its surfaces: the skin, the hair, the features of the face, and so forth. Unfortunately, all of those features are lost when we consider the remains of ancient Maya bodies. We are left with the skeleton, a once vital tissue that in life we only indirectly perceive as the shape of our heads, our stature, or the aches in our joints. Despite being largely unseen, bones are essential to who we are, and the study of this tissue has much to tell us about the body in life and in death.

Bodies are historically contingent and thus are of immediate interest to anthropologists who study past societies. The bodies of today are the cumulative product not only of our own lives but also of those of our ancestors, both our immediate kin and our most distant evolutionary forebears. Bodies are also dynamic: even our skeletons continue to change as they are remodeled over the course of our life. Although the body's basic form is dictated by the genes that we have inherited, it is also heavily subject to exogenous factors, from the food we eat to the microorganisms that invade and make us ill.

The past few decades have produced a veritable flood of research on the body in the humanities and social sciences. One of the primary concerns has been the relationship between the body and the self. On one hand, the self is an ideological construct of the body; seated in our brains, our concept of self is created by the firing of neurons and other biochemical reactions. On the other hand, it could also be said that the body is the physical manifestation of the self. For archaeologists, the works of Michel Foucault and Judith Butler have been highly influential and there has been a recent move to view the human body as a form of material culture, a social construct that is made and remade through performance and social discourse.[21]

The challenge, of course, is that most of the biological processes that shape the human body are beyond the domain of human agency: they occur without any conscious effort, even if we may wish otherwise. Instead much physiology and anatomy is rooted at the molecular level, ultimately in our DNA, itself the product of millennia of evolutionary forces (selection, drift, mutation, and gene flow). But to say that the form and substance of the human body are beyond the influence of our own will is obviously incorrect. Evolutionary processes have left us with a mind that allows us to reflect on our bodies and, in some instances, intervene directly to shape the course of its anatomy and physiology. Moreover, we are social creatures, so our bodies must contend, for better or worse, with norms of behaviors, power structures, and the general influence of others.

Constructivist theories of the body have recently been extended to the study of the ancient human skeleton, leading some to suggest that even the skeleton is socially constructed, that it is "material culture."[22] Such approaches to the skeleton are heavily influenced by recent approaches to "materiality" in material culture studies. Although the term evades easy definition, the general sense is that meaning is constituted in our material objects (artifacts) and that those objects in turn shape us; humans and their objects perpetually make and remake one another. Some scholars suggest that the inanimate world of "things" is imbued with agency.[23] Trying to describe the body as material culture, however, involves a problematic circularity of logic. Take, for example, this passage by Joanna Sofaer:

> The material qualities of the human body are key to its materiality. The materiality of the archaeological body, in other words the particular form that a body may take, can be understood as the material outcomes of human plasticity at a given point in time. The materiality of the archaeological body is not therefore given and immutable, but follows from the specific material qualities of the human skeleton that permit or constrain its change and development. The materiality of each body is context dependent, temporally described, produced and unique. Nonetheless, because the plasticity of the body is not limitless, and people have common experiences, or are situated in contexts with common social values, bodies may have common expressions (albeit to different degrees).[24]

If materiality is a theory of the agency of objects, why would such a theory need to be applied to bodies which, by virtue of being human, are the very definition of physical forms with agency? Applying a theory of materiality to the body is akin to applying Darwinian evolutionary theory to human psychology and then trying to use evolutionary psychology to explain the origins and extinction of dinosaurs. The circularity of attempted reason just does not make sense.

Tim Ingold provides a more succinct and lucid explanation of how the body may express aspects of materiality: "The acquisition of culturally specific skills is part and parcel of the overall developmental process of the human organism, and through this process they come to be literally embodied in the organism, in its neurology, its musculature, even in features of its anatomy."[25] Indeed, if we scrape away the obfuscating jargon and pay attention to prior scholarship, we find that osteologists have been making such observations about the body for decades now. The analysis of the human skeleton is like a textual exegesis where the analyst reads the skeleton and disentangles the various processes that contributed to its final form: the skeleton at death. Even the staunchest osteological positivists use narrative metaphors in describing their work, likening the skeleton to a person's biography (or, perhaps better, obituary), as evidenced by popular books with titles such as *Written in Bone* or *Dead Men Do Tell Tales*.[26] In fact, Frank Saul coined the term "osteobiography" decades ago, which he described as studying "skeletons as life histories recorded in bone."[27] John Robb provides a recent update to Saul's approach and suggests that an aim of skeletal analysis should be to reconstruct a person's "biography as a cultural narrative." Robb's goal is "to focus on the cultural understanding of life events and to encompass the history of human remains after death." In a sense his approach is a comparative biography: by integrating archaeology, osteology, and taphonomy, he explores how different aspects of a person's identity (such as age) shaped responses to illness, injury, and even death.[28]

I would expand Robb's view and argue that the study of Maya skeletons must take an integrated approach that is centered on a blend of laboratory osteology and field archaeology, coupled with reference to social theory, ethnography, ethnohistory, epigraphy, and iconography. Skeletons must be interpreted within their broader archaeological, taphonomic, social, and historic contexts. We must also be explicit about the nature of the human body—it is a biological organism that has been uniquely shaped by social action. Unlike the constructivists, we cannot pretend that the body is entirely constituted through social action. We must also recognize, however, that the body is more than a composite of biological functions.

The problem of the Maya lived body is the subject of chapter 1, which establishes a framework for Classic Maya souls upon which the rest of this study rests. I introduce ethnographic concepts of Maya self and souls and explore how those beliefs are reflected in Classic Maya treatments of the body, particularly as evidenced in the human skeleton. I also discuss how these concepts may have informed Classic period understandings of health and illness.

RITUAL AND THE MORTUARY LANDSCAPE

The remaining chapters focus on the dead body within the mortuary context and explore the broader significance of funerary rites and postinterment ritual practice. Ritual is familiar and exists universally across human societies, yet it also escapes easy definition. At its most basic, ritual is defined in opposition to other activities that are assumed to be nonritual. As Catherine Bell explains, "ritualization is fundamentally a way of doing things to trigger the perception that these practices are distinct and the associations that they engender are special."[29] Ritual is established by the context in which it is enacted. This study focuses on those actions that were enacted in the context of a dying person, especially those that pertained to the disposal of a dead body and the management of that body's souls. It also considers the importance of later rites that incorporate the bodies and memories of the dead, in some cases decades after the death event. Ultimately I explore how these rituals of death and veneration relate to the broader social fabric of the Classic Maya community and kingdom.

Rituals are inherently divisible; smaller ritual acts may combine in different and meaningful ways to establish the greater ritual performance. At the same time, rituals are not always easily bounded and how we frame ritual is to some degree arbitrary. For example, we can parse Classic Maya mortuary ritual into its smaller components, such as the lighting of an incense burner or the wrapping of a body. The repetition of action helps to frame ritual and is of great benefit to archaeological study. Although the idiosyncratic can also be ritual, repeated acts are more likely to be identifiable and interpretable in archaeological contexts. In this volume I focus on both the repeated elements of Maya mortuary rites and how those practices reflect elements of deeper Maya ideology.

Rituals are inherently "bodycentric": costumes are worn, gestures are repeated, feasting and fasting occurs, and so forth. Mortuary ritual is distinct in that it involves the performance of both living and dead bodies. For the most part the ritual is largely directed by those living bodies, though we should not ignore the profound influence of the bloating, stinking, decomposing corpse, which forcibly and unpleasantly inserts itself into the ritual process and must be carefully managed. We can never hope to reconstruct all aspects of the mortuary ritual from the archaeological record. What we can do, however, is explore the final product of ritual action as evidenced in the placement of the body, the objects arranged with it, and the landscape in which it was interred.

People have a tendency to perceive ritual as particularly concerned with supernatural forces. Even Bell, the eminent specialist on ritual, suggested that archaeologists should just stick with a simple definition: "namely, rituals are those activities that address the gods or other supernatural forces," as proposed by Colin Renfrew.[30] Such a definition of ritual is problematic, however: it assumes that people make a distinction between the natural and supernatural when they engage in ritual. In this definition any activity that is perceived as secular or atheistic could not then be ritual. Such a distinction is problematic for describing ritual in our own lives and certainly is problematic when applied to the Classic Maya, who had no such binary division between the natural and supernatural. Ethnography suggests that contemporary Maya mortuary rituals pertain most immediately to the management of dead souls and more abstractly to the reparation of the social rupture caused by the death of the individual.[31] In other words, Maya death ritual reflects what we would consider both sacred and secular concerns.

Ritual is also meaningful. What can Classic Maya mortuary rituals tell us about understandings of the soul? Why were certain practices repeated in certain places and times? How might an understanding of mortuary rites help us understand the nature of Classic period communities and kingdoms? As Bell suggests, "ritual acts must be understood within a semantic framework whereby the significance of an action is dependent upon its place and relationship within a context of all other ways of acting: what it echoes, what it inverts, what it alludes to, what it denies."[32] Each aspect of a mortuary ritual, from the wrapping of the corpse to the placing of a vase, may have had its own specific meaning. Yet it is really the composite of all of these actions that is most important. At the same time, the meaning of each action may have differed from person to person, community to community, or century to century. Our capacity to reconstruct meaning in a particular mortuary context is contingent on our ability to find tendrils of that meaning in other lines of archaeological, iconographic, and epigraphic data.

In order to reconstruct meaning in the past, we must turn to ethnohistoric and ethnographic clues,[33] but not without caution. Such efforts not only can lead to historical anachronism but can also risk treating the Maya as an essentialized, fossilized, unchanging people. Yet to disregard the existence of deep historical continuity of certain aspects of Maya belief and practice would deny the powerful strength of tradition, which has undoubtedly been one of the Maya's greatest assets. If we ignore the larger body of ethnographic literature on Maya concepts of body, self, and soul we also risk transposing our own worldviews to the Classic Maya. At least some of the pitfalls of anachronism and essentialism can be avoided if we look for patterns in the ethnographic literature that encompass many Maya communities and ethnic groups, such as the Yucatec in the north, the Cho'rti' in the east, the Tz'utujil in the south, and the Tzeltal in the west. Ethnographic analogy to the past is strengthened if similar practices and beliefs can be traced back to the period of European contact. When commonalities emerge among geographically and temporally distant groups, we can tentatively argue that such patterns of thought and action may also be found in the Classic period, assuming that we can marshal evidence to demonstrate this. Nevertheless, we must also be sensitive to variability in practice and belief in order to illuminate the places and times in which different approaches to ritual action emerged, even among neighboring, contemporary polities of the Classic period. The goal here is to identify the basic grammar of Classic Maya mortuary practices, while exploring the unique variants of belief and practice that developed in different parts of the Maya lowlands.

Chapter 2 explores how the treatment of the corpse provides a useful window into Classic Maya understandings of bodies and souls. I first introduce evidence that different Maya kingdoms formed distinct communities of ritual practice. It is unlikely that each polity had wildly different concepts of souls. Instead, distinctive traditions presumably reflect different approaches to how souls were managed. Chapter 3 considers the mortuary space, loci that contained the bones of the Maya. I focus especially on evidence of liminality and show that burial spaces were understood as axes mundi, places of communication and travel among the earthly, underworld, and celestial realms.

The final chapter of this book is concerned with the relationship of burials to the greater social landscape. In the United States we are surrounded by the memory of the deceased: statues of historically prominent men and women fill our parks, and pictures of dead presidents line our wallets. Yet we are careful to contain the actual remains of our dead within cemeteries, separated from the quotidian world of the living. For the Maya, however, the living and mortuary landscapes were one and the same. Burials were located below the floor of houses, within pyramids in the monumental core of the city, and in caves and other places throughout the natural landscape. The placement of the dead also reflects fundamental Maya concepts of time and space. Both were ordered by understandings of the daily and annual movements of the sun over the surface of the earth. The symbolic parallels are especially meaningful, as the movement of the soul was likened to the ascent of the sun.

This daily proximity of Maya graves had important implications for the way the Maya related to their dead, particularly those who became "ancestors" through memory and practice. The final chapter of this book devotes specific attention to the Classic Maya practice of domestic inhumation and how the management of ancestral souls was essential to the negotiation of inheritance within lineages. Specifically, access to the bones and burials of prominent lineage members during rites of veneration aided claims to resources against other members of the same lineage. In a similar fashion, the prominent location of dead kings and queens shows how access to royal souls was vital for accession to the Classic Maya throne. Both situations echo David Kertzer's observation that "far from simply propping up the status quo, ritual provides an important weapon in political struggle, a weapon used both by contestants for power within stable political systems and by those who seek to protect or to overthrow unstable systems."[34] For the Maya, ancestor veneration was ritual as negotiation.

CHAPTER 1

LIVED BODIES

It is unfortunate that many books that are ostensibly about ancient graves say very little about the bodies they contain. Perhaps this is due to a mistaken assumption that the remains of ancient people have little to tell us about complex issues such as ritual, identity, and belief. That could not be further from the truth. There is a long anthropological legacy for the study of human skeletal remains, affording insight into factors such as sex, age at death, stature, health, diet, ancestry, and many other facets of an ancient person's lived identity. Yet a divide persists between what is generally understood as "mortuary archaeology" and "bioarchaeology."[1] There is no intrinsic reason why the study of the mortuary context should be divorced from the study of the skeleton. Rather, this arrangement is merely an artificial divide that reflects the training, methodologies, and research questions of the practitioners of these two fields.

In this chapter I integrate osteological data with insight gained from archaeology, epigraphy, iconography, ethnohistory, and ethnography regarding Maya concepts of self, the body, and the soul, establishing the framework upon which the rest of the book is built. How did the Maya understand the self to be constituted? What is the relationship between body and soul? What happened to the self after death? How are the answers to these questions relevant for understanding meaning in Maya burial practices? Here and throughout much of the book I rely on data from burials that I have studied along the Usumacinta River. This includes 122 burials from Piedras Negras and another 11 burials from its subordinate sites of El Porvenir and Budsilha. I reference data from 20 burials excavated at three sites in the Yaxchilan kingdom: Tecolote, El Kinel, and Zancudero. In later chapters I also draw from my recent research on the skeletons from El Zotz and Yaxha. Complete descriptions of all of these burials and subsequent skeletal analysis can be found in the original research reports.[2]

BODY AND SELF, HEART AND SOUL

Decades of comparative ethnography have shown that humans have no universal understanding of the relationship between the individual, self, or person and the body.[3] We can define "self" as an agent of behavior that is capable of reflecting on those actions; it is linked to the individual (though it is not always the same) and is constructed through social interaction.[4] Western, particularly American, thought treats the self as the unique and indivisible person (the individual) and the body as its physical manifestation. Yet our bodies do not always obey the will of the self. I may prefer not to grow old, yet I cannot stop my body from doing so. I would rather not develop cancer, but I might. This basic point has very important implications for understanding the relationship between the body and self, particularly among the Maya.

Most readers of this book likely have a view of the self and body that incorporates rational scientific knowledge regarding the working of the mind, perhaps mixed with varying degrees of Judeo-Christian belief about souls. We accept that the thinking, perceiving, and experiencing self is largely centered around the activities of the brain. Yet most of us would prefer not to reduce our sense of self to the biochemical firings of neurons within our heads. Rather, we recognize a partnership of sorts between self and body, an arrangement that ends in death and (depending on personal beliefs) results in either oblivion or an afterlife for the self as a soul, detached from the body. The self as soul inhabits the body but is not contingent on it; they are not inalienable. In Western thought, the self is greater than the body; this perspective, among other things, informs constructivist theories of the body.[5]

As is evident from a large body of ethnography, the contemporary Maya have very different understandings of the relationship of the self, individual, body, and soul. As Inga Calvin notes, these can be roughly summarized into four categories: the predetermined life span, souls related to personality and consciousness, essences related to blood and vitality, and animal and supernatural co-essences.[6] Pedro Pitarch offers a thorough exegesis of the Maya self, body, and souls based on his ethnographic work with the Tzeltal of Cancuc, Chiapas.[7] The self, the "I," is the body for the Tzeltal. As Pitarch writes:

> From the Indian [Tzeltal] viewpoint, appearance is the locus of personal identity and morality. One is what one shows to others. The birth of the body is what initiates the process of differentiation, that is, humanization and, ultimately, Indianization. The body gradually takes its shape through nurture and the development of bodily *gestus*. If the Christian moral imperative is the cultivation of the soul, then the indigenous imperative is the cultivation of the body.[8]

For the Tzeltal, the body is inhabited by a number of sacred/spiritual entities known as *ch'ulel*. The word *ch'ulel*, unlike the Western notion of soul, does not mean the self but "the body's other."[9] *Ch'ulel* exist both outside and inside of the human body. Moreover, as a Tzeltal archaeologist explained to me during our fieldwork, you are not born with your *ch'ulel*, you accumulate them over the course of your childhood and adolescence.

For the Tzeltal, three classes of soul inhabit the body. The first is the "bird of our heart," a sacred entity that lives in each human heart. The bird of our heart is essential to life but skittish and is the favorite food of other supernatural beings. The second is the *bats'il ch'ulel* (the genuine *ch'ulel*), an essence that "[a]lso lodges in the heart and is needed to continue living; but in addition, it plays a role in shaping each person's individual character. The *ch'ulel* is the seat of memory, feelings, and emotions; it is responsible for dreams; and it is the source of language. The different nature of each *ch'ulel* is, in a nutshell, what gives each human being a unique 'temperament.'"[10]

This second *ch'ulel* is most like the Western concept of soul in that it is linked to the essence of each person and is anthropomorphic. Yet it differs in that it exists in a double form: one part inhabits the body and the other inhabits a layered, sacred flower mountain. A person's well-being in the real world is directly linked both to the state of the *ch'ulel* in the mountain and to the status of the *ch'ulel* in the heart. The *ch'ulel* in the mountain live alongside other *ch'ulel* of the same lineage in a society that is parallel, yet different from that of the real world. The *ch'ulel* of the heart is able to leave the body, especially as an individual sleeps, which is particularly dangerous; harm caused to the *ch'ulel* in its wanderings can be the source of illness and even death.

The third type of sacred essence linked to the body of the Tzeltal is the *lab*. Much like the second *ch'ulel*, the *lab* has variants that exist both within the heart and outside the body. The *lab* are most often animals but also include *lab* of the water (usually freshwater snakes, linked to metallic objects), celestial *lab* (winds, lightning, rainbows, shooting stars; Pitarch calls these "meteor *lab*"), and the dreaded illness-giving *lab*, which, for Pitarch's informants, include a being depicted as a diminutive Catholic priest, his companion dog, a dwarf-scribe, a Castilian, sheep, goats, owls, and evangelical music. As presented by Pitarch, the Tzeltal concept of *lab* is extremely complex and at times contradictory. In essence, people can have a various number of *lab* (thirteen is ideal and has implications for what happens after death). Certain *lab* are associated with power, such as the jaguar.

Variations of these concepts are held across the Maya world.[11] Evon Vogt reports that for the Tzotzil, like the Tzeltal, *ch'ulel* "is placed in the body of an unborn embryo by the ancestral deities who live in the sacred mountains." The Tzotzil *ch'ulel* resides in the heart and blood. After death it "rejoins the 'pool' or 'supply' of *ch'uleletik* that are kept by the ancestral deities. It is later utilized for another person."[12] Vogt's description of Tzotzil souls is important in a number of regards. First, it highlights the concept of soul reuse, a resurrection of sorts. Moreover, it underscores that, after death, souls will eventually travel to an otherworld place where they dwell with other souls and supernatural beings. Both of these concepts are pervasive among contemporary Maya ethnic groups. I would suggest that their ubiquity indicates that they are core beliefs in Maya thought that likely date to Precolumbian times. These observations are critical for much of the analysis that follows.

Modern Maya spiritual essences are closely linked with the heart, blood, and breath (the lungs being functionally and anatomically close to the heart). Classic period texts reveal a close connection of blood, the sacred (*k'uhul*), and wind (*ik'*). The heart, the focus of the sacred within the body, was the part of humans most coveted by supernatural beings in ancient times and was the target of sacrifice. In the Postclassic Dresden Codex, the Sun God is shown seated before a plate of human hearts, likened to a plate of tamales.[13] Offerings of human hearts thus served to appease capricious supernatural beings. In a modern variant, the Yucatec offer tamales as a substitution sacrifice (*k'ex*) to placate spirits, including dangerous winds that may cause illness.[14]

Souls and co-essences are also essential for understanding concepts of health and well-being among the contemporary Maya. The Tzotzil, Tzeltal, Lacandon, Yucatec, Q'eqchi, Mam, and Mopan recognize that souls can wander at night while people sleep and that illness may relate to the loss of an individual's spirit or an attack by a malicious spirit.[15] Among the Tzeltal, Tzotzil, Lacandon, Q'eqchi, and Mam the status of a soul can be checked through the pulse in the wrists and ankles: the soul is closely linked to the heart, the flow of blood, and also the lungs and breathing.[16] Pulsing is thus an important component of health care in many Maya communities. As Ann Fink notes, as a Mopan child's shadow-soul (*pixan*) becomes better attached to the body, the circulation system becomes more complete and the diagnosis of illness by "reading the blood" becomes easier.[17]

In the Classic period the close linkage between the sacred essence of royal beings (*k'uh*) and their blood suggests that ancient Maya souls were also found in the heart and blood.[18] *Ch'ulel* is linguistically tied to the Classic period concepts of *k'uh* (supernatural being) and *k'uhul* (sacred, divine, of supernatural character) and is a loose cognate, capturing the concept of a spirit force that dwells in a person's heart and flows through the veins. The glyph for *k'uh* appears as an unusual head with a prognathic face and small, circular, vacant eyes (figs. 1.1a and 1.1b). This being corresponds to Paul Schellhas's God C and seems to be a general concept of god or supernatural being.[19] The *k'uh* glyph bears a superficial resemblance to the "*ajaw*-face" sign employed in a range of contexts, including the day sign *ajaw* (figs. 1.1c and 1.1d). Many of the uses of the "*ajaw*-face" glyph have little do with lordship. The glyph must have a broader meaning, perhaps relating to animate or vital seed (see chapter 2). This spirit or supernatural quality may explain why the *k'uh* and the "*ajaw*-face" glyph are superficially similar, including their vacant circular eyes, mouths that occasionally have *ik'*-shaped teeth, and a prognathic face (obvious in the case of *k'uh* and suggested by the circle around the "mouth" of the "*ajaw*-face"). The origins and original meanings of these vacant face signs are murky. They could be masks, relating to the Mesoamerican practice of using such devices to embody supernatural beings, though other possibilities are equally plausible.[20]

In the Classic period *k'uh* was especially linked to royalty, indicating that they were possessed of a particular character or abundance of spirit that distinguished them from other humans. The adjective *k'uhul* is the *k'uh* head with a prefix that most commonly includes the *k'an* cross (precious) and a spray of dots (fig. 1.2). The full *k'uhul* glyph is not common. Rather, its most frequent occurrence is as a component in emblem glyphs, where the *k'uh* head is usually covered by a sign that most likely corresponds to the respective names of different Maya polities or perhaps different dynasties or lineages. In his recent discussion of emblem glyphs, Alexandre Tokovinine suggests that one of the functions of the emblem glyphs was to link individuals of the distinct *ajaw* status to a mythical place of origin that is identified in each emblem glyph's name component (such as *mutal*, *kaan*, and *baak*).[21] Hence the emblem glyphs may be toponymic, associated with a sacred place (real

FIGURE 1.1. Glyphic examples of sacred essence: (a) Early Classic *k'uh* from Tikal Stela 26; (b) Late Classic *k'uh* from Yaxchilan Lintel 25; (c) Early Classic "*ajaw*-face" in day sign from Tikal Stela 31; (d) Late Classic "*ajaw*-face" in day sign from Yaxchilan Lintel 3 (all drawings by author).

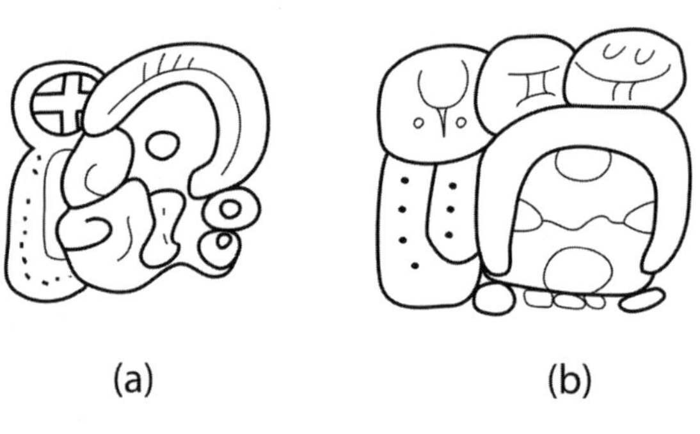

FIGURE 1.2. Classic period *k'uhul*: (a) as a complete glyphic expression from the Palenque Tablet of the Foliated Cross; and (b) as part of the Palenque emblem glyph expression on the Tablet of the Sun (drawings by author).

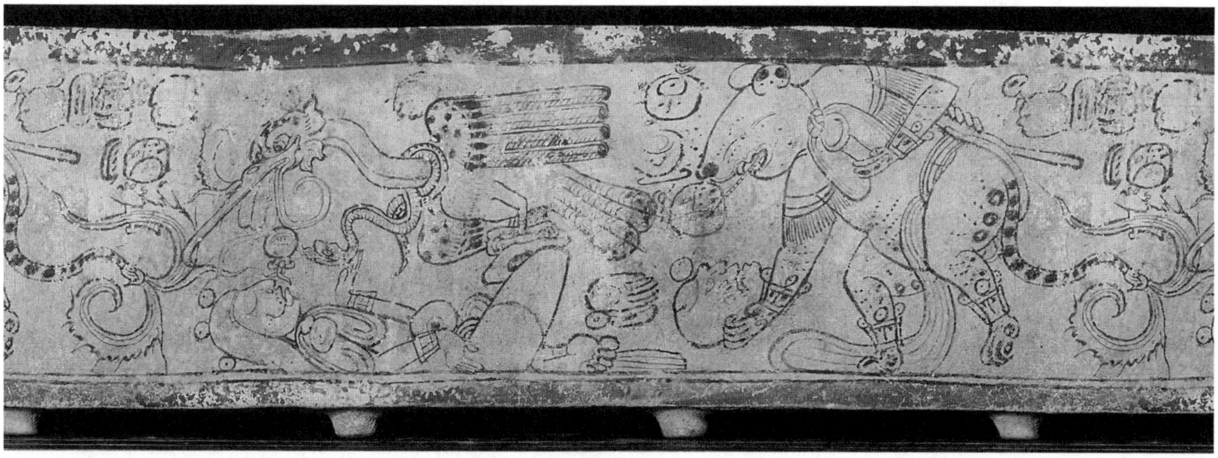

FIGURE 1.4. *Wahy* in what seems to be a death scene (K1228 © Justin Kerr).

or mythic), and perhaps also associated with particular lineages that received their legitimacy to govern through their associations with these places.

WAHY

If *k'uh* is a rough cognate for the Tzeltal *ch'ulel*, then the Classic period *wahy* are superficially equivalent to the Tzeltal *lab*.[22] Other Tzeltal and Tzotzil communities refer to these animal companion spirits as *chanul*, *wayhel*, *holomal*, or *lab*.[23] Other Maya groups, such as the Yucatec, use the Nahuatl loan word *nawal* to refer to these sometimes malicious entities. The Classic period term *wahy* is closely linked to the widespread modern verb *way* (to sleep or dream), and the Tzotzil refer to the animal companions of witches as *wayhel*.[24]

Our understanding of the *wahy* of the Classic period is largely derived from their depiction and textual description on polychrome bowls and cylinder vases of the southern lowlands. *Wahy* are shown as zoomorphic (jaguar, bird, snake) and anthropomorphic (especially skeletal) beings. The *wahy* glyph consists of the "*ajaw*-face" partially covered in a jaguar pelt (fig. 1.3). They inhabit underworld or otherworldly places and interact with other supernatural beings, including other *wahy*, in a fashion similar to modern Maya spiritual co-essences (fig. 1.4). Stephen Houston and David Stuart suggested that "many of the supernatural figures, once described as 'gods,' 'underworld denizens,' or 'deities,' are instead co-essences of supernaturals or humans. . . . It appears now that much of the imagery on ceramics relates to Maya perceptions of self."[25] The Maya concept of *wahy* is part of a broader animistic worldview whereby all humans, animals, plants, portable objects, buildings, mountains, rivers, and so forth have spiritual essences. In parallel to the Tzotzil concept of *wayhel*, Stuart suggests that Classic period kings used knowledge of their own *wahy* and those of their enemies as a form of royal witchcraft during the Classic period.[26] These Classic period *wahy* are depicted on portable objects that could be exchanged, which suggests some degree of shared belief and understanding of these beings, at least among the ruling elites of the southern lowlands. This contrasts with the more diverse and varied understandings of Maya co-essences in modern times.

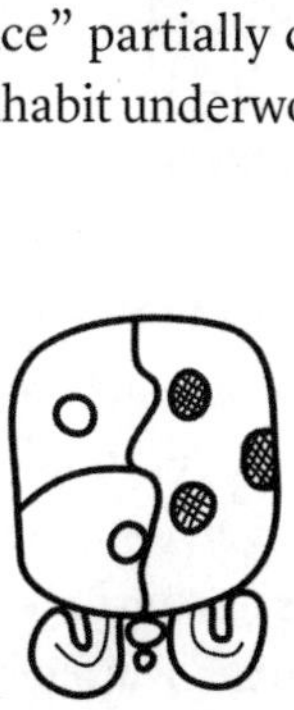

FIGURE 1.3. *Wahy* glyph from Yaxchilan Lintel 15 (drawing by author).

According to Vogt, *chanul*, the Zinacantán Tzotzil equivalent, are said to confer a certain personality type.[27] For example, jaguars are the most powerful *chanul* and are possessed by political leaders and shamans. Small animals like opossums are the spirit companions of the meekest people. Similar symbolism seems to be at work with Classic period *wahy*.[28] Jaguars and serpents are among the most powerful spiritual entities and are frequent protagonists in Classic period supernatural scenes. In contrast, small mammals, particularly rodents, seem to scurry about doing tasks for other beings.

GODS AND OTHER CLASSIC PERIOD SUPERNATURALS

Classic Maya souls and co-essences existed in a complex supernatural world that we are only now beginning to understand. Much of this work can be traced back to the pioneering efforts of Paul Schellhas, who compiled the first list of Precolumbian Maya supernaturals.[29] Schellhas recognized repetition of figures across the three contact period codices and classified the anthropomorphic beings as "gods" and the zoomorphic beings as "mythological animals." Many of the entities identified by Schellhas have since been linked to beings in Classic period imagery. Among those beings identified by Schellhas, the most important for this volume are:

God A: The Death God
God B: The God with the Long Nose and Lolling Tongue (now recognized as Chahk)
God C: The God with the Ornamented Face (now believed to be a general representation for god or supernatural)
God D: The Moon and Night God (now recognized as a creator deity or patron of the gods, Itzamnaaj in colonial Yucatec sources)
God E: The Maize God
God G: The Sun God (today frequently referred to as K'inich Ajaw)
God K: The God with the Ornamented Nose (now known as K'awiil)
God L: The Old, Black God (now recognized as an important lord of the underworld)
God F: Schellhas conflated three distinct entities as his God F. Karl Taube has disentangled these entities, one of which includes a being known in the literature as God A' or Akan.[30]

A number of other important Classic period supernatural beings are not referenced in the codices but are relevant for this book. This includes two aspects of the sun, the Jaguar God of the Underworld and an aquatic being generally referred to as GI of the Palenque Triad. Scenes of the Maize God often include other actors, including twins and maidens. The supernatural embodiment of paper has been known for quite some time as the Jester God.[31] An unusual jawless being associated with trees serves as the patron of the month Pax. Two aged beings are frequently shown navigating a canoe and are known in the literature as the Paddler Gods.

Schellhas assumed that any supernatural anthropomorphic being represented in the codices was a god. Now that we have a better understanding of Classic Maya belief thanks to advances in epigraphy and iconography, this is perhaps an oversimplification. It ignores the subtleties of Maya cosmology, the diverse roles of these beings, their relationship to humanity, and their relative importance to one another. As an analogy, it would be incorrect to suggest that all ancient Greek supernatural beings (the twelve Olympians, Titans, immortal heroes, monsters, nymphs, satyrs, centaurs, and so forth) are gods, a concept that implies some sort of equivalency or commonality. Unfortunately, unlike the Greek situation, the Classic Maya texts are not explicit in regard to the nature of most supernatural beings. Nevertheless, we can at least consider how different supernatural entities were depicted in Classic period imagery, how they are shown to interact with humans, and how they relate more broadly to aspects of Maya cosmology.

Certain supernaturals seem to have held tremendous power over the lives of humans, judging by the frequency and diversity of contexts in which they appear in Maya texts and images. These tend to be entities that were petitioned, placated, or even summoned in ritual performances. They are frequently imitated or embodied by Classic period kings, many of whom even took their names as their own. Many of these entities remain important protagonists in contemporary Maya cosmology.

As beings of supreme import and action, this group of supernaturals best corresponds to what is traditionally thought of as a "god" or "deity" in other societies. This includes, for example, the Sun God (and his variants), Chahk, God D, and, perhaps, God L. Some of these entities had their own courts filled with supernatural servants, including messengers that they could send to the world of the living. Moreover, the Sun God and Chahk (a deity of rain and storms) represent forces of nature that were fundamental to the survival of humanity.

Other anthropomorphic and zoomorphic supernaturals accord better with the Maya's animistic understanding of the world. Not only people but also places and things have a spiritual essence or, as Houston writes, a "life within."[32] This includes the Jester God, the Patron of Pax, and the spiritual embodiments of *witz* (mountains). But even this division between gods and animistic forces is problematic and forces a binary that does not accord well with the Maya worldview. For example, K'awiil was the embodiment of lightning and was wielded as an axe by Chahk, yet he was also one of three supernatural patrons at Palenque.

Other supernatural beings seem to have been the embodiment of the very forces of nature, both good and bad. These entities find a parallel in contemporary Maya conceptualizations of beings that lurk in the forest and along paths, move through the skies as celestial phenomena, or are manifest in the wind.[33] One such class of immoral animistic beings are the skeletal Death Gods, a broad range of vile named beings that include both *wahy* co-essences and other malicious denizens of the underworld.[34] In this sense they are not really gods at all but the embodiment of things that are bad or evil, especially disease. Even though the Maya Death Gods bear some iconographic and conceptual similarities to the Aztec death god, Mictlantecuhtli, they are not perfect cognates. Because of the very complicated and imprecise nature of Maya supernaturals, I find the term "Death God" a misleading reference to this class of Maya entities. Nevertheless, in accord with the current literature, I refer to these beings as such throughout this book.

Classic Maya godlike entities and named spirits appear as protagonists in an impressively rich and diverse corpus of Classic period myths, fragments of which are depicted on polychrome painted vases.[35] These entities are complemented in these depictions by other important named characters in Maya mythology, such as the Maize God, the Hero Twins, or the Paddler Gods. Some of these entities seem to have had little or no impact on human life but were instead mythic charters, metaphors for ideal ways of being, and characters that demonstrate a moral way of acting. In some sense they are functionally like the supernatural heroes, villains, and monsters of Greek mythology (Hercules, Perseus, the Gorgons). Yet the Maya also embodied these entities in some of their ritual performances and even in daily life, in what can be understood as mythology made real.

This is perhaps most evident in myths pertaining to the Maize God that, among other things, ordered some Classic Maya practices pertaining to treatment of the dead and the management of their souls. Souls were at peril of being trapped or harmed by nefarious underworld beings, but acts could be performed in emulation of the Maize God (and also the sun) that replicated their own triumph over the forces of immorality, death, and the underworld. By enacting these myths in mortuary rites, the Maya sought to ensure ascent of souls to ancestral mountains and other celestial paradises. For this and other reasons, maize deities and the sun become especially linked to Maya kings and the continuation of their dynasties.

Some of these tropes are evident on an impressive group of ceramic vessels known as the "Holmul dancer vases."[36] Although most of these vases are from unprovenanced contexts, one was fortuitously excavated by Jennifer Taschek and Joseph Ball at Buenavista del Cayo, Belize (fig. 1.5).[37] These vessels show a series of dancing Maize Gods that sport complex backracks consisting of a supernatural animal or creature covered by a sky-band

FIGURE 1.5. The Buenavista Vase showing paired Maize God dancers with the *wahy* of two royal lineages in their backracks (K4464 © Justin Kerr).

and seated atop personified *witz*. Monuments at Tikal, Dos Pilas, and La Corona show kings dressed in similar garb. What is especially notable about this corpus of vases is that specific Maize Gods and supernatural creatures are linked to particular royal dynasties or polities, including Kaan (Calakmul) and Mutal (Tikal). Some vases describe these maize beings as "ascending" to their polity name (Mutal, Kaan, etc.) and others describe them as "inside/ before the *ch'e'n* [cave]."[38] This may be in reference to the mountains that they wear on their backs.

The association of mountains or caves, specific supernaturals, and particular polities (or dynasties) parallels contemporary views of Maya souls and lineages. For example, in the Tzeltal village of Cancuc, each of the four lineages has its own mountain (*ch'iibal*) in which only the *ch'ulel* of each respective lineage dwell. In a similar fashion, the Tzeltal animal co-essences are also gathered together in a particular location. Among the Tzotzil of Chamula, the junior aspects of all of the town's *chanul* are corralled at night by the town's patron, St. Jerome, in the sacred mountain, Tzontevitz.[39] Within the corral they are safe from harm, especially from the dangers of witchcraft. The Tzotzil of neighboring Zinacantán also envision a cave near the summit of Tzontevitz as the entrance to the abode of their patron, San Juan. Vogt and Stuart see this as a contemporary parallel to the sky caves of the Classic Maya (see chapter 4).[40]

The distinctly local character of Maya patron saints in modern times and the plurality of Maize Gods of the Holmul vases remind us that supernaturals were not universal and singular but existed in many forms that reflected local understandings of these beings. For example, Chahk was widely recognized by the Maya as a being of storms and warfare and was consistently shown as an axe-wielding, anthropomorphic reptilian being. He first appears in the iconography of the Preclassic period and is still understood to be responsible for rain in some contemporary Maya communities.[41] Although we often speak of Chahk in the singular, he exists in many forms, much like the multiple Maize Gods of the Holmul style vase. In the Classic period multiple Chahks occasionally appear acting in concert with one another in the same image.[42] By the time of the conquest the Yucatec Maya identified different forms of Chahk with the cardinal directions.[43] In Classic period texts various forms of Chahk are described, based on their associations with meteorological events, such as Yax Ha'al Chahk (Green [First] Rain Chahk).[44] Some of these variants of Chahk relate to his adoption as a local patron by various royal lineages. For example, Stuart identifies Aj K'ahk O'Chahk as the probable patron of the royal throne at Yaxchilan. This form of Chahk is frequently shown as part of the costuming of the Yaxchilan lords.[45]

BODY AND SELF, SOULS AND OTHERS

According to Pitarch, the Tzeltal soul "is first and foremost an 'other.' The souls of the Tzeltal are made up of beings that are the antitheses of their native selves . . . souls represent, in their maximum expression, that which is alien." This concept is in almost direct opposition to Western concepts of the soul. For the Tzeltal and other Maya ethnic groups, individuals are not responsible for the shape of their souls. Rather, it is the body that is subject to human action; morality is defined by how people comport themselves within society. As Pitarch explains, "the body is thought of as belonging to the realm of the 'cultural,' of what human beings can and should morally do, and thus to be *fabricated through human intervention*, while the soul belongs to the realm of the 'sacred,' and *comes to this world as something already given*" (emphasis added).[46] Similarly, Gary Gossen observes that, because of this particular relationship between souls and the body, among the Tzotzil: "Individual fate and fortune are always to be understood and reflected on as phenomena that are *predestined* but also, secondarily, *subject to the agency and will of others*, both human and supernatural. . . . The healthy body carries this constellation of influences in equilibrium; it is the passive bearer of forces over which it really has no control" (emphasis in the original).[47]

The Maya soul is in effect the inscrutable and generally inaccessible other that lies within. Judgment of a person's character is thus not based on what that individual might be feeling or thinking but on what the person is doing and projecting to the external world. As Kevin Groark explains, the result is that the Maya contend with a world of social opacity where "accurate knowing of a person's inner states is extremely difficult—almost impossible—yet, at the same time, indispensable for navigating the social world. . . . The problem of accessing individual intentionality and feeling, paired with a widespread awareness of potentially occluded aspects of self and hidden potencies of others, creates a marked preoccupation with questions of surface and depth, inner and outer, and public and private."[48] For the Tzotzil, dreaming and the discussion of those dreams offer some understanding of these otherwise inaccessible co-essences, recalling again the etymological linkage between dreaming and co-essences (*wahy*).[49]

The inner soul is in effect managed through the head. For the Tzeltal, as Pitarch observes, "the inside of someone's heart is, as we have seen, his *talel*, 'that which comes as given,' in other words, the character bestowed upon him before birth and therefore formed a priori. On the other hand, reasoning and discernment are a function of the head."[50] As one Tzotzil informant succinctly explained to Groark, "first we think in our heart, second we think in our head."[51] Both the head and heart are engaged in emotion and memory, but the penetration of both is much deeper in the heart than it is in the head. Calixta Guiteras-Holmes elaborates: "[S]ometimes the mind is overpowered by the feeling in the heart; and although decisions are made in the mind, they are not followed if the heart refuses to cooperate."[52] Hearts are associated with the spiritual others that are at the essence of who a person is, yet the head is the outward aspect of the body and most signifies an individual's self, identity, and proper socialization.

The importance of the head as signifier of self and window into a person's inner nature is well exemplified in the fifteenth-century Achi Maya dance drama the *Rabinal Achi*, which recounts the final encounter between a K'iche' prisoner and his Achi captors. In his decision to meet the captive, the Achi lord proclaims: "Make haste, then bring him in here, sir, in my teeth, in my face. I have yet to look him in the teeth, yet to look him in the face to see how brave he may be, how manly."[53] The characters regularly use the expression "in my teeth, in my face" throughout the play to describe when people meet or take action against one another.

The linkage of self, identity, and the head is evident in the Classic period textual use of the glyph *baah* (head, face, body, self) and the *u-baah* expression (himself, herself, itself) (fig. 1.6).[54] As Houston, Stuart, and Taube observe for the Classic Maya, "the bridge between notions of 'body' and 'head' is crucial. It is the front surface or top of the head that facilitates individual recognition and receives reflexive acts. As the locus of identity, the face or head established individual difference and serves logically as the recipient of reflexive action."[55] The convention of the head as a unique signifier is fundamental to Maya writing systems. A significant portion of logographs were drawn as the heads of various entities. A range of subjects from humans, supernatural beings, animals, and even numbers were represented in text as portrait "head variants" (figs. 1.6c and 1.6d).[56] The Maya understood glyphs to have vitality, prompting artists of the Late Classic period to render them as full-bodied (for example, on stelae from Quirigua). The head-as-glyph is the logograph's body reduced to its unique signifier.

Because the head was the key to individuation, humans could also be depicted simply by their heads. Decapitation and the display of those severed heads, worn as trophies on the Bonampak murals, was a means of demonstrating permanent possession of a captive (fig. 1.7a). But isolated heads do not always have the same negative connotation as they do in Western convention. Maya kings regularly wore depictions of ancestral heads as belt ornaments, especially during dance and other ritual performances (fig. 1.7b). In a similar fashion, the Olmec represented apparent ancestors, presumably deceased rulers, as carved colossal heads (fig. 1.7c). Shortly before the conquest, Cocom lords (the rulers of Mayapan) reconstructed the heads of their ancestors from their skulls and placed them "in oratories of their houses with their idols, holding them in very great reverence and respect."[57] In all instances, such heads should not be understood as simply reproductions of the likeness of ancestors but instead as objects that embodied these beings: an ancestral head was not the representation of an ancestor, it was the ancestor.[58]

In Western convention we associate the chest with the person, as when we gesture or wear name tags at a public event. For the Maya, naming was associated with the head, the seat of identity. In painted and sculptural scenes the Maya invariably placed names next to a person's head, ideally in front of their gaze. Some depictions

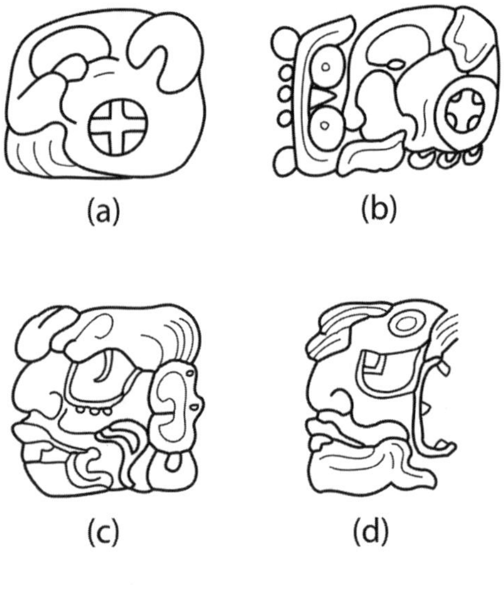

FIGURE 1.6. Heads in Classic Maya writing: (a) *baah* from Palenque Tablet of the Slaves; (b) *u-baah* from Yaxchilan Lintel 26; (c) the Palenque Supernatural Patron GI written as his head on the Tablet of the Foliated Cross; (d) the head of the Sun God as the number four on Palenque Tablet of the Cross (all drawings by author).

of Maya kings show them wearing their names as their headdress.[59] In contrast, names are rarely shown associated with other body parts. The important exception to this rule are captives, who usually do not have their names associated with their heads but instead are frequently depicted with the glyphs written across various body parts (fig. 1.8). This convention not only belittles the prisoners but accords with the dismemberment and disembodiment of captives (discussed later in this book).

FIGURE 1.7. Heads as embodiments of people: (a) worn as trophies at Bonampak (reconstruction painting of Bonampak, Mexico: Yale University Art Gallery, Gift of Bonampak Documentation Project, illustrated by Heather Hurst and Leonard Ashby); (b) as an ancestral belt ornament on La Pasadita Lintel 1 (drawing by Linda Schele, © David Schele, courtesy Foundation for the Advancement of Mesoamerican Studies, Inc.); (c) Olmec colossal head, Monument 1, La Venta (photograph by author).

Individuation by head is all the more evident in Classic period conventions for depicting supernatural beings. Each being (or class of being) is distinguished by aspects of its head that are distinct from other supernatural beings and from humans (fig. 1.9). Eyes are the most important distinction: crossed eyes for celestial beings and swirls for spirits of the underworld (compare fig. 1.9 to fig. 1.6d).[60] But other variables of the head were used to identify different supernaturals, such as the smoking celt in the forehead of K'awiil. From the neck down, most Classic period supernaturals are quite anthropomorphic, generally distinguishable from humans only by their skin markings. Only a few supernaturals, such as K'awiil and the Death Gods, have characteristics that easily distinguish them from humans. The Maya situation is in sharp contrast to the gods of ancient Greece, beings that are largely indistinguishable in body attributes from one another and from humans. Instead, they are sorted by their possessions; a trident for Poseidon, a shield

FIGURE 1.8. K'inich K'an Joy Chitam as captive on Monument 122 from Tonina. The Palenque lord's title is inscribed on his thigh (drawing by Linda Schele, © David Schele, courtesy Foundation for the Advancement of Mesoamerican Studies, Inc.).

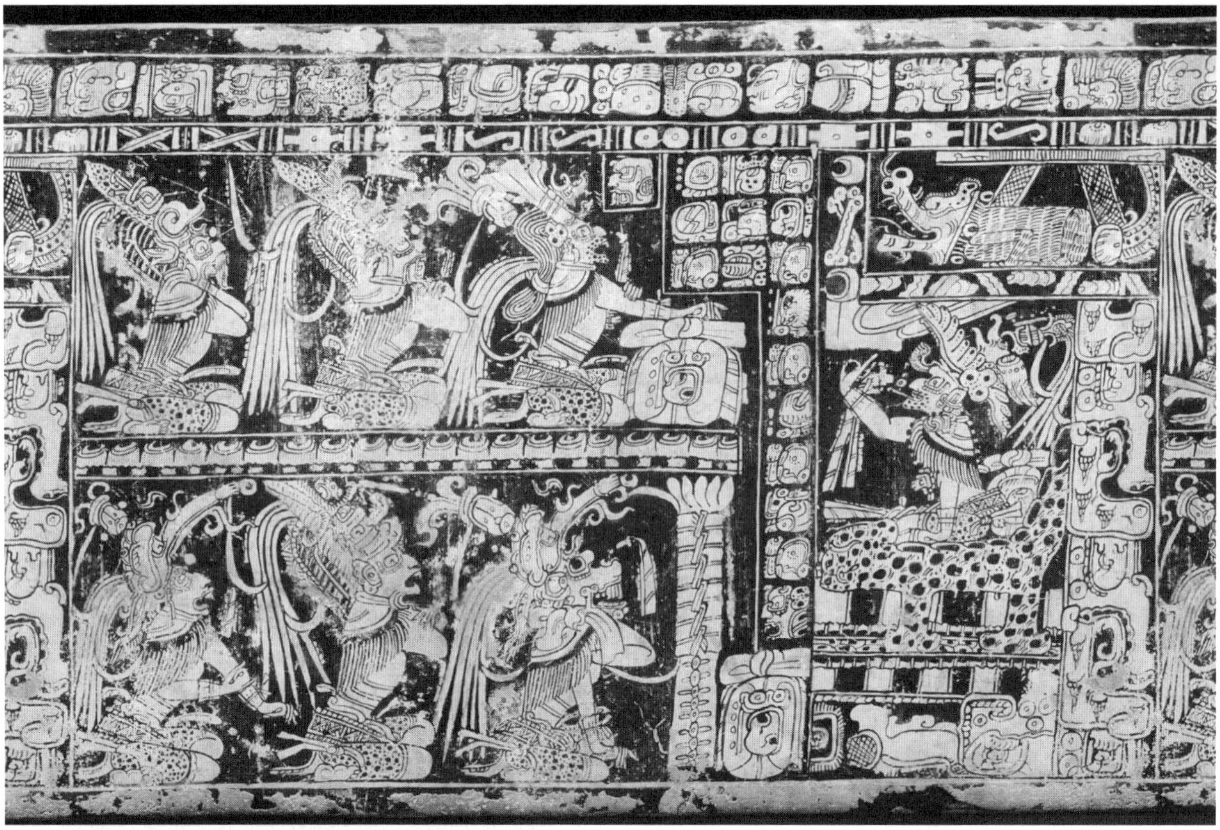

FIGURE 1.9. (*above*) Vase showing seven underworld supernaturals. These beings are nearly identical from the neck down but are distinguished by their heads and headdresses (K2796 © Justin Kerr).

FIGURE 1.10. (*left*) A K'iche' boy embodies the monkey during a dance performance at Momostenango, Guatemala, July 2008 (photograph by author).

and spear for Athena, a lightning bolt for Zeus, and so forth. The divine Greek identities are tethered to what they carry, what they can do, or what they can produce: things external to the body.

The importance of the head as signifier relates closely to the Mesoamerican tradition of masking. Ancient and contemporary Maya ritual practitioners who seek to become supernatural beings in their performances must adopt their visages; internal or otherworldly essences are manifested by covering the face with the likeness of that being. Today these masks are generally of wood, though in the past they may have been of other materials (fig. 1.10).

HEADS OF FRUIT, BODIES OF MAIZE

Heads and skulls are frequently conflated with cultivars by both the ancient and modern Maya, underscoring a fundamental metaphor that compares the human life cycle to that of maize, fruits, and other plants. For example, seeds are likened to skulls and other bones, and vice versa, as when the Tz'utujil refer to seeds as *jolooma* (little skulls) or *muk* (interred ones).[61] Conceptually, the next generation is born from the remains of the ancestors.

Among the contemporary Maya, childhood socialization is likened to the development of fruit. As Groark explains, Tzotzil "young children have 'unripe heads' (*unin sjol*), and don't feel ashamed when they act badly. As the child ages, his head 'matures' or 'ripens' (*-yijub*), reason and a knowledge of right and wrong enter (*-och rasonal*), and the ability to feel shame and embarrassment develops. . . . A good person, one who is thoroughly socialized, 'knows how to feel shame/embarrassment' (*ta sna' xk'exav*)."[62] A fully developed head is able to manage the desires of the heart and allows people to comport themselves properly in society. As Pitarch describes, the Tzeltal "head is not formed at the moment of birth, and children learn slowly because their brains (*chinam*) are still tender. Only with time, by means of learning and experience, is the brain gradually toughened up, filled out, and given shape, a process that is enabled by the working of the senses."[63]

In Classic period Maya iconography the Maize God's head was shown to embody both maize and cacao (fig. 1.11). In the sixteenth-century K'iché *Popol Vuh*, the severed head of Hun Hunahpu is mistaken for a calabash when it is placed in a tree and becomes lost among its fruit.[64] Later in the myth a squash is made to look like the head of the Hero Twin Hunahpu in order to trick the lords of the underworld. In the *Rabinal Achi* the K'iche captive initially confuses his grandfather's skull for a calabash drinking cup.[65] The metaphor of heads as gourds has an obvious logic. Gourds, especially calabashes, are

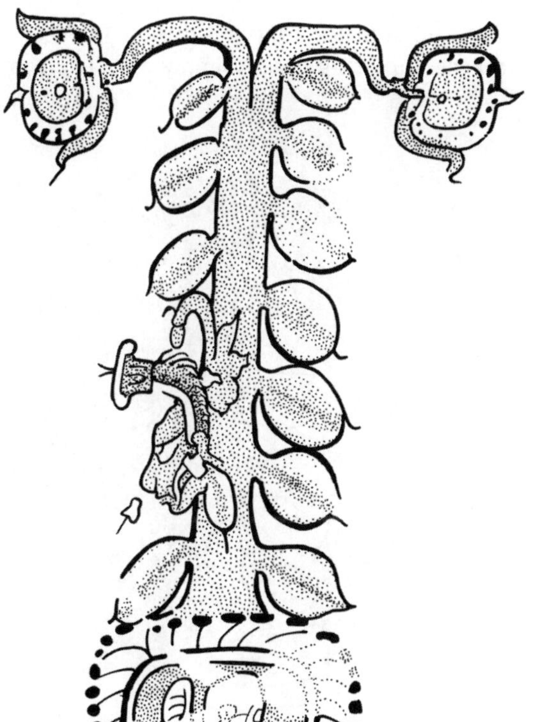

FIGURE 1.11. The head of the Maize God as a cacao pod from an unprovenanced vase (drawing by Simon Martin based on K5615).

similar in size, shape, and heft to a human head. Their smooth external surfaces conceal and protect a viscous mess of inner gore, venous stringy material, and tiny white elements (seeds/teeth). The reproductive potential of gourd-heads is suggested by the impregnation of the maiden by the gourd-head of One Hunahpu in the K'iche' *Popol Vuh* and is evident in the San Bartolo murals and the imagery at Teotihuacan, where beings are born from gourds.[66]

As their heads were likened to fruit, the Maya also perceived themselves as crafted of maize. Today the consumption of maize is understood to be essential for childhood development and for the maintenance of adult health. While conducting fieldwork along the Usumacinta River, one of my excavation assistants commented that he was worried about his brother's health because he lived in New York City and might not have had access to an adequate supply of maize tortillas.

FIGURE 1.12. Resurrection of the Maize God on a Classic period dish in the Museum of Fine Arts, Boston (drawing by Linda Schele, © David Schele, courtesy Foundation for the Advancement of Mesoamerican Studies, Inc.).

One of Allen Christenson's informants explained that mothers in the Tz'utujil village of Santiago Atitlán "give this water mixed with maize [from making tortillas] to their young children to drink so they will learn how to speak and learn the customs of their ancestors."[67] As Christenson notes, maize is incorporated into many rites of passage. Both the Classic and contemporary Maya speak of adolescents as *ch'ok* (unripe, a sprout).

Maize was personified in the myth of the Maize God in antiquity, a story of death, rebirth, and renewal.[68] For the Classic Maya, the Maize God was the embodiment of beauty, fertility, and youthfulness and was emulated by humans, especially kings. The Maize God is depicted as young and handsome with distinctly human features that stand in marked opposition to the imagery of most of the other major deities. Based on Preclassic and Classic period images on painted vessels and murals, his mythic cycle seems to involve death, dressing and rebirth in the underworld, and a resurrection that is likened to a maize plant erupting from the earth (fig. 1.12).[69] A later form of this myth is evident in the K'iche' *Popol Vuh*, where One Hunahpu (a figure who in some ways parallels the Classic period Maize God) is killed by the lords of the underworld. His twin sons undergo a series of trials, including their own death and return to life. Ultimately they triumph, restoring the face of their father and comforting his heart by ensuring that later peoples would call upon him and worship him. They then ascended into the sky as the sun and moon.[70] The K'iche' myth includes two important themes revisited throughout this book: death and renewal and the cycle of underworld descent and celestial ascent.

Today the most important foods made from maize are tortillas, tamales, and atole. Although no evidence for tortilla making in the Maya region prior to the conquest has been found, atole and tamales are well attested in Classic period imagery.[71] Atole is especially ubiquitous today, made in the home, and sold in markets. Classic period murals from Calakmul show atole being sold in an apparent market.[72] Also important is pozol, a beverage

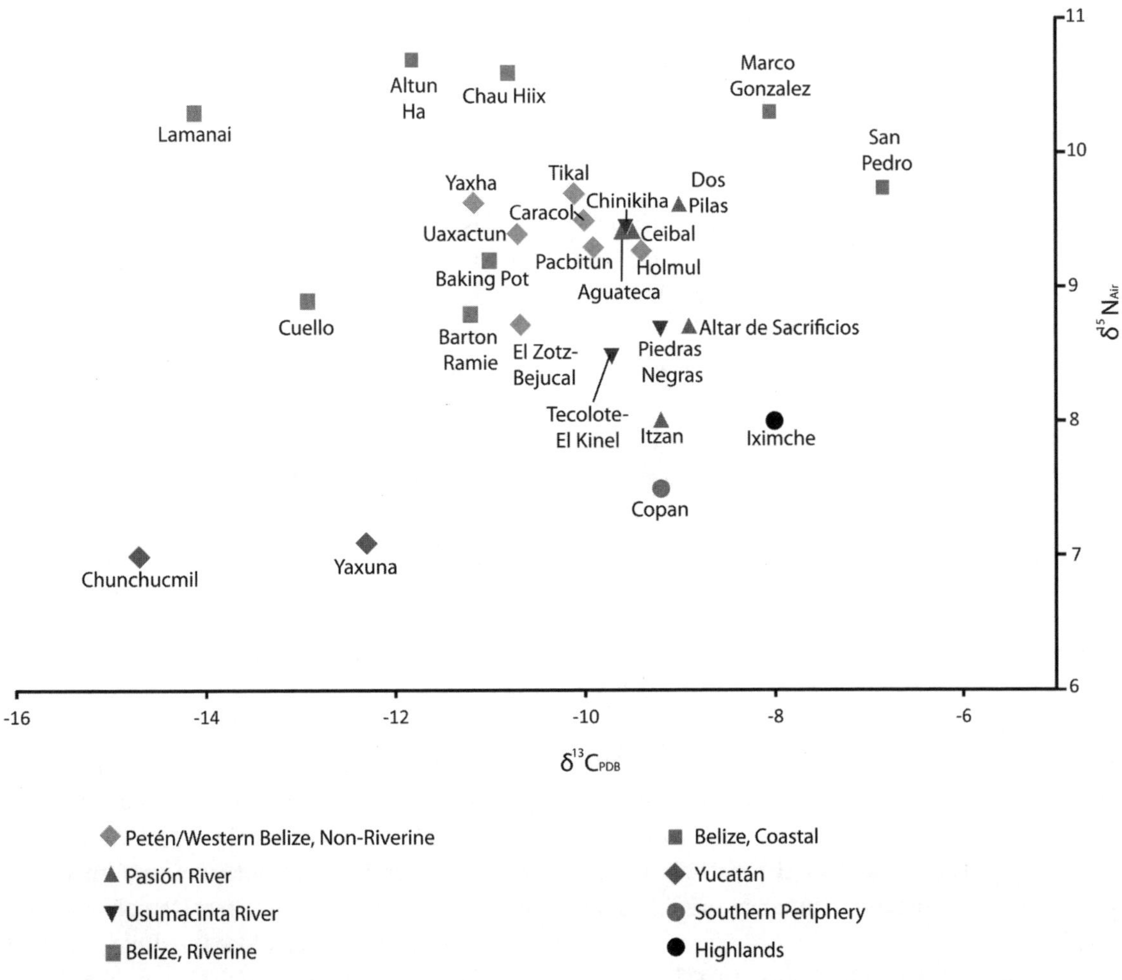

FIGURE 1.13. Mean stable carbon and nitrogen isotope ratios from across the Maya area (figure by author based on published data).

made by mixing lightly fermented maize dough with water. Both atole and pozol can be easily transported in plastic bottles, so it is the most common food consumed when people travel, especially to work in the fields (or on an archaeological excavation). Gourds could have substituted for plastic bottles, making atole and pozol the perfect ancient on-the-go comestibles.

Isotopic analysis of human bone provides a further line of evidence regarding the importance of maize for the Classic Maya. As maize consumption increases, the body's tissues generally are more heavily enriched in stable isotope ^{13}C, which is reported relative to the more common ^{12}C and in comparison to a reference standard as $\delta^{13}C$.[73] A summary perspective of mean isotopic data from across the Maya area reveals the primacy of maize in Classic period diets, but with some distinct geographic variability (fig. 1.13).[74] The Maya mean $\delta^{13}C$ range from –14.7 to –6.85‰. As a point of comparison, skeletons from a cemetery outside of Rome dating from around AD 0 to AD 200 demonstrate a mean $\delta^{13}C$ of –18.8‰, a value so low that it would not even appear on the Classic Maya isotopic plot.[75] In other words, the Maya belief that their bodies are crafted from maize is well attested in their bone chemistry.

It is notable that the inland sites of the southern lowlands tend to have the highest $\delta^{13}C$ ratios in the Maya sample, indicating greater levels of maize consumption. In contrast, data from sites of the Yucatán peninsula and sites along the various rivers in Belize indicate slightly less consumption of maize. Surprisingly, two coastal Belize sites show very high $\delta^{13}C$ ratios that the original study's authors suggest was due to these populations' dependence on marine resources, as sea grass and algae are also elevated in $\delta^{13}C$. Elevated $\delta^{13}C$ ratios in humans living by the coast may be due to consumption of such plants or more likely fish and other marine animals linked to that food web. The remarkably low isotopic values for the two Yucatan sites is enigmatic and points to less consumption of maize in the northern lowlands. This may reflect the greater difficulty of growing maize in this relatively dry region. Robert Redfield and Alfonso Villa Rojas report that one to four harvests were yielded each year in the Yucatec village of Chan Kom, though two harvests was the most common yield.[76] A farmer who lives near Mérida informed me that one maize crop per year was most common. In contrast, three harvests are typical in the communities where I work along the Usumacinta River because of the heavy rainfall in that region. This potential for greater maize yield in the wetter, southern inland areas accords well with the enriched carbon values noted for the Usumacinta and Pasión River sites.

Figure 1.13 also provides information regarding meat consumption among the Maya. As the amount of meat increases, so too does the ratio of ^{15}N relative to ^{14}N (measured relative to the standard air and reported as $\delta^{15}N$).[77] The increase in ^{15}N is due to a trophic effect. Eating the meat of carnivores or omnivores, especially those that have eaten other carnivores or omnivores, will result in greater $\delta^{15}N$ enrichment than eating the meat of herbivores. The Maya typically did not eat terrestrial carnivores. They did eat aquatic carnivores, however, such as fish. Hence the food web is generally more complex in riverine, lacustrine, and especially marine contexts and is reflected in the elevated nitrogen values of people who consume shellfish, fish, and other aquatic animals, as compared to those who eat primarily terrestrial animals. The Petén sites and some of the riverine sites cluster together, suggesting similar patterns of carnivory. Sites such as Tikal and Caracol may have been importing dried fish from riverine and lacustrine locales. This seems likely considering that their mean nitrogen value is comparable to that of Yaxha, a Petén site located on the shore of a lake. Other riverine sites, including Piedras Negras, Altar de Sacrificios, Tecolote, and El Kinel, have lower nitrogen ratios, indicating less access to meat or perhaps differences in the types of fish consumed. All four of these sites are located on the fast-moving Usumacinta River, where bottom feeders (lower trophic level) are easier to catch than carnivorous fish (higher trophic levels). The low values at Chunchucmil and Yaxuna suggest less access to meat resources overall, perhaps because of the lack of rivers and deep forest coverage in the Yucatán peninsula.

MAYA HEADS AND THE SHAPING OF SELF

In Precolumbian times the cultivation of the person was manifest in the Maya practice of cranial remodeling.[78] From Preclassic times forward, Maya infants' heads were reshaped by their mothers into forms that were retained throughout their lives. Osteologists describe the most common ancient Maya skull shape as the tabular oblique form, indicating the use of boards or other firm implements as the binding device that produced oblique elongation of the skull (fig. 1.14). Schellhas was among the first scholars to identify the elongate head as distinctive of the Maize God (his God E) and to suggest that this form replicated the shape of an ear of maize (see figs. 1.11 and 1.12).[79] The Maize God and Maya men and women are depicted with their hair bound in locks reminiscent of the silk that flows from a mature maize plant (fig. 1.14a). Some depictions show that the altered cranial shape was further accentuated by tonsuring parts of the scalp.

In the description of cranial remodeling in Diego de Landa's *Relación de las Cosas de Yucatan*, boards were placed on the front and back of the newborn infant's head, presumably held in place by cloth wrappings. This method would have resulted in frontal and posterior depressions that are characteristic of tabular forms of modification, including the tabular oblique shape favored

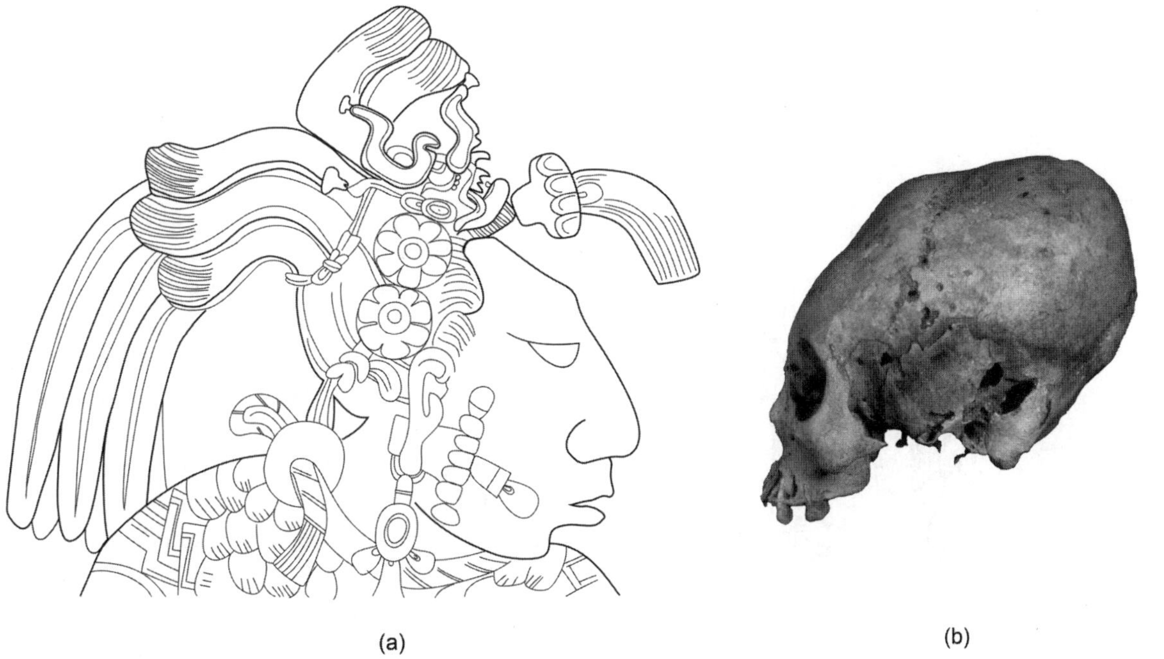

FIGURE 1.14. Tabular oblique cranial modification: (a) Itzamnaaj Bahlam III from Yaxchilan Lintel 26 (drawing by author); and (b) left lateral view of the cranium from Piedras Negras Burial 32 (photograph by Lori Wright).

in the western Maya lowlands (shown in fig. 1.14). The shape includes a long, concave forehead, accentuating the locus of identity and creating a large surface that could be adorned with a variety of ornaments. The shape may also have been useful when tumplines were used to carry goods. Less common among the Classic period Maya was the tabular erect form, a vertical head shape distinctive of Olmec iconography.[80] Other modified head shapes have been documented, though most pertain to subtle differences in the execution of the oblique and erect forms.[81]

A common misconception is that cranial modification was an elite-only practice.[82] I have tabulated the presence or absence of cranial modification for 55 individuals along the Usumacinta River: 47 skulls (85.5 percent) were modified. Although all of the elite crania were modified (8⁄8), most of the nonelite crania were also modified (39⁄47; 83.0 percent). There is a slight bias by sex, with more females (11⁄12; 91.7 percent) demonstrating cranial modification than males (8⁄12; 66.7 percent). This trend may have more to do with temporal variation, however: the female sample only includes one Early Classic skull, while the male sample includes four individuals.

Considering the marked change that a modified skull made to a person's appearance, some scholars have suggested that variability in head shape among the Maya may reflect ethnic differences.[83] Yet the data are ambiguous, in part because of the lack of consensus as to what constituted different ethnicities among the Classic Maya, but also because no supporting evidence indicates that the ancient Maya conceptualized group identity based on head shape. Head form is strongly uniform across the competing kingdoms of the western lowlands, including Palenque, Piedras Negras, and Yaxchilan. At Late Classic period Piedras Negras and its secondary centers, 18⁄23 (78.3 percent) of observable individuals exhibit tabular oblique modification.[84] Though data are still forthcoming from Yaxchilan, 3⁄3 (100 percent) of the observable skulls at its secondary center of El Kinel have tabular oblique deformation. Similarly, at Palenque 90 percent of the skeletal population demonstrates some variant of tabular oblique deformation.[85] Members of these royal courts distinguished themselves from other courts through royal regalia and aspects of material culture. But head shape was not distinctive for any single Maya kingdom.

FIGURE 1.15. Supernatural being crafting a human head or mask from an unprovenanced vessel (drawing by author after K8457).

Contemporary Maya belief holds that the inner self is inscrutable to outsiders and that morality is performed through bodily action. Parents are responsible for preparing the bodies of their children and teaching them how to comport themselves properly. For the Classic Maya, one of the first steps in this process was the reshaping of their children's heads. This notion of crafting Maya heads is shown on a number of polychrome vessels, where supernatural artisans shape what are ambiguously masks or human heads. Erik Boot places these particular images in mythological time, near the point of creation (fig. 1.15).[86] According to Landa's *Relación*, the Yucatec of the conquest era placed the binding devices on their infants four or five days after they were born.[87] All of the nine infants' skulls that I have examined from the western Maya lowlands demonstrate evidence of modification, confirming that the Classic Maya reshaped their children's skulls shortly after birth.

Although we cannot fully understand the context and significance of infant head binding, the act should be considered part of a broader tradition of head wrapping that still persists among the Maya. Ritual specialists and officeholders today wrap their head in cloth, and royal accession in the Classic period was marked by the binding of a paper headband around the head of the new king.[88] Stuart likens the act of wrapping the headband of rulership around the heads of Maya kings to the wrapping of stelae.[89] In this sense rulers were literally bound into the duty of kingship. F. Kent Reilly argues that a similar act of elite head wrapping was employed by the Olmec, as evidenced by the caps and chords worn on the carved colossal heads (fig. 1.7c).[90] The wrapping of heads can also be treated as part of a broader tradition of wrapping and bundling spiritually charged objects, including not only heads and stelae but bloodletters, sacrificial knives, royal regalia, and even corpses (as explored in chapter 2).[91] Bundling protects and contains the divine essence of sacred objects. Today special care is given to the bundling and wrapping of infants in caps, blankets, and shawls in order to protect them from dangerous animate forces and winds.[92]

A very early example of infant head wrapping is evident in the gourd birth scene in the San Bartolo murals (fig. 1.16).[93] Five infants are shown emerging from a gourd, an important vegetal symbol of reproduction. Only one of the infants, the central and apparently final infant to emerge, has a wrapped head. This same infant is also the only one to wear ornaments, including earflares and a beaded belt. Also, as William Saturno and colleagues deftly point out, the central infant is the only one both to spill blood and to exhale breath volutes, including a nose bead, a quintessential symbol of animate life. The other four infants all have umbilical cords, one spills blood (but has no breath), and another

FIGURE 1.16. Central infant with wrapped head from the gourd birth scene from the north wall of the murals at San Bartolo (drawing by Heather Hurst).

has breath volutes (but no blood). The meaning is murky. Are the other four infants incomplete or just not fully developed, as implied by their attached umbilici? Are they failed attempts at earlier birth-creations? Whatever the case, the central infant is the focus: its humanity or perhaps even royalty is demonstrated by the wrapping of its head and the adorning of its body.

Head shaping can also be understood as a rite of passage. According to Landa's *Relación*, the head was bound for "several days" in order to alter the shape of the head, after which the device was removed and the child was brought to an indigenous priest to receive a name.[94] The road to Maya personhood began once the head was given a new shape in accord with cultural norms. Like the development of Tzeltal and Tzotzil infantile heads discussed earlier, the heads of Classic period infants may have been shaped not only to achieve a desired aesthetic but to ensure that the intellect would grow and that the child would be properly socialized.

Landa describes other rites of passage that focus on the head. One is *caput sihil* (to be born anew), a childhood coming-of-age ceremony that he misidentifies as baptism.[95] The rite was performed when children were somewhere between the ages of three and twelve. A bead was attached to the hair on the top of boys' heads. Girls had a thin chord with a shell tied over their genital area. These objects were worn until the time of the ceremony, in which a white cloth was placed on the heads of the children who underwent the rite. The ceremony ended with the cloth being removed, the bead cut from the heads of the boys, and the chord and shell removed from the girls.

THE FINISHED SELF

The development of the body was a stepwise process, much as the Tzeltal acquire *ch'ulel* in three stages: during infancy, in adolescence, and at the transition to adulthood.[96] Binding of the head began the making of the Maya self. Filing the teeth was another important milestone, if not the completion of the process for many individuals. The ancient Maya modified their anterior teeth by cutting notches (primarily on the distal and incisal surfaces), by filing away the entire incisal surface, and by drilling holes on the labial surfaces and inserting stone encrustations (fig. 1.17). Only the anterior teeth were modified, partly because the modifications were meant to be seen but also because it would have been exceptionally difficult to work the posterior teeth. Modifications generally were limited to incisors and canines, though the first premolars were also worked on occasion. Both the maxillary (upper) and mandibular (lower) teeth were targeted, though in some instances the mandibular teeth were ignored or received more simple modifications relative to their maxillary counterparts.

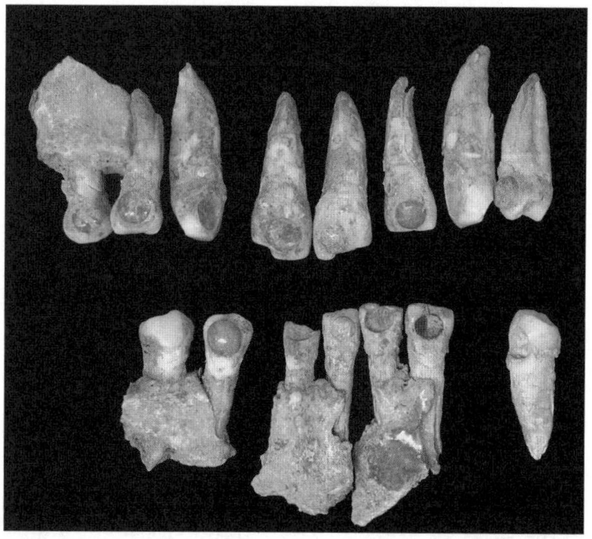

FIGURE 1.17. The modified teeth of the primary individual from Burial 5 of Piedras Negras, identified as the remains of Ruler 3 (photograph by author).

Today scholars record dental modification using Javier Romero's typology, which recognizes seven basic types (A–G), each with a series of numerical variants.[97]

Development and eruption of the permanent anterior teeth is complete by about the age of twelve or thirteen. The central maxillary incisors, the focal teeth for Maya dental modification, are in occlusion a few years earlier. Thus teeth could have been modified in late childhood or early adolescence (deciduous teeth were never modified). Yet, of all the skeletons I have studied from along the Usumacinta River, the youngest people to have modified teeth were aged osteologically between eighteen and twenty years old at the time of their death. No adolescents in the Usumacinta sample have modified teeth. Notably, a Piedras Negras juvenile buried with a stingray spine that identifies him as a *ch'ok k'in ajaw* (a young or unripe sun lord) did not have modified teeth. I established his age osteologically to be fifteen to sixteen at the time of death. In contrast, all adult royal skeletons from Piedras Negras have dental modification (see fig. 1.17).[98]

This age-related pattern of dental modification indicates that teeth were shaped at the transition to adulthood. It may be that teeth were modified at the completion of the first *k'atun*, an important milestone in the Maya life span that is roughly equivalent to twenty years. Undoubtedly, filing was conducted in a ritual context, perhaps accompanied by some combination of fasting, sweating in a bathhouse, and consumption of intoxicants. As Pamela Geller points out, dental filing is painful and requires stamina.[99] Endurance and the capacity to manage pain are fundamental to many indigenous American adolescent rites of passage, such as the Apache Sunrise Ceremony.[100] For the Classic period Maya, the first bloodletting was another painful rite of passage. A panel from Dos Pilas shows a young *ch'ok ajaw* engaged in what may be his first bloodletting ritual, sometime near the onset of puberty.[101] Dental filing probably was the final rite in a sequence of painful acts of transformation that marked the body's evolution

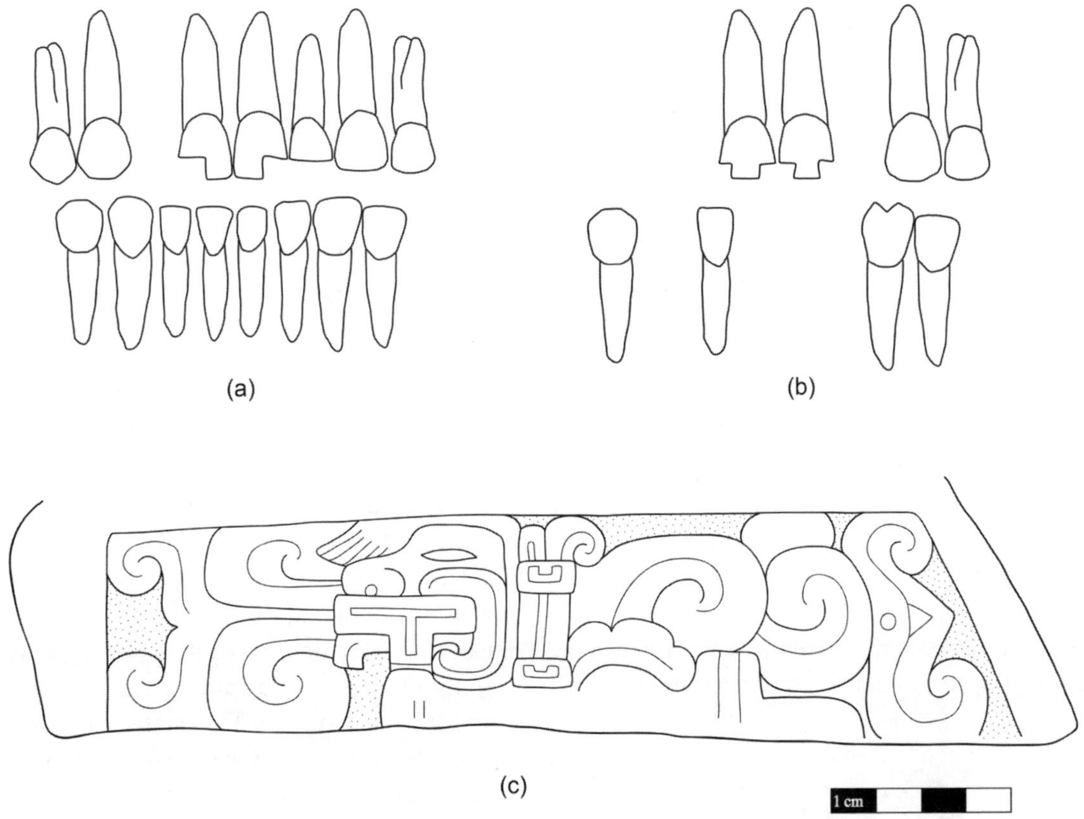

FIGURE 1.18. *Ik'* pattern of dental modification: (a) Piedras Negras Burial 92 (drawing by author); (b) Piedras Negras Burial 44 (drawing by author); (c) Kaminaljuyu Sculpture 14 showing wind supernatural with *ik'* emblem over the mouth: his "buckteeth" emerge from the left side of the emblem (drawing by Lucia Henderson).

from infancy to adulthood, a process that began with head shaping at birth.[102] Although similar acts of painful transformation are no longer practiced among the contemporary Maya, ritual payments to supernatural beings and other rites are performed by the Lacandon to mark a child's passage from infancy into adulthood and to celebrate incorporation into society.[103]

Analysis of tooth microwear and experimentation with modern teeth indicates that a combination of cutting with stone tools (chert or obsidian) and abrasion, presumably grit on a string, was used to reshape the teeth.[104] According to Landa's *Relación*, dental filing among the Yucatec Maya was conducted by older women, who worked with water and stone to file younger women's teeth.[105] It is unclear whether women were also responsible for filing teeth during the Classic period or whether the gendered nature of this role relates to the general decline in male dental modification noted in skeletons in the Yucatan during the Postclassic period.[106] Whatever the case, dental modification requires some skill, especially the drilling and insertion of inlays.

In Southeast Asia, where dental filing is still practiced, the anterior teeth can be reshaped in roughly fifteen minutes to half an hour with the aid of metal implements (files and machete-like blades).[107] Among Balinese Hindus, tooth filing is an important rite of passage into adulthood. As Clare Fischer observes, Balinese tooth filing is conducted under a mythic charter whereby the act of tooth filing makes a person more human: "the mouth is basically reshaped in an effort to reduce animality."[108] Moreover, Southeast Asian dental filing is also an act of beautification. As one Indonesian woman explained in a recent *National Geographic* documentary, "[N]ow that my teeth are sharp, I look more beautiful for my husband so he won't leave me."[109] The Maya likely considered filed teeth beautiful and, like cranial modification, a means to make the body more human.

Different combinations of filed teeth had specific meanings for the Maya. As Vera Tiesler suggests, it is more useful to understand filed teeth as part of a composite mouth rather than simply to quantify modification types by individual tooth.[110] In other words, simply to consider each modified tooth as an isolate is akin to studying Roman mosaics by tabulating tesserae by color or shape with no consideration of the overall designs. In the Usumacinta River skeletal sample, 64 adult individuals have sufficiently intact anterior dentition to observe for dental modification. Of those, 37 (57.8 percent) demonstrate dental modification. Of the observable maxillary dentitions with incisors ($n = 32$), the most common pattern in the Usumacinta sample involved the notching of the central incisors (Romero Type B4) so that the combined teeth formed a tau ($^{12}/_{32}$; 37.5 percent) (fig. 1.18a). Frans Blom was among the first scholars to recognize that these tau-shaped teeth correspond to the Maya sign for *ik'*, denoting wind or breath (fig. 1.18c).[111] Typically, the lateral incisors were filed down (Romero Type A4) to allow the central incisors to stand out (fig. 1.18a). In a Late Preclassic burial from Zancudero, Guatemala, and a Late Classic burial from Piedras Negras each tooth was individually transformed into a tau-shaped *ik'* sign (fig. 1.18b). By transforming the dentition into an *ik'* sign the Maya not only alluded to the breath that emanates from the mouth but more specifically denoted the body's vital essence. The cessation of the body's winds was understood to mark death, as evidenced by the Late Classic period death hieroglyphic expression "it is finished, his flower breath."[112] The concept of spiritual essences as embodied by both winds and human breath is widely held among the contemporary Maya.

FIGURE 1.19. Dental modification on woman from SE-S1 scene of the murals from Structure Sub 1-4, Calakmul (drawing by author after Martin, "Hieroglyphs from the Painted Pyramid," fig. 26).

Teeth are rarely seen in depictions of humans in Maya art, and modified teeth even less so. It may be that the concealment of teeth was part of a deeper convention in Maya art where emphasis was placed on the concealment of emotion and the inner self. Expressed emotion is generally negative in Maya art, reserved for drunkards and captives who lack control over their lust, fear, and other feelings.[113] Although teeth are occasionally shown in sculpture and some polychrome vase paintings, the recently reported Calakmul murals offer a rare example where many of the protagonists are shown with open mouths and exposed upper teeth (fig. 1.19). It may be that such depictions were meant to suggest the

FIGURE 1.20. The prominent and projecting anterior teeth of Classic Maya supernaturals: (a) Sun God stucco façade from El Diablo, Zotz (image courtesy of Stephen Houston); (b) unprovenanced GI mask, likely from Río Azul (drawing by author); (c) Chahk architectural ornament from Copan (photograph by author); (d) "Death God" stucco façade from the fifth terrace of the Tonina Acropolis (drawing by Linda Schele, © David Schele, courtesy Foundation for the Advancement of Mesoamerican Studies, Inc.).

liveliness or expressiveness of these scenes of quotidian life, contrasted with more somber and formalized stately affairs and ritual performances that otherwise dominate the bulk of Classic period art. The Calakmul murals are especially expressive: women blush, clothing is diaphanous, food is consumed, and overall we see a dimension of Maya life otherwise missing from the imagery of the Classic period.

In contrast to the hidden teeth of humans, supernatural beings invariably reveal their anterior teeth. Most Classic Maya supernaturals are shown with large, projecting central teeth (fig. 1.20). The Sun God is consistently depicted with a central tooth in the shape of the *ik'* sign (fig. 1.20a). The *ik'* symbolism indicates not only his vital essence but more generally that he is the source of healthy winds and essences vital for human life. This concept is well illustrated by the Late Preclassic supernatural being whose mouth is covered by an *ik'* sign as he expels a gush of wind (fig. 1.18c). Another solar being, GI from the Palenque triad, also has a single projecting tooth that is either a shark's tooth or a stingray spine, not only underscoring his aquatic nature but linking him to acts of bloodletting and perhaps even suggesting his own thirst for blood (fig. 1.20b). Chahk and K'awiil have particularly long and exaggerated central

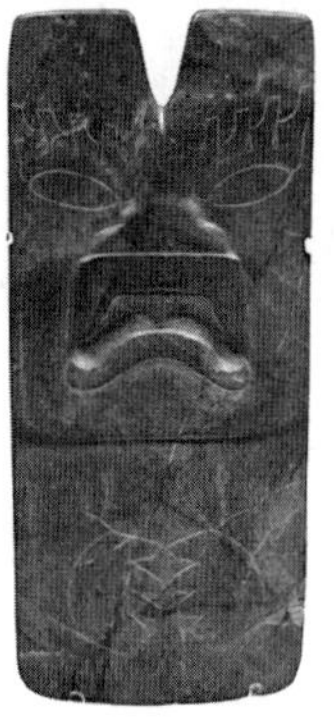

(a)

(b)

(a)

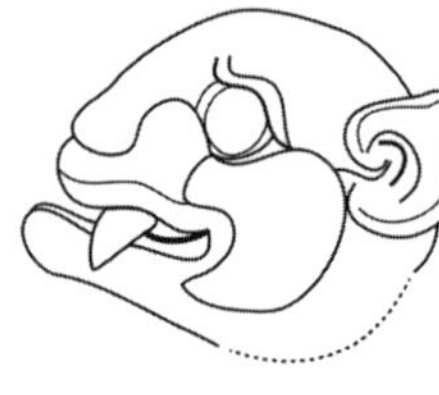

(b)

FIGURE 1.21. Preclassic center teeth: (a) unprovenanced Olmec axe-figurine depicting a maize or lightning being with a central fang (photograph by author); (b) the Maize God (center figure) with projecting buckteeth dancing inside of the earth-turtle from the west wall of the murals of San Bartolo. Also note the long central fang on this early depiction of Chahk (left figure) (illustration by Heather Hurst).

FIGURE 1.22. Central fang: (a) of the supernatural Jaguar God of the Underworld from the Tablet of the Sun at Palenque (drawing by Linda Schele, © David Schele, courtesy Foundation for the Advancement of Mesoamerican Studies, Inc.); (b) as compared to the normal canines of an earthly jaguar, shown here as a helmet on Yaxchilan Lintel 26 (drawing by author).

fangs, though the significance of this fang is unclear (fig. 1.20c). Their central fang may also be a reference to bloodletting. Alternatively, it may relate to the serpentine or reptilian character of these beings, perhaps even likening them to the quick-striking vipers of the Maya lowlands. Among the few supernatural beings without projecting central teeth are the Death Gods. With skulls for heads, they are consistently shown with unmodified human incisors of normal size (fig. 1.20d).

Looking back to the Preclassic period, we see that Olmec imagery is rife with beings with projecting fang-like incisors (fig. 1.21a). Similar incisor-fangs are evident in early depictions of the Maya Maize God (fig. 1.21b). In the Early Classic period the Maize God's central teeth are more typically shown as unshaped projecting incisors or "buckteeth," similar to those shown on the monument from Kaminaljuyu (see fig. 1.18c).[114] By the Late Classic period he is generally shown with a closed mouth, stoic and emotionless like kings and other humans, very much in contrast to the more expressive supernaturals (see figs. 1.11 and 1.12).[115] Even supernatural jaguars can be distinguished from "normal" jaguars by their teeth. The Jaguar God of the Underworld and his youthful variant, the Baby Jaguar, are depicted with either a prominent *ik'* tooth or a central fang, whereas other jaguars have prominent canines located in their proper position in the dental arcade, roughly one-third of the way back from the front of the mouth (fig. 1.22). Considering that incisors are the first teeth to appear in infancy and often the first lost with old age, the incisor fangs of most supernaturals may be meant specifically to call attention to their vitality. The most aged of all Maya supernaturals, God D or Itzamnaaj, is a toothless old being.

As in the case of the supernaturals, it is likely that the modification of human teeth was meant to convey or allude to a person's nature or inner essence. Thus *ik'* style modification was favored not because the Maya wanted to inscribe "wind or breath" inside their mouths (which would be rather banal) but instead because they sought to demonstrate that their essence was like the warm

FIGURE 1.23. Teotihuacan butterfly souls: (a) maxillary left incisor from Piedras Negras Burial 13 (photograph by author); (b) anterior maxillary teeth from Yaxha Burial 4 (photograph by author); (c) butterfly in Tepantitla murals of Teotihuacan (photograph by author); (d) Teotihuacan *incensario* with stylized butterfly nose ornament (photograph by author); (e) butterfly (black) conflated with a stylized nose ornament that is suspended in front of a floral element from a Teotihuacan ceramic fragment (drawing by author after von Winning, *La iconografía de Teotihuacan*, fig. X-3b).

and healthy winds of the sun, akin to that of the Sun God and other celestial beings. In this regard the Death Gods are shown to be fundamentally different from most supernaturals and humans. Not only do they lack dental modification, but their skulls are unshaped.[116] Perhaps this convention was merely intended to reinforce their deathly visage. More likely, however, the unmodified state of their teeth was meant to demonstrate an absence of healthy vitality: their essence was different from that of other supernaturals and was antithetical to humanity.

A variant of the *ik'* maxillary central incisors consists of angled notching on the distal edges of the maxillary central incisors (Romero's Type B5), as evident in an adult woman from Yaxha (fig. 1.23b). This pattern is quite rare among the Classic Maya but is reported with some regularity among the skeletal remains at the great central Mexican city of Teotihuacan.[117] Despite its infrequency among the Maya, it is notable that at least three individuals from royal tombs have this type of modification: Ruler 4 from Piedras Negras (fig. 1.23a), a skeleton within the tomb of Yax Nuun Ayiin (Burial 10) from Tikal, and the human remains attributed to K'inich Yax Kuk Mo' from Copan.[118] Epigraphy, iconography, and material culture indicate that the Tikal and Copan kings, who

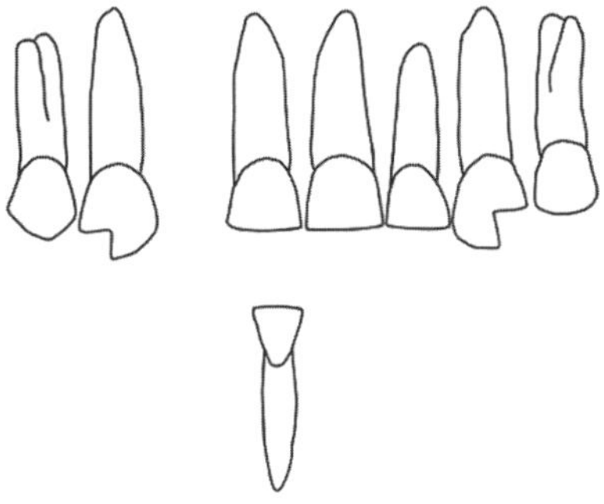

(a)

(b)

FIGURE 1.24. Dental modification emphasizing the canines: (a) Burial 55 from Piedras Negras (drawing by author); (b) central Mexican Storm God from the Tepantitla Murals of Teotihuacan (photograph by author).

reigned during the Early Classic period, had metaphoric if not actual connections to Teotihuacan.[119] The kings of Piedras Negras employed Teotihuacan imagery throughout the Classic period, such as the central Mexican fire serpent headdress that they wear on a number of stelae from the site.[120] At Teotihuacan, souls—particularly those of warriors—are represented as butterflies.[121] Butterfly-souls are depicted in the Tepantitla murals, and stylized butterflies are commonly shown as nose ornaments in imagery both from Teotihuacan and across Classic period Mesoamerica (fig. 1.23c). Such ornaments are especially common on ceramic anthropomorphic beings from *incensarios* that are commonly interpreted as representations of the Teotihuacan dead (fig. 1.23d). The connection between the nose ornaments and butterflies is securely established on a ceramic fragment that conflates the nose ornament and the butterfly (fig. 1.23e). Moreover, skeletons excavated from the Temple of the Feathered Serpent at Teotihuacan wore nose ornaments that are unambiguously butterflies.[122]

Hanging from the nose and in front of the mouth, the butterfly ornament illustrates how the concepts of breath and soul were conflated at Teotihuacan much as they were for the Maya. In that sense the Maya *ik'* sign and the central Mexican butterfly emblem can be understood as like-in-kind concepts. The triangular side-notched incisors from Yaxha, Piedras Negras, Tikal, Copan, and other sites seem to replicate the stylized version of the butterfly-soul, offering a Teotihuacan-influenced variant of the more common *ik'* style modification. We can only speculate on the meaning; such modifications may relate to the more general adoption of Teotihuacan imagery in the Maya area as a means to legitimize power and authority by establishing a connection to this distant place that achieved mythological significance in later times after its own demise at the end of the Early Classic period.

The focus thus far has been on variation of the central incisors. However, 11/34 (32.4 percent) of the individuals from the Usumacinta region sample have dentitions with notches on the distal aspect of the canines. Of these, nine are triangular wedges (Romero type B5) similar to the butterfly incisors just discussed. The other two are similar to the *ik'* style incisors (Romero type B4). Of these eleven individuals, eight have no apparent modification of the incisors. It is likely, however, that dental wear obscured the file work. In the case of an individual eighteen to twenty-one years old who demonstrated little dental wear (Burial 55 from Piedras Negras), it was quite clear that the height of the incisors had been reduced by filing (Romero Type A4) as a means to accentuate the notched canines (fig. 1.24). It is possible that the notching of the canines was intended to create one large *ik'* or butterfly sign. Alternatively, this modification may have been meant to replicate the prominent canines that are widely associated with mammals in Maya iconography, especially jaguars (see fig. 1.22b). Such teeth are also typical of the central Mexican Storm God and other supernaturals from that region (fig. 1.24b).

Contrary to popular conception, dental modification (including stone inlays) was not limited to Maya elites. In the Usumacinta sample, 5/8 (62.5 percent) elites and 32/56 (57.1 percent) of nonelites exhibit some form of tooth modification. The only real exceptions are the jade encrustations that seem largely limited to the Maya elite.[123] In addition to jade, the Maya encrusted their teeth with pyrite, hematite, and other greenstones. In the Usumacinta River sample 8/34 (23.5 percent) of the individuals with modified teeth demonstrate inlays, all from Piedras Negras. These stone encrustations must have been a sign of both beauty and prestige for the Maya. These concepts are illustrated in the story of Seven Macaw of the K'iche' *Popol Vuh*, which provides a mythic admonition against vanity and pride. Seven Macaw proclaimed that "my teeth, along with my eyes, are my finery." However, when "all of the blue/green jewels in his mouth were removed" by two elderly supernatural beings, "no longer did he appear as a lord."[124]

It is important to remember that dental modification, unlike cranial shaping, was enacted at an age where at least the potential for personal choice and self-expression was present. This is evident in the occasional idiosyncrasies of Maya dental filing and encrustation, such as the case of a young nobleman from Piedras Negras (Burial 45) who lived and died in the Early Classic period (fig. 1.25). Perhaps to enhance the unusualness of his supernumerary teeth, five inlays were set into one of his central incisors. The rest of his teeth were left untouched, creating what was undoubtedly one of the most unique smiles in the Maya world.

At its source in the Motagua River valley, jade is not particularly rare. Its value derives from the labor of transporting and working this extremely hard stone. Furthermore, the drill work necessary to inlay the teeth would have required a degree of skill, particularly the delicate work necessary to set five inlays in a single tooth, as shown in figure 1.25. It is possible that the craftspeople responsible for placing the inlays were lapidarists, which may in part explain the lack of inlay work outside of Piedras Negras in the Usumacinta sample. Stone-tipped drills were used to drill the holes for inlays, and the inlays were likely set with resin and other cements.[125]

Aside from its value as an object of prestige, jade (and blue-green stones in general) was a durable symbol of fertility and vitality. As Taube shows, jade is closely affiliated with wind and breath, so jade-encrusted teeth reinforced the same concept of vitality as *ik'*-modified teeth (see fig. 1.17).[126] As noted earlier, teeth are even conflated with seeds in modern Maya thought. This is also evident in the *Popol Vuh* when Seven Macaw's jeweled teeth are extracted and replaced by maize grains.[127]

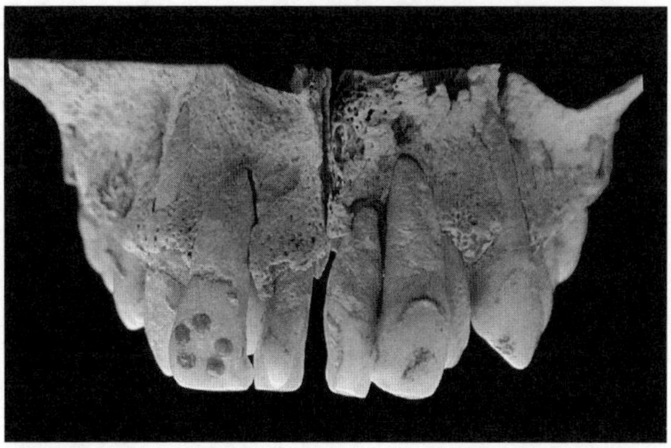

FIGURE 1.25. Jade inlays and supernumerary incisors of a young adult male from Piedras Negras Burial 45 (photograph by author).

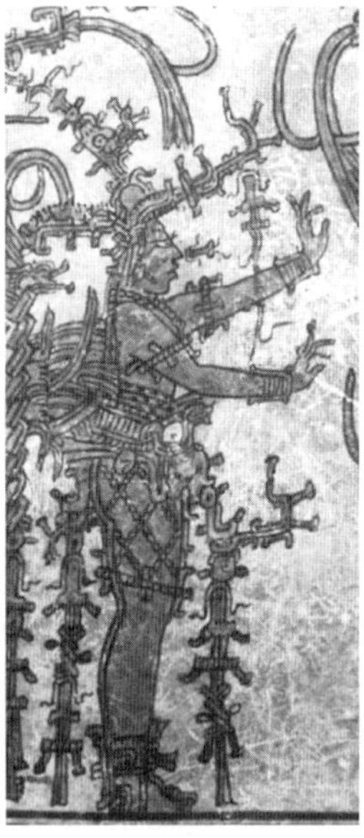

(a)

(b)

(c)

(d)

SEX AND GENDER

The analysis of burials and the bodies they contain provides one of the tantalizing yet problematic lines of evidence for the study of sex and gender in antiquity. Social constructivists have argued that gender is constructed, performed, and reified through daily practice, regardless of biological sex. This position has gained some traction among archaeologists in recent years.[128] For archaeologists, the concern is primarily how material culture can be used to identify gender (usually male and female) and understand the actions and meanings associated with those genders. Social constructivists and nonconstructivists alike have a dangerous tendency to see objects as gendered. In the case of the Maya, stingray spines have frequently been perceived as male and skirts as female, resulting in many problematic interpretations of the material record. In one recent example, an ancient Maya female skeleton buried with a stingray spine was suggested to be an individual "who engaged in gender-bending ritual performances while alive."[129] This assertion is based entirely on the assumption that bloodletting by stingray spine was a male prerogative, despite abundant evidence that Classic Maya women also pierced themselves with stingray spines (for example, Yaxchilan Lintel 25). Moreover, people possess objects that belong to other genders for a variety of reasons that say nothing about their own genders. Medieval knights carried bits of women's clothing with them (scarves, veils, sleeves) not because they wanted to challenge the boundaries of gender in the Middle Ages but because they were given them as favors by admiring ladies.

FIGURE 1.26. (*opposite*) Skirts and dresses of the Maya: (a) the Maize God's jade skirt from the Buenavista vase (K4464 © Justin Kerr); (b) a jade skirt from a stucco figure from the Temple of the Inscriptions (drawings by Linda Schele, © David Schele, courtesy Foundation for the Advancement of Mesoamerican Studies, Inc.); (c) full-bodied netted jade huipil worn by Lady Sak K'uk' and a loincloth-skirt worn by Pakal on the Palenque Oval Tablet (drawings by Linda Schele, © David Schele, courtesy Foundation for the Advancement of Mesoamerican Studies, Inc.); (d) Copan Stela H (photograph by author).

Osteologists can reconstruct sex with a reasonable degree of accuracy, based on morphological differences of the os coxae (pelvis) and cranium. Identification of sex in the Maya area is significantly hampered by poor skeletal preservation. Readers should be dubious of skeletal reports from the region in which the majority of burials have an identified sex. For example, of the 149 skeletal individuals at Piedras Negras, I was able definitively to identify sex for only thirty individuals (20.1 percent). Part of the problem is that osteologists cannot reliably determine sex for anyone less than about eighteen years of age. The other issue is that the most reliable bone for sex reconstruction, the pubis, is rarely well preserved.

Although the determination of sex from the skeleton is relatively straightforward, biological sex is not the same as socially constructed gender.[130] Gender differences among the contemporary Maya are well codified along biological lines and involve proscriptions regarding clothing, work, comportment in public and domestic space, and modes of speech.[131] Although some scholars see a degree of gender fluidity among the Classic period Maya, the perfect concordance between visual manifestations of gender and epigraphic references to male and female would suggest otherwise.[132]

Much of the confusion rests on the perceived femininity of the Maize God, a being frequently impersonated by Maya kings. The Maize God is often depicted wearing the netted jade skirt that some scholars perceive as feminine (fig. 1.26a). The skirt usually hangs mid-thigh, which is typical of male dress. On women, however, the shirt generally reaches the ankle. In one unusual case, the Copan king Waxaklajuun Ubaah K'awiil (18 Rabbit) wears an ankle-length netted skirt on Copan Stela H (fig. 1.26d). This depiction in particular has been cited as evidence of the feminine quality of the Maize God and his impersonators. Yet there really is no evidence to suggest that wearing skirts, or even skirt length, was a

FIGURE 1.27. Pakal (right figure) as a deceased old man depicted with youthful vigor on an unprovenanced panel from the Palenque area, Dumbarton Oaks (drawing by Alexandre Tokovinine).

gender signal among the ancient Maya. Rather, a better signal of gender in depictions of the human figure seems to be the overall extent of bodily coverage, especially in depictions of Maya kings and queens. In a sense Classic Maya norms of bodily coverage were reversed from those of our own society. Maya men were allowed if not expected to show skin, perhaps in part because it was more comfortable in the hot and humid environment but also because Classic Maya society valued a youthful, vigorous body. Take, for example, the figure depicted on Pier C of the Palenque Temple of the Inscriptions. Some scholars have assumed that the ankle-length skirt indicates this person was female. However, the lack of a top signals this person is a male as all other topless figures at Palenque are male; compare the attire of Lady Sak K'uk' to that of her son K'inich Janaab Pakal I (1.26c).

The actual significance of the beaded skirt is poorly understood. It may have been meant to imply the turtle shell (a metaphor for the earth) from which the Maize God rises in his resurrection (see fig. 1.12). Alternatively, the beaded skirt may be suggestive of the nets that the Maya used to harvest maize, as referenced in the K'iche' *Popol Vuh*.[133] Even today the Maya bring their produce to market in nets. Landa's *Relación* also notes that adornments and other things of value were carried in "purses of net."[134] Whatever the specific meaning, the skirts are simply the appropriate dance costume worn by anyone performing in emulation of the quintessential supernatural dancer, the Maize God. Although no such beaded skirt has been found archaeologically, a netted jade collar was recovered at Calakmul.[135] Most other netted skirts were quite likely inherited and not interred with the dead.

Unfortunately, the cotton garments of the Maya are not preserved in the archaeological record. This establishes a particular problem for the study of gender in Maya mortuary contexts. One of the few key lines of evidence for gendering—the extent of clothing worn—is largely undetectable in archaeological contexts. Other objects that may seem to have gendered significance (such as weapons or weaving tools) may enter the mortuary record for other reasons unrelated to gender or even identity. Rather, as this book demonstrates, funerary objects often reflect beliefs and practices pertaining to the management of the dead and their souls and may say little about personal identity.

AGING BODIES

Ancient skeletons provide access to a wealth of information regarding different age grades within Maya society. As I like to tell my students, the story of the skeleton for the first two decades of life is one of growth and development; thereafter it is a history of steady bodily degradation. Skeletal growth, development, and, to a lesser degree, degradation happen at a roughly comparable rate and sequence for all people, which is the key to the determination of age at death. In a typical skeletal sample from a preindustrial society, children and adolescents should constitute approximately one-third of the sample. In that regard the skeletal sample at Piedras Negras is quite typical, with 30.9 percent of the sample being children or adolescents. Life is particularly dangerous in the first three years, when the immune system is still developing and children are confronted with a host of microscopic and macroscopic hazards.

Epigraphic and iconographic evidence from the Classic period suggests that the Maya were deeply concerned with the human life cycle and especially prized youthfulness and vitality. One of the most elderly of kings, K'inich Janaab Pakal of Palenque, died at the ripe age of eighty. Posthumous depictions of the king show him as vigorous and youthful, even in chronologically impossible scenes where a long dead Pakal meets with his adult grandchildren (fig. 1.27).[136] When the Maya did choose to depict old age, they focused on key aspects of body deterioration: tooth loss, wrinkled skin, and a stooped back (fig. 1.28). The Classic Maya worldview reveals some ambivalence about old age. Among the most prominent old-aged supernatural beings are God L, God D (Itzamnaaj), and Goddess O. God L is an underworld being of ill-repute and is in thematic opposition to the youthful Maize God. In contrast, God D, a creator being, and Goddess O, the spiritual embodiment of postmenopausal women, both exude power and seem to have commanded respect.

The lives of Classic Maya children are remarkably difficult to access. Children rarely appear in Classic Maya art and when they do are often shown as diminutive adults. They are, however, unfortunately common in the mortuary record (fig. 1.29). Notably, children's graves are found among those of adults and in many respects mirror adult burials in general funerary treatment, though with fewer grave goods. Although emotion is difficult to wrest from the archaeological record, the

FIGURE 1.28. The "Old Man" sculpture from Copan (photograph by author).

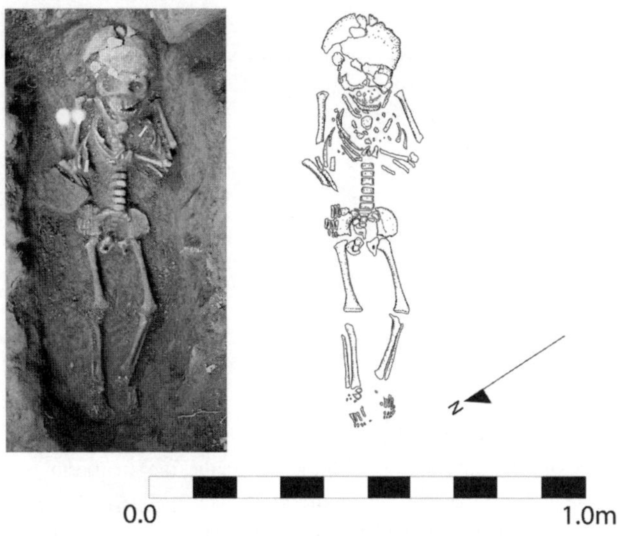

FIGURE 1.29. Photograph and drawing of a two- to four-year-old child from El Kinel (Burial 4) (photograph and drawing by author).

general equality of treatment of children and adult graves seems to suggest a comparable sense of loss and respect, suggesting that a child's death was equally troubling as that of someone who lived a full life. The preciousness of children likely also explains their presence in more grisly contexts, as sacrificial offerings, especially in and around Early Classic period tombs (see the discussion of child sacrifices in chapter 3).

From the perspective of the skeleton, the Maya aged remarkably well relative to other ancient populations. They suffered from few severe infectious diseases, at least not those that leave their mark on bone; nor was their dental health particularly bad.[137] For example, when Lori Wright and I compared mean dental wear among 297 Maya individuals to a global sample of hunter-gatherers and agriculturalists using data and methods developed by Holly Smith, the Maya mean was one standard deviation below Smith's comparative samples (fig. 1.30).[138] The relative lack of dental wear among the Maya was due to the soft foods that dominated their diet, which included atole and tamales, both made from processed maize. Wright and I did find that dental wear was slightly higher at the highland site of Kaminaljuyu, which we attribute to the use of basalt and granite grinding stones in the highlands versus softer limestone grinding stones in the lowlands. Both grinding stones gradually erode with use, but limestone tends to pulverize, whereas the mineral clasts in igneous rock will retain their crystalline structure and enter the ground-maize as grit.

Dental caries were a greater concern for the Maya. More commonly known as cavities, these pits in the enamel surface form as a result of the bacterial fermentation of simple carbohydrates (fig. 1.31). Though lacking in sugar, the Classic Maya diet included maize, honey, and other sticky carbohydrates that can promote caries formation when lodged in the mouth. In general there seems to be a correlation between frequency of dental caries and levels of maize consumption in the Maya area, as indicated by carbon stable isotope analysis, though many other factors can also contribute to caries

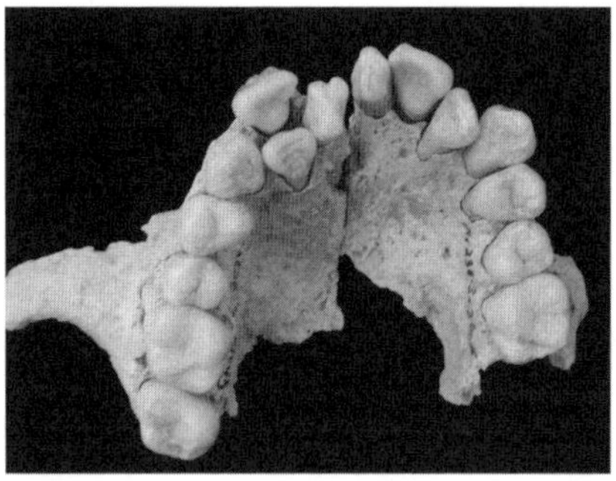

(a)

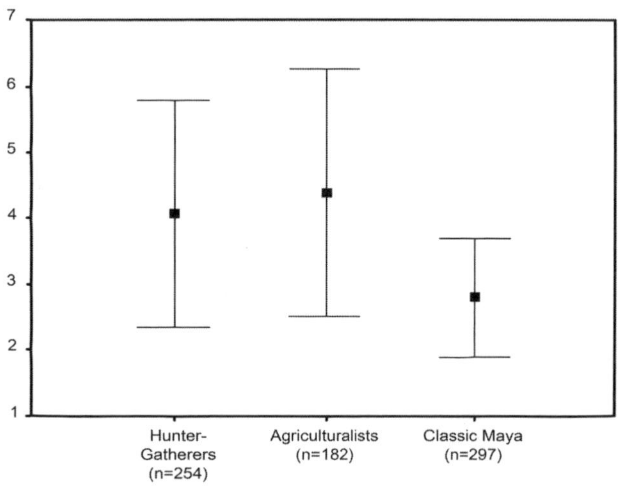

(b)

FIGURE 1.30. Maya dental wear: (a) occlusal view of Piedras Negras Burial 45 showing minimal wear: note the supernumerary incisors (photograph by author; see also fig. 1.25); and (b) mean and first standard deviation of Classic Maya dental wear compared to B. H. Smith's ("Patterns of Molar Wear in Hunter-Gathers and Agriculturalists") global data on hunter-gatherers and agriculturalists (figure by author).

prevalence.[139] If significant enough, dental caries can expose the pulp chamber, causing significant pain and possibly infection. Both will lead to tooth loss, either when the tooth is mechanically pulled because the pain has become too intense or when the bone that holds the tooth in place is destroyed by abscess.

At Piedras Negras 294 tooth sockets from 32 adult individuals were observable. Of those, only 17 tooth sockets (5.8 percent) exhibit antemortem tooth loss. This figure is low relative to other samples from around the world and may partly relate to the lack of dental wear at Piedras Negras and other Maya communities.[140] The relatively good dental health of the ancient Maya may reflect a general concern with bad odors of the body. Tooth decay, periodontal disease, and calculus (calcified plaque) all lead to bad breath (halitosis) (fig. 1.32). Mesoamerican worldview sees a strong dichotomy between the flowery breath that represents the essences and speech of nobility as compared to the foul breath of deathly underworld beings.[141] How the Maya inhibited halitosis, removed calculus, and generally combated oral pathology is not well known. According to Bernardino de Sahagún, the Mexica had a number of solutions, including a mix of *tlatlauhcaptatli* root (possibly geranium), salt, and chili, followed by scraping of the teeth.[142] Sahagún also indicates that chicle (gum) was chewed to produce saliva and keep the teeth clean.[143] Chicle is gathered from the trees of the genus *Manilkara* (sapodilla or *zapote*), which are indigenous to the forests of the Maya lowlands. Modern studies have shown that sugarless gum reduces the risk of tooth decay, so chewing chicle may have been one of a variety of means by which the Maya minimized tooth decay.[144]

The skeleton of an adult male from Piedras Negras Burial 119 has the most pathological mouth that I have encountered in the Maya area, including a combination of dental caries, abscesses, and antemortem tooth loss (fig. 1.31). The most dramatic feature of the mouth, however, was the malformation of the maxillae. The floor of the nasal aperture is uneven and the maxillary

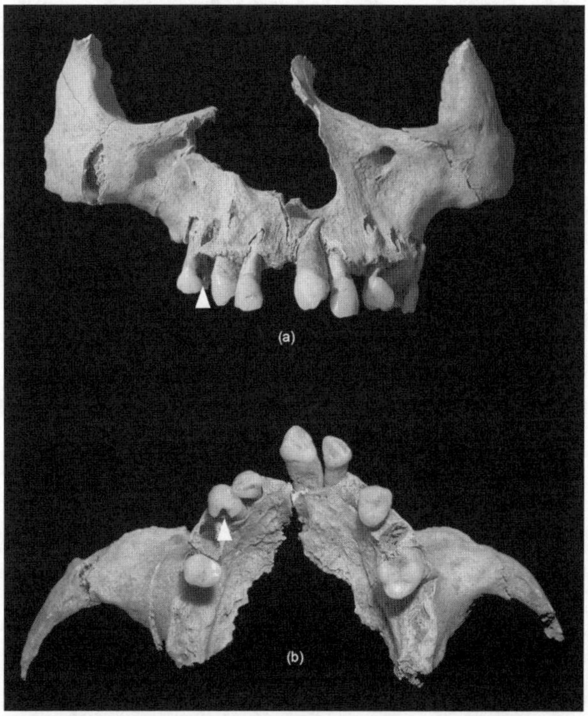

FIGURE 1.31. Dental caries, periodontal disease, and antemortem tooth loss in Piedras Negras Burial 119: (a) anterior view; and (b) occlusal view (photographs by author). The arrows indicate dental caries. All missing teeth were lost antemortem and the bone (alveolus) was remodeled. Abscesses are visible as perforations at the root. The left canine is located in the position of the left central incisor.

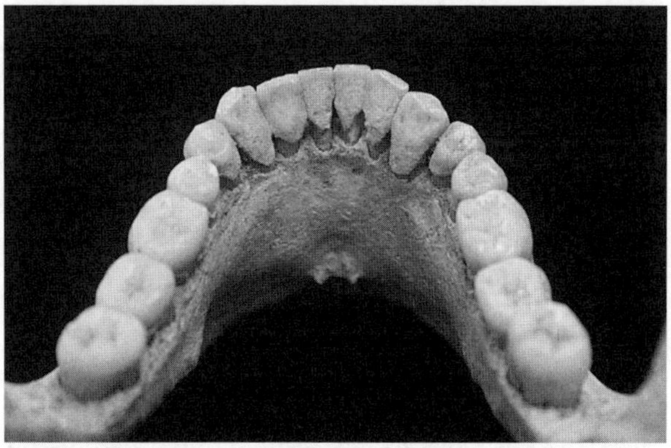

FIGURE 1.32. Dental calculus on the lingual aspect of the incisors and canines of Yaxha Burial 14 (photograph by author).

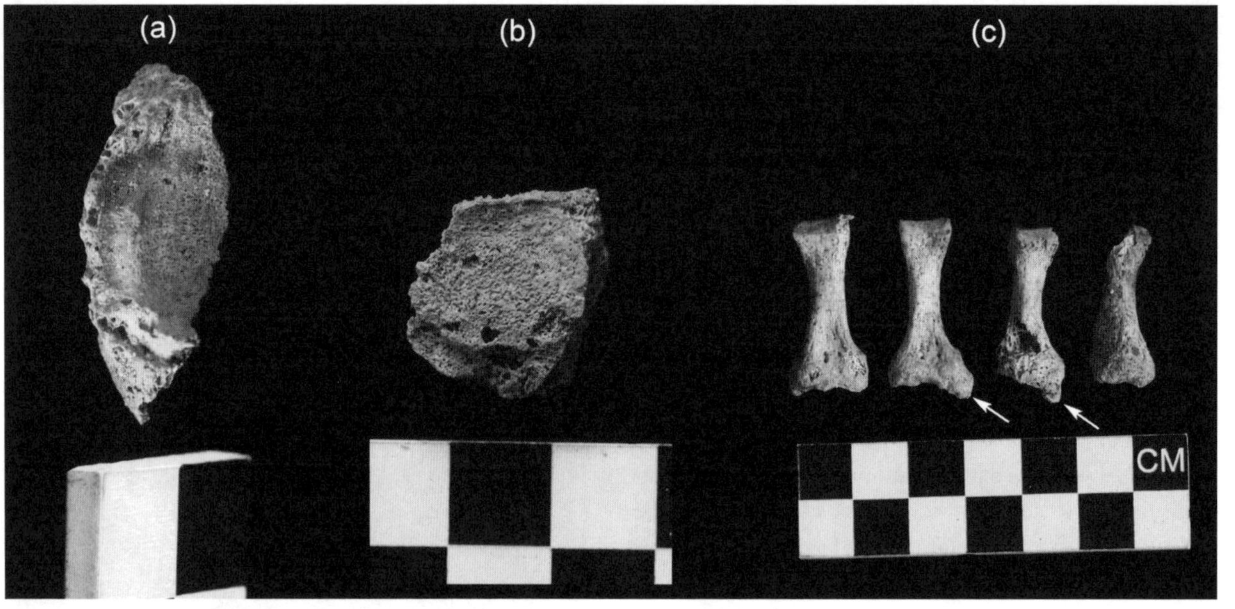

FIGURE 1.33. Osteoarthritis of an Early Classic King (El Zotz Burial 9): (a) right glenoid fossa (shoulder); (b) superior view of upper thoracic vertebra; (c) plantar view of proximal foot phalanges, with osteophytes indicated by arrows (photographs by author).

canine is misplaced and would have protruded from the center of the mouth. This person was found in a modest burial near the foot of the K-5 pyramid where a number of other simple burials had been placed.[145] It is impossible to understand how this individual's unusual appearance would have been perceived by the Maya or why he developed such terrible dental pathology in a society that seems otherwise concerned with oral hygiene, but whatever the case, it is striking that he possessed a canine that protruded from his mouth, not unlike the incisor-fangs of Maya supernaturals.

Osteoarthritis—the degeneration of the articular surfaces of the bones—is one of the most telltale signs of aging and is directly related to the overall lifestyle of the individual. This inescapable wear and tear of the body affected Maya elites and commoners alike. An Early Classic king from Piedras Negras (Burial 110) exhibits arthritis throughout his vertebral column, and a recently excavated Early Classic lord from El Zotz, Guatemala, demonstrates bony spurs (osteophytes) and porosity on nearly all of his observable joint surfaces (fig. 1.33). Jane Buikstra noted that the skeleton attributed to the founder of the Copan dynasty also exhibits osteoarthritis throughout much of the skeleton, some of which can be linked to a series of antemortem fractures that he sustained many years before his death.[146] Tiesler observes that K'inich Janaab Pakal of Palenque is similarly affected by osteoarthritis on much of his skeleton.[147] Although Maya kings undoubtedly did not toil in the fields or spend hours carrying and placing stone blocks in construction efforts, it is perhaps incorrect to assume that they entirely escaped physical activity and hardship. The antemortem trauma of the Copan skeleton suggests that Maya lords went to war and fought in battles, among other things.

Advanced age is also associated with loss in bone mineral density: osteopenia and osteoporosis. An elderly woman from Piedras Negras suffered from an overall loss of trabeculae and a reduction in her cortical bone thickness, as evident in the broken cross section of one of her femora (fig. 1.34).[148] Similarly, Tiesler notes that Pakal also suffered from bone density loss as a result of his old age. What is intriguing about the elderly woman from Piedras Negras is that she was buried with a ceramic vessel decorated with eyeballs and crossbones. Although such symbols are generally tied to underworld beings like the Death Gods, they are also linked to the elderly Goddess O, a supernatural being connected with midwifery who is often depicted with crossbones and eyeballs on her skirt (fig. 1.35).[149] The selection of this particular vessel is unlikely to be coincidental and suggests that this elderly woman may have performed such duties at Piedras Negras in the years before her death. The cross-bone and eyeball motif also appears in some forms of the glyph *ch'e'n* (cave) and is often shown on the wings of bats, denizens of such dark places.[150] Thus this symbol probably is not narrowly associated with death but more broadly with the underworld. As a midwife and general administrator of health care, Goddess O and her real world counterparts may have been linked to crossbones and eyeballs because of their role as intermediaries between earthly and underworld places, where diseases and illness were likely understood to originate.

FIGURE 1.34. Femur cross section and vase from the grave of an elderly woman from Piedras Negras Burial 105 (photograph by author; drawing by A. René Muñoz).

THE SPIRITUALITY OF DISEASE

Beyond the natural deterioration of the body, the Maya also confronted a battery of infectious, metabolic, and congenital illnesses. Ethnolinguists studying contemporary understandings of health suggest that the Maya classify illness into two categories: naturalistic and personal.[151] A better way of conceptualizing the dichotomy, more in line with the Maya worldview, is that some disease represents pathology of the body, whereas other illnesses are afflictions of the soul. Curing depends on assessing the source of the illness and prescribing a proper treatment. The modern Maya have a spectrum of health care specialists, ranging from those that focus on the body to those that deal almost exclusively with the soul, including herbalists, bonesetters, midwives, and pulse readers/ritual healers. Illnesses of the body are expressed by a range of symptoms, such as diarrhea, headaches, tooth decay, and dandruff. Healing the body is accomplished by directly addressing the symptom, often with herbs, with a particular concern for restoring the hot-cold balance.[152]

FIGURE 1.35. Goddess O assisting a vomiting supernatural (drawing by author after K6020).

Heat is generally associated with good health. The most powerful members of contemporary Maya society have souls that are hot and fiery, likening them to the sun. As Gossen notes for the Tzotzil of Chamula, some individuals "become politically, economically or spiritually dominant over others; they have hotter hearts (*mas k'isin yo ?nton yu ?un*). Others are less successful; they have colder hearts (*mas sikil yo ?nton yu ?un*)."[153] Heat is accumulated over the span of life and can be made unpredictable through alcohol, anger, or witchcraft. Heat is associated with being well spoken and is linked to fertility. Illness is often viewed as a loss of heat. Gastrointestinal illnesses, aches, and pains are all understood to be cold and are combated with herbs perceived as hot.[154] There is good evidence that some of the modern Maya perceptions of health and illness are rooted in antiquity. For example, both the contemporary Tzeltal and the sixteenth-century K'iche' ascribe tooth decay to a worm that enters the tooth.[155] Furthermore, a concern with the balance of hot and cold is an indigenous concept embraced across Mesoamerica. For the modern Maya, cures for diseases of the body are widely known. Anyone can prescribe a remedy, though herbal specialists may have more detailed knowledge than the average person.

In contrast, illnesses of the soul require the intervention of a ritual specialist who has a spiritual calling for working such cures. Illnesses of the soul are triggered by some sort of spiritual disruption: the person has inadvertently angered a spiritual being, the individual's spiritual co-essence has become lost or has endangered itself, or someone else has sent a disease to attack the person.[156] Curing illnesses of the soul usually begins with a reading of the person's pulse coupled with a ceremony to ascertain the nature of the spiritual disruption. A solution is sought or exchange offerings (*k'ex*) are made in an attempt to placate the offended supernatural being. Illnesses that have a spiritual etiology are typically dangerous and life threatening. The boundary between illnesses of the body and of the soul is not always clear. For example, an illness of the body that fails to respond to herbal cures may be reassessed as something more dangerous, a symptom of an illness of the soul.

It is important, however, not to confuse this distinction between illnesses of the body and diseases of the soul as a dichotomy between the secular and the sacred. For example, modern Maya bone setting is not so much a profession but a spiritual calling; its practitioners are summoned to the art through dreams and heal not merely by physically manipulating the body but also with prayer and the aid of sacred animate objects.[157] Although contemporary Maya curers are usually specialized in a single task, occasionally they do work on multiple dimensions of the body, as in the case of a bonesetter who was also a midwife in the Tz'utujil community of San Pedro Laguna, Guatemala.[158] Although it is difficult to know how specialized body health care was in the past, it is likely that curers treated a variety of ailments and conditions. For example, in the K'iche' *Popol Vuh*, the first Grandfather and Grandmother proclaim that they "remove worms from teeth, cure eye ailments, and set bones."[159] These pathologies would all be understood as ailments of the body in contemporary Maya thinking, and the healing of the soul may have been seen as a separate practice even in antiquity, a task best handled by a shaman-priest, who was generally more knowledgeable and experienced in spirits and otherworldly places.

Epigraphic analysis of Classic period descriptions of *wahy* suggests that the notion of soul illness has very deep roots and was of particular interest for the Classic Maya. Current interpretation holds that some (if not the majority) of Classic period *wahy* represent diseases.[160] Most obvious are the Death Gods. These foul beings are hieroglyphically tagged with richly descriptive names such as "One Death," "Fire Center Death," "Red Bile Death," "Stinking Death," and "Knot Mouth." Christophe Helmke and Jesper Nielsen identify additional creatures and motifs in Classic period texts and images that are possibly associated with particular symptoms of disease. For example, they suggest that avian creatures are linked with asthma and seizures and fire-tailed beings with ulcers.[161] Such a proposition is interesting but may be too narrow a reading of the symbolism of Classic period vases.

For the Tzeltal, to be ill is understood to be literally on death's door. The term for illness (*chamel*) is derived from the verb to die (*cham*) and means "death-like state."[162] The Tzeltal describe illness as something received from an external agent: contracting an illness is expressed as "illness grabs one" (*ya stzak chamel*) or "it gives one illness" (*ya yak' chamel*).[163] The descriptive names of Classic period *wahy* are quite similar to the names of the illnesses of the Tzeltal, which are based on their most obvious symptom and include "bloody diarrhea" (*ch'ich tza'nel*), "anus comes out" (*lok' itil*), "choke-choke cough" (*jik'jik' obal*), "major boil" (*apon chakal*), and "white cough" (*sak obal*; tuberculosis).[164] The sixteenth-century Yucatecan *Chilam Balam of Tzimin* explicitly describes disease and plagues as "blood vomit" (*xe kik*), "scab plague" (*mitan kuch*), "fever of the nose and limbs" (*kak tun i u kakob*), "runny asses" (*cucul it*), and "deer death" (*cim cehil*).[165] Similarly, the Hero Twins of the K'iche' *Popol Vuh* defeat such underworld beings as "Flying Scab," "Jaundice Demon," "Pus Demon," and "Bloody Teeth" in a series of struggles that seem to be metaphors for overcoming illness and death. Moreover, the Hero Twins must survive the House of Cold and House of Fire, two of the most important symptoms in Maya diagnosis of illness.

Classic period illness *wahy* are shown primarily on polychrome painted vessels. An obvious question arises: why would the Maya decorate their drinking vessels (most of which are textually identified as used for preparing and serving cacao) with such hideous content? Houston suggests that these vessels may be instructional.[166] Some of the *wahy* are linked to particular royal lineages, so it may be that such beings were used by Maya kings against their enemies in a form of spiritual warfare.[167] The suggestion is not preposterous when we consider that the Tzeltal and Tzotzil term for what scholars describe as witchcraft is literally "giving illness" (*ak' chamel*), a practice that is well attested in the ethnographic literature.[168] The Maya may have painted these particularly nasty *wahy* on vessels that were traded and circulated among Maya courts as a means to boast about their supernatural prowess or to demonstrate their knowledge of the *wahy* of other lineages.

We should be careful, however, not to equate all Classic period *wahy* with illness.[169] Certainly the Tzeltal recognize that only a portion of *lab* are empowered to be *ak' chamel*.[170] In most Mayan languages the word *wahy* and its cognates are linguistically linked to sleep and dreams, not disease. As Guiteras-Holmes explains for the Tzotzil, "the word for *wahyel* has the same root as the verb for *to sleep* and *to dream*, which is the same thing. All Pedranos [in reference to the community in which she worked] declare that oneiric experiences relate to the *wayhel*, and dreams are interpreted according to this belief. Therefore, communication with the deities and the dead, which occurs in dreams, is effected through the *wahyel*" (emphasis in the original).[171] Where some *wahy* are clearly illnesses (such as the Death Gods), others are not and may have served other purposes as, for example, patrons or emblems of powerful dynasties.

BURNING BONES

The topics of heat, fire, and the human skeleton are revisited throughout this volume. As noted, a balance between hot and cold was fundamental to Maya health. In sixteenth-century K'iche' lore the Hero Twins had to confront these temperature extremes during their underworld trials. The contemporary Tzotzil and Tzeltal refer to the underworld as *k'atinbak* (place of burning bones).[172] According to Pitarch, the spirits of those who have died light their way using bones that they have pulled from graves.[173] The Tzotzil *k'atinbak* is described as "terribly hot—all of the bones of sinners are burning 'like firewood.'"[174] Perhaps the notion of *k'atinbak* merely reflects the influence of Christian notions of fiery hell. But nothing in Christian theology specifically references the notion of burning bones. Yet a scene from a Classic period vessel shows a deathly being named "stinking death" seated before burning crossed bones (fig. 1.36).[175]

The extremes of temperature in the living body—feverish chills or hot and inflamed tissue—are symptomatic of illness. The viral culprits of many common and not so common diseases are unfortunately unknowable to us in the archaeological record. Bacteria, however, especially the sort that cause fiery inflammation of the body's tissues, are detectable when they affect the skeleton. Inflammation of the bone is most commonly evident as periostitis, the ossification of the periosteum (the membrane that covers bone). Periostitis forms when the periosteum is inflamed by trauma, localized bacterial infection, or systemic bacterial infection (fig. 1.37). When bacteria infest the bone itself, the pathological response is more dramatic, resulting in a condition known as osteomyelitis. A variety of bacterial infections can trigger periostitis and osteomyelitis, including the common staph infection (*Staphylococcus aureus*) and the more frightful treponematoses (*Treponematosis* sp.). The latter is a group of bacterial infections include venereal syphilis. Fortunately for the Classic Maya, they only had to worry about a more benign form of the treponematoses that was likely yaws or a close equivalent (probably the source of the periostitis shown in fig. 1.37). For the most part, mild periostitis is unnoticed by the sufferer. But the more severe cases of periostitis and especially osteomyelitis are reflective of disease processes that would have involved swollen inflammation of the limbs and even the drainage of pus on the skin's surface.

FIGURE 1.36. Death God warming himself on burning crossbones (drawing by author after K927).

The study of the human skeleton provides a useful though imperfect opportunity to access the disease conditions of the Classic Maya. Conditions such as periostitis and osteomyelitis usually cannot be linked to a specific disease but instead can be used to assess general levels of morbidity. Moreover, not all disease manifests itself as bone pathology. At Piedras Negras 26/34 (76.5 percent) of adult tibiae demonstrate periostitis.[176] The prevalence of periostitis is comparable to that observed for the Pasión Region (68.0 percent of tibiae) and Copan (55.3 percent). Although this may seem like an inordinately high level of morbidity, Lori Wright and Christine White have shown that the prevalence of periostitis among the Maya is comparable to that in other global populations, particularly those with endemic treponematosis.[177] The Maya had to contend with disease, but no more so than other preindustrial societies. Wright and White's synthetic study also shows no clear trend in Maya health over time, contrary to what is often assumed. At some sites health conditions worsened over the course of the Classic period, at other places morbidity declined, and at others the level of disease never changed.

What is perhaps most sobering is a comparison of the health of the ancient Maya with the health of the Maya following the Spanish conquest. Adult attained stature is a robust measure of general health (disease burden and malnutrition), especially when traced diachronically in populations of common genetic descent.[178] Though the

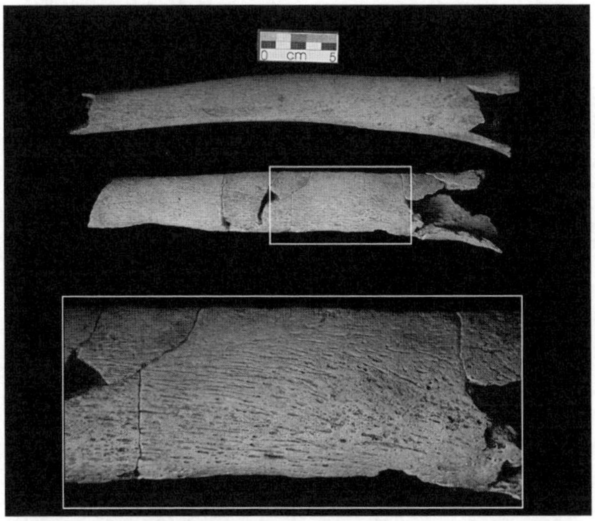

FIGURE 1.37. Severe periostitis of the right tibia from Piedras Negras Burial 111 (lower bone) compared to a typical right tibia. The inset photo shows the irregular, woven appearance of rapidly deposited new bone (photographs by author).

sample size is small, the ancient inhabitants of Piedras Negras were tall by Maya standards. The mean stature for men is 166.3 cm ($n = 5$) and for women 150.4 cm ($n = 8$); both are above the means reported by Marie Danforth for her composite pool of Precolumbian Maya stature (males = 160.1 cm; females = 147.8). By the twentieth century, mean Maya height was only 155 cm for males and 143 cm for females, making the Maya among the shortest populations in the world.[179] The Classic period Maya of Piedras Negras were taller than their modern descendants because they enjoyed less disease burden and better nutrition. Fortunately, there is hope for future generations. From my own experience, current youths in the Petén are noticeably taller than their parents' generation. Barry Bogin has shown that Maya people who have had the opportunity to live and work in the United States and have access to better nutrition and health care are substantially taller than previous generations.[180] Unfortunately, at a modest height of 170 cm, I still tower above most Maya people in the more impoverished parts of the Guatemalan highlands, despite being shorter than the average man in the United States.

THE LIVED BODY

Bodies are the cumulative histories of our biological and social lives. Accessed through the skeleton, ancient Maya bodies provide a wealth of information regarding diet, health, and life history. Yet it is also important to consider how the Maya understood the body. The body was the self: a host to a variety of souls that in combination ascribed the character of a person. For the contemporary Maya, understanding the soul is relevant for a range of issues, from childhood development to health care and ultimately to the way an individual behaves and is understood by society. As this book shows, Maya mortuary practices were organized not only to deal with the body after death but to manage those souls long after mortal life ended.

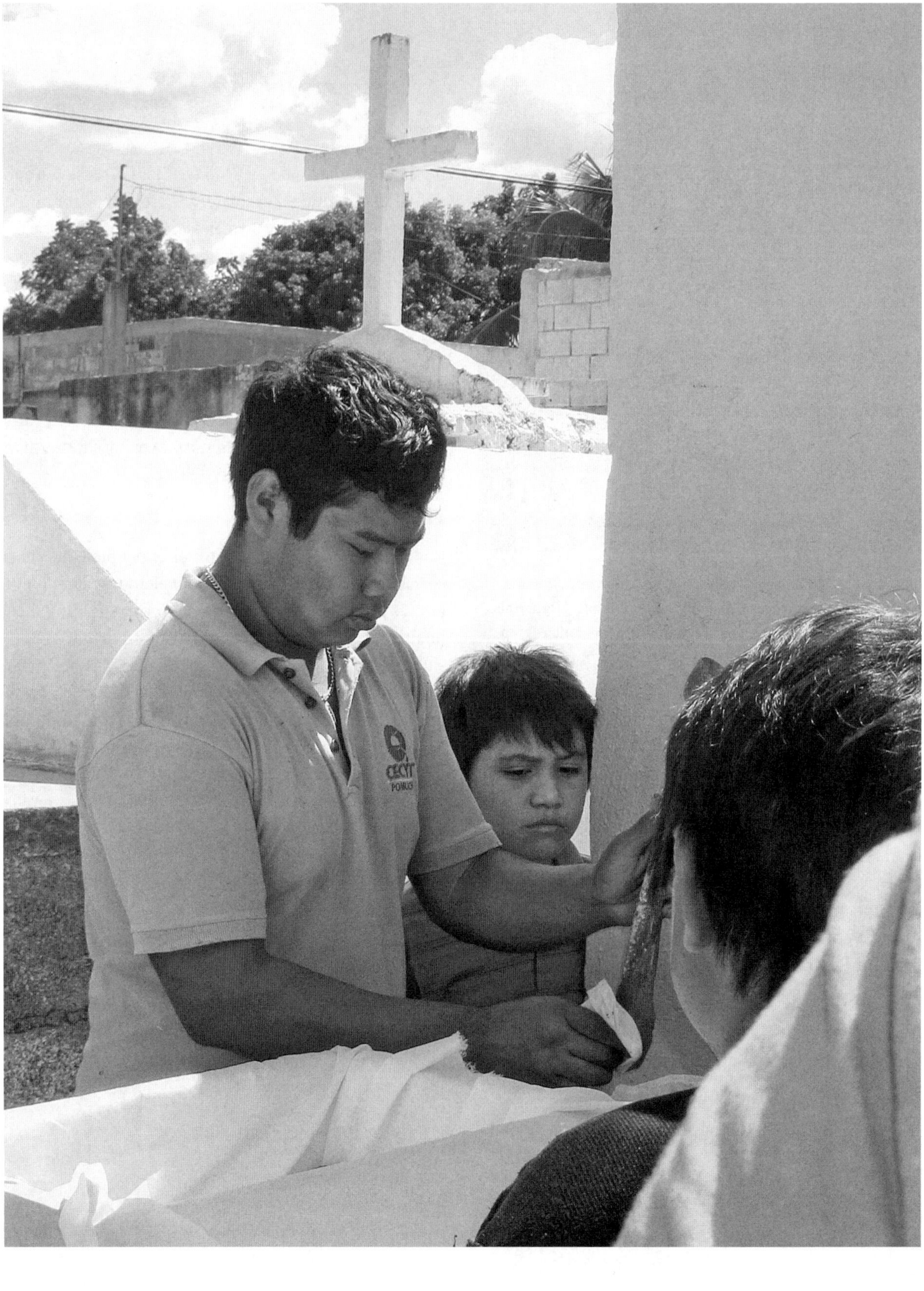

CHAPTER 2

DEAD BODIES

FIGURE 2.1. (*opposite*) A long bone is cleaned in a Yucatec community in Campeche as part of Día de los Muertos (photograph by author).

Since the inception of the field, anthropologists have studied mortuary beliefs and practices in order to infer deeper aspects of social behavior.[1] How different societies prepare, manipulate, inter, exhume, and reinter the bodies of their dead reveals not only how they grapple with the rupture of death but ultimately how society organizes itself and how its members perceive themselves within the greater universe. This chapter explores Maya practices pertaining to the treatment of the dead body. What objects adorn the corpse? How was the body dressed and prepared for interment? How was it arranged within the funerary space? In what contexts were human remains removed from burials? Why did some body parts never make it into the grave? And, ultimately, what does all of this tell us more broadly about the Maya's relationship to one another, to their ancestors, and to the greater realm of the supernatural?

Funerary dress and mortuary furniture may reflect aspects of the decedent's lived identity. Such connections must be demonstrated, however, and not simply assumed. In life we consciously select our wardrobe to conform to our sense of self and to project that identity to the world around us. But the dead do not dress themselves. In that sense the costumes of the dead are better understood as reflecting the intentions of the living and may tell us more about broader norms regarding the proper treatment of the corpse than about anything specific to the identity of the deceased. As this chapter shows, Maya funerary costumes do indeed reveal practices of individuation yet also demonstrate deeper customs that reflect both local tradition and universal aspects of Maya mortuary practice. Fundamentally, mortuary archaeology reveals that the treatment of the dead not only was determined by beliefs about dead bodies but also reflects the need to manage postmortem souls within a religious system that emphasized rebirth and regeneration.

SOULS AT BIRTH AND DEATH

For the contemporary Maya, souls are joined to the body at birth. Death occurs when injury or illness dislodge or destroy those souls. Maya souls persist when separated

FIGURE 2.2. Bird Jaguar IV's parents look down from a celestial place as ancestors on Yaxchilan Stela 11 (drawing by Linda Schele, © David Schele, courtesy Foundation for the Advancement of Mesoamerican Studies, Inc.).

from the body, and some may eventually join the ambiguous pool of ancestors that hold influence (good and bad) over the world of the living.[2] For both the ancient and modern Maya, the relationship between the living and the dead was one of reciprocal need. In many modern Maya languages, the ancestors are referred to collectively as *mam* ("mother-fathers" or "grandfathers").[3] This term can be traced back to Classic period inscriptions, underscoring the antiquity of the concept.[4] *Mam* may be quite specific, referring to named individuals. Yet the term may also refer to the collective body of ancestors that are linked to a particular place or lineage. In many contemporary Maya communities such ancestors are conflated with other spirits that dwell within the mountains.[5] The days near the end of the agricultural cycle, corresponding today to Día de los Muertos, mark the time when such ancestors must be honored with food and other offerings. People in the Yucatán have a long tradition of removing the skeletal remains of the dead to be cleaned and fed in association with this day (fig. 2.1).[6]

Abundant iconographic evidence shows the importance of ancestors among the Classic Maya. Funerary pyramids were constructed and monuments were erected in order to commemorate especially important and powerful royal ancestors who were also the object of intensive and extensive rituals of veneration. In Classic period imagery ancestral spirits are shown looking down on the activities of the living, with numerous descriptions of the invocation of royal ancestors in rituals conducted by Maya kings and queens (fig. 2.2). The presence of ancestral shrines within nonroyal house compounds indicates that all sectors of Maya society venerated revered members of their family.[7]

The linkage between the spirits of the dead and their descendants is well attested in the ethnographies of the Maya highlands.[8] For the Tzotzil of Zinacantán, as Vogt reports, the soul may remain associated with its grave for as many years as it inhabited the body, after which it "enters the pool of innate souls kept by the ancestors. It may then again be implanted in another embryo."[9]

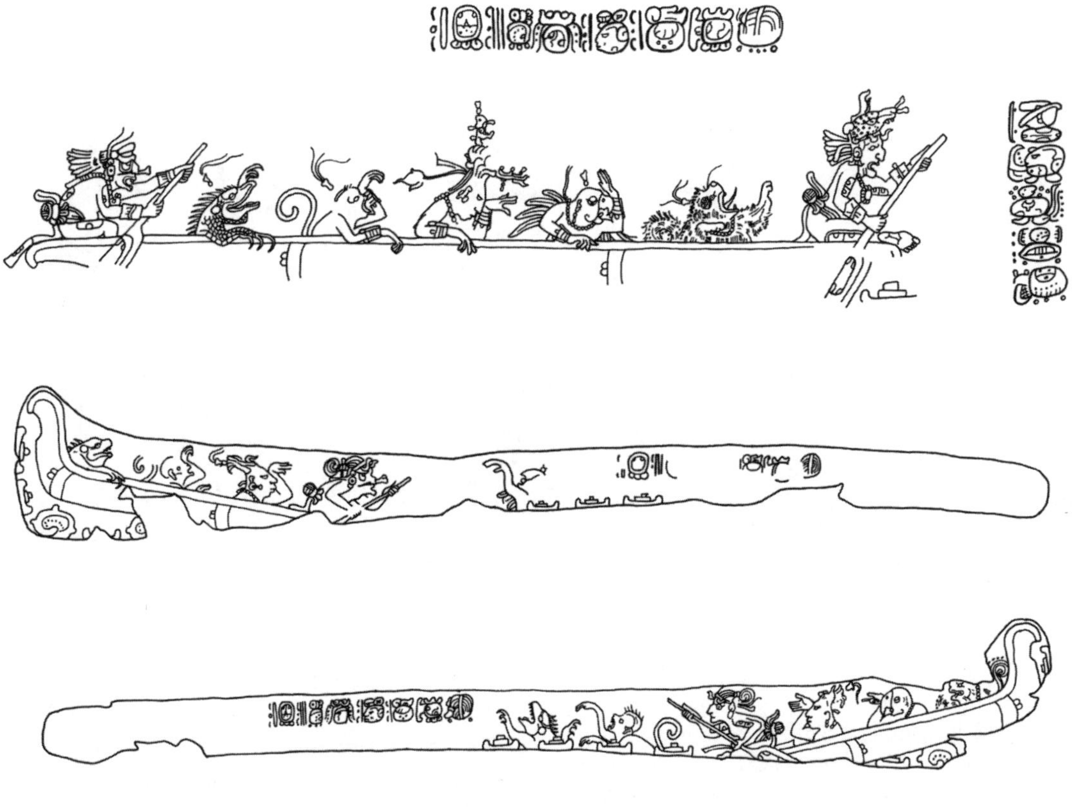

FIGURE 2.3. Inscribed bones from Tikal Burial 116. The Maize God descends into the underworld in a canoe, possibly accompanied by *wahy* (drawing by Linda Schele, © David Schele, courtesy Foundation for the Advancement of Mesoamerican Studies, Inc.).

As a Yucatec informant from Chan Kom explained to Robert Redfield and Alfonso Villa Rojas, souls return to the earth because there are "not enough souls to keep forever repopulating the earth."[10] The Maya liken the concept of soul rebirth to the life cycle of plants, particularly maize. According to Allen Christenson, "once placed in the grave, Atitecos [the Tz'utujil of Santiago Atitlán] raise a small mound of earth over the body and plant a tree on top which represents the soul of the dead reborn to new life. The community cemetery has long rows of graves bearing trees, giving the appearance of a great orchard or grove."[11] Francisco Ximénez described a similar practice in highland Guatemala at the beginning of the eighteenth century and noted that people were frequently buried in the maize fields, suggesting that the dead were reborn as maize.[12]

Today the Maya see a special linkage between grandparents and their grandchildren, who serve as replacements (*k'ex* in many Maya languages) for them. As Robert Carlsen and Martín Prechtel note for the Tz'utujil, "because of high infant mortality, until recently it was common in [Santiago] Atitlán for parents to give many of their children the same (grandparent's) name, thus insuring a *k'exel* to pass on the ancestral life form."[13] Scholars suggest that similar beliefs operated among the Classic Maya.[14] Certainly the act of adopting grandparents' names is evident in the naming conventions of some Classic period dynasties, stressing the continuity of the royal line.[15]

As noted in chapter 1, the cycle of life, death, and rebirth was manifest in the myth of the Maize God.[16] In life he is a being of beauty, vitality, and fertility.[17] In death the Maize God travels to the underworld, often shown as a journey in a canoe (fig. 2.3).

Other imagery shows not only the Maize God's canoe voyage but his underworld or underwater rebirth and subsequent adornment in his regalia, assisted by maidens (fig. 2.4). After his rebirth he resurrects, breaking

FIGURE 2.4. The death, rebirth, and dressing of the Maize God (drawing by Linda Schele after K3033, © David Schele, courtesy Foundation for the Advancement of Mesoamerican Studies, Inc.). Note the three depictions of the Maize God, one of which is in the "birth position," apparently disgorged from the mouth of an underwater serpent.

through the hard shell of the earth, usually with the assistance of the Hero Twins or aided by the rain deity Chahk (see fig. 1.12). The antiquity of elements of this myth is testified by the murals of San Bartolo, which date somewhere in the first century before Christ.[18] A late version of the myth comes to us as the sixteenth-century K'iche' *Popol Vuh*.[19] It is important to keep in mind, however, that no definitive version of the Maize God myth exists. The story has much geographic and especially temporal variability in its details. Yet the fundamental parable of death, journey, and renewal is intrinsic to Maya spirituality and essential for deciphering aspects of Classic period mortuary practices.

CROSSROADS

Death of the body marks the beginning of travel for the soul for the contemporary Maya. Yet many souls prefer to linger, so the living must take action to encourage their departure and to ensure that the souls of the dead do not attempt to take any living souls with them. For the Tzotzil of Zinacantán, this may involve opening a hole in the roof of the house, kicking the coffin, and spitting saltwater on the location where the coffin was placed during the wake.[20] In one account of the souls' journey, the Tzotzil spirits must cross a river and may be aided on their journey by a large black dog.[21] In the Classic period the canoe voyage of the Maize God clearly points to the notion of a postmortem soul's journey and may be suggestive of a broader trope relating to the watery descent of souls.[22] Other evidence for the soul's journey comes from Classic period inscriptions: one common euphemism for death is *och bih* (he enters the road), a reference to a journey, perhaps along the great celestial path of the sun (see chapter 4).[23]

The Milky Way is another celestial path that may be implied. The Tzotzil speak of the summer Milky Way as the *be vo'* (road of water) and the winter Milky Way as *be taiv* (road of frost).[24] Similarly, the modern K'iche' call the summer Milky Way the *sac bey* (white road) and the winter Milky Way *xibal bey* (road of fright).[25] For the Cho'rti', the Milky Way is a road upon which Santiago, lord of thunder and lightning, rides.[26] It is conceptualized as a white serpent that runs north-south across the sky, perpendicular to the path of the sun. At midnight of the feast of Santiago, the sun's path and the Milky Way are understood to form a cross, marking the beginning of the *canícula* (a late July break in the rainy season), when Santiago holds off the rains to allow the second sowing of maize.

Both contemporary and Classic Maya thought recognize important connections among journeying souls, their routes of travel, and crosses.[27] The modern Maya landscape is populated with wooden crosses that, according to ethnographers, have nothing to do with Christianity but instead are related to concepts of the world tree (the axis mundi) and general conduits that allow movement between earthly and supernatural places.[28] For example, the contemporary Tzeltal use crosses to mark routes of travel and both the Tzeltal and Tzotzil believe that crosses are the conduit by which souls travel during dreams or in death.[29] As Pedro Pitarch notes for the Tzeltal, crosses operate as "channels allowing communication that breaks the traditional boundaries of time and space, connecting the world of the flesh-and-blood Indians with their soul world."[30] As Vogt reports, the Tzotzil of Apas removed their crosses from their homes after a loved one died, which he believed was an effort to dissuade the deceased souls from using the cross as a conduit to return.[31] On the Palenque sarcophagus lid, Pakal's ascent from the underworld occurs along the length of a shiny wooden cross-tree (fig. 2.5).

In some ancient and modern Maya contexts, the cross is understood to be a stylized representation of the great world tree, a *ceiba*. Such a cross centers a place as an axis mundi and the heart of creation. This belief resonates throughout the indigenous Americas, where world trees are understood as centers of a quadripartite world. In Mesoamerica such motifs date well into the Preclassic period.[32] On the Palenque sarcophagus lid, the cross-tree likely represents both a supernatural conduit of travel and the vitality of the Palenque kingdom.[33] The roots of the tree are in the underworld, and the trunk rises toward the heavens, where its branches spread. The vertical up-down axis of world trees can also be conflated with north-south. The perpendicular direction, east-west, is the path of the sun and is the principal axis in Maya conceptions of the earth. Such cosmology is well attested among the contemporary Maya. As Gossen explains for the Tzotzil, "the sun and its life-giving heat determine the basic categories of temporal and spatial order."[34]

Today crosses are used by the Tzeltal for communicating with "otherworldly powers" and are employed to direct food (conceptualized as incense) and other offerings to supernatural beings.[35] For the contemporary Yucatec, crosses are understood to be alive, embodied with animate spirits, and capable of communication with the living. They are carefully tended and are often wrapped or given clothing.[36] Spiritual embodiment and

FIGURE 2.5. The sarcophagus lid of K'inich Janaab Pakal of Palenque (drawing by Linda Schele, © David Schele, courtesy Foundation for the Advancement of Mesoamerican Studies, Inc.).

FIGURE 2.6. Palenque Tablet of the Cross (drawing by Linda Schele, © David Schele, courtesy Foundation for the Advancement of Mesoamerican Studies, Inc.).

otherworldly interconnectedness were depicted in a variety of means in Classic period contexts. On the sarcophagus lid a face on the tree trunk shows the animate spirit of the object.[37] On the tablet from the Temple of Cross at Palenque, solar square-nosed breath serpents (see the following section) emerge from flowers on the tree's branches (fig. 2.6). The Tablet of the Cross shows both a youthful and adult Kan Bahlam atop a sky-band, employing the cross to access ancestors and otherworldly beings that dwell in such celestial places. As David Stuart and George Stuart suggest, the Palenque cross-trees "served to animate the larger theme of ancestral resurrection through the sun's eastern ascent."[38]

The regenerative power of trees is evident throughout Maya history. In some instances fruit trees substitute for the *ceiba* as world tree, especially gourd or cacao trees (see fig. 1.11). An early example is the calabash tree from the San Bartolo murals.[39] In the K'iche' *Popol Vuh* the head of One Hunahpu (the equivalent of the Maize God) is likened to a gourd and placed in a tree, where he impregnates a daughter of an underworld lord. Maya rulers and their supernatural counterpart, the Maize God, were understood to be fertile axes mundi and were likened to sacred trees.[40] The inverted, acrobatic form of the Maize God in particular is the embodiment of the world tree and is closely associated with scenes of mythic rebirth, especially in the Early Classic period. He appears on a remarkable effigy vessel excavated from Burial 10 at Tikal, with cacao or gourds hanging from his legs and a cross-shaped tree emerging from between his legs (fig. 2.7a).[41] A similar floral tree was incised on a funerary vase at Oxkintok (fig. 2.7b). A series of diving maize-being (or perhaps ancestor) trees is shown on the famous Berlin Vase (see fig. 2.48). On this vessel the head seems to operate as the seed from which the trunk-body rises and the arms and hands are the roots.

IMAGES OF THE SOUL

Animate breath—a visual cue for life or the soul—is typically shown in Classic period art as beads or tiny flowers suspended in front of the noses and mouths of humans and supernatural beings (see, for example, the bead located in front of the nose of Pakal in fig. 2.5).[42] These floral bead exhalations reference the flowery nature of the celestial supernaturals and Maya nobility that populate Classic period art. In general jade beads, respiration, and the animate life force are closely associated.[43] To some extent these floral exhalations also reference the pervasive trope that the human life cycle is like-in-kind to the cycle of plants. Equating human life with flowers underscores its preciousness and fragility. Flowers are a surprising rarity in the jungles of the Maya lowlands. When they do appear they offer a sharp visual and aromatic contrast to the monotony of the green forest, with its subtle yet pervasive smell of pungent vegetal decay. As Taube shows, the sweet and fragrant floral soul is associated with the Maya celestial paradise, a place that he refers to as "Flower Mountain."[44] Among other things, Flower Mountain is the destination for the flowery soul after death.

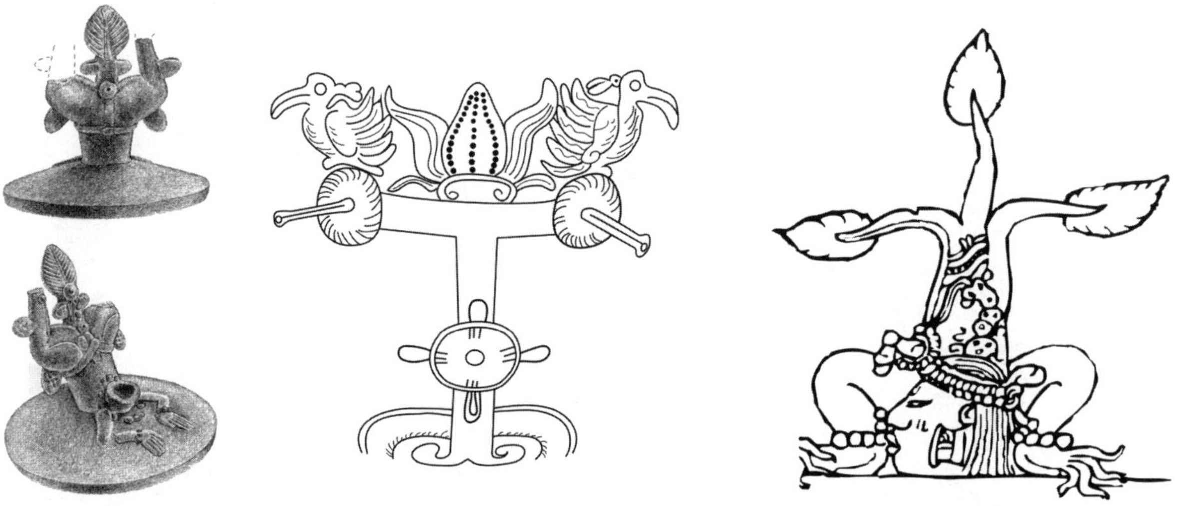

(a) (b) (c)

FIGURE 2.7. Classic period crosses: (a) the Diving Maize God from the lid of a ceramic vessel from Tikal Burial 10 (drawing courtesy of the University of Pennsylvania Museum of Archaeology and Anthropology); (b) an incised ceramic vessel from Tomb 8 at Oxkintok (drawing by author after Martínez del Campo, *Rostros de la divinidad*, fig. Oxkintok.15); (c) ancestral tree for the Berlin Vase (modified from drawing by Nikolai Grube; see fig. 2.48). The top of the Oxkintok cross is a maize plant. The other three arms of the cross are marked by symbols for wind or soul.

As noted in chapter 1, the "*ajaw*-face" sign and the glyphs for *k'uh* and *k'uhul* relate to Classic period understandings of spirit, souls, or internal essence of beings (see fig. 1.1). The "*ajaw*-face" was inserted into the glyph **SAK** (white, clean, or clear) to form a glyph that some epigraphers read as *sak nik* (white flower), a metaphor for the vital spirit. A common death euphemism has been translated as *k'a'ay u-sak nik ik'il* (it ends, the white flower breath) (fig. 2.8).[45] The reading of the "*ajaw*-face" as *nik* is not secure, however, and the meaning of **SAK** is ambiguous in this context. Whatever the actual reading of this sign, it undoubtedly relates to the vital essence of people (though further analysis is needed to elucidate the correct reading of the glyph).

The **SAK** "*ajaw*-face" expression is prominent on a series of altars from Tonina that celebrate postinterment funerary rites (fig. 2.9). The text on both monuments includes the expiring breath euphemism to describe the death of a Tonina lord, followed by reentry into his tomb 260 days later.[46] On each monument the dead lord holds a bar capped on either end by the **SAK** "*ajaw*-face" soul-essence. A third such emblem is located between the lord's legs on Monument 69. The span of 260 days is an obvious reference to the short count calendar, which itself likely recalls the nine months of human gestation. This also happens to be the span of time between first planting (March) and last harvest (November) in the agricultural calendar of the Chiapan highlands.[47] Thus,

FIGURE 2.8. **SAK** "*ajaw*-face" as vital essence in a death expression from Yaxchilan Lintel 27 (drawing by author).

(a) (b)

FIGURE 2.9. SAK "*ajaw*-face" signs on mortuary sculptures at Tonina: (a) Monument 69; and (b) Monument 149 (drawing of Monument 69, disc, by Ian Graham and Peter Mathews, from *Corpus of Maya Hieroglyphic Inscriptions*, vol. 6, part 2, *Tonina*, 6:103 © 1996 by the President and Fellows of Harvard College, Peabody Museum of Archaeology and Ethnology, Harvard University, PM# 2004.15.6.15.60 [digital file# 125870001]; drawing of Monument 149, by Ian Graham; inked by Lucia Henderson, from Graham et al., *Corpus of Maya Hieroglyphic Inscriptions*, vol. 9, part 2, Tonina, 9:82 © 2006 by the President and Fellows of Harvard College, Peabody Museum of Archaeology and Ethnology, Harvard University, PM# 2004.15.6.20.9 [digital file# 99040064]).

like a human fetus, the Tonina lord required nine months to prepare in the earth before he was ready for rebirth.[48]

The genesis, original meaning, and evolution of the "*ajaw*-face" sign are poorly understood, despite the sign's ubiquity in Maya text and imagery. What seems to be a very early form of it is shown as part of the exhalation or vocalizations of a bird from the Olmec site of San Andrés, La Venta (ca. 650 BC), yet its meaning in this context is not clear (fig. 2.10a).[49] The sign appears on at least two ceramic stamps from Chiapa de Corzo, where it seems to sprout vegetal curls, foreshadowing the foliated *ajaw*-face head ornaments of the Early Classic period (fig. 2.10b; compare with fig. 2.11a). What may be a variant is depicted within the emanation of a lord on La Mojarra Stela 1 (Veracruz, 2nd century AD), where it seems to signify his vital essence or flowery speech (fig. 2.10c).

An Early Classic period version of the "*ajaw*-face" that sprouts three leaves also appears on the foreheads of supernaturals or located atop their heads (fig. 2.12). As Stuart has recently shown, this composite sign seems to have the value of Ux Yop Huun (three leaf paper), the name given to the headbands worn by Maya rulers.[50] Ux Yop Huun is most regularly embodied by a raptor that is one of three distinct beings frequently glossed as the "Jester God" in Mayanist literature.[51] The jade cobble from Altun Ha illustrated in figure 2.11b is a rendering of this avian creature with the "*ajaw*-face" sprouting three leaves in its forehead. In some contexts, especially in the Early Classic period, the raptor and the "*ajaw*-face" merge in what seems to be shorthand for the headband or, more specifically, the idea or value that the headband is meant to symbolize. As Stuart points out, the ancestors on the sides of Pakal's sarcophagus offer a good example.[52] The foliated "*ajaw*-face" and the foliated bird substitute for one another in the ancestors' headdresses; five ancestors have the "*ajaw*-face" and the other five have the raptor.[53] Both Maya royalty and various celestial

(a)

(b)

(c)

FIGURE 2.10. (*above*) Preclassic signs similar to the Classic period "*ajaw*-face" signs: (a) bird vocalization from Olmec stamp; (b) ceramic stamps from Chiapa de Corzo; (c) breath volute from La Mojarra Stela 1 (drawings by author).

FIGURE 2.11. (*right*) Early Classic period "*ajaw*-face" as part of a supernatural head ornament: (a) atop the head of the Patron of the month of Pax (drawing by Karl Taube); (b) in the forehead of the avian Ux Yop Huun (Jester God) (drawing by author).

(a)

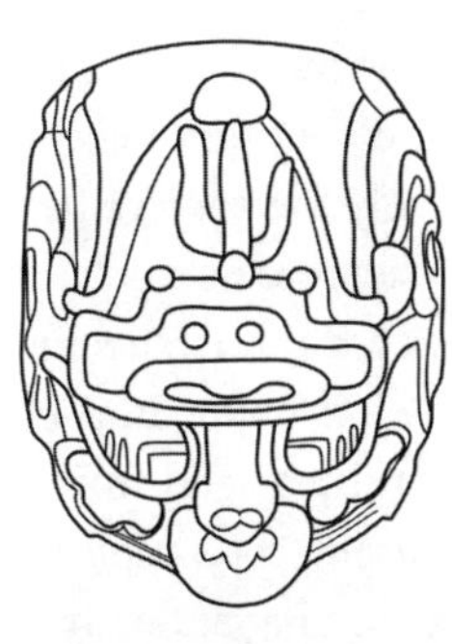

(b)

FIGURE 2.12. Ux Yop Huun avian Jester Gods from Tomb 19 at Río Azul. The eastern (central) head is marked by the "*ajaw*-face" (drawing by author after Hall, "Realm of Death," fig. 15 and Stuart, "The Name of Paper," fig. 18).

supernaturals (identifiable by their square-shaped pupils) wear the foliated "*ajaw*-face," which may simply imply the like-in-kind nature of these beings. Yet it is probable that something more complex is at work. As Stuart suggests, the Ux Yop Huun headband and ornament may also have been worn in reference to some concept manifest in the poorly understood myth of the slaying of a celestial bird by Juun Ajaw (the headband-wearing member of the Hero Twins), who is depicted on Classic period vases shooting a raptor from a tree with a blowgun.[54] Overall, the significance of the "*ajaw*-face" in the Ux Yop Huun headband remains murky but seems to tie into the concepts of spirit, trees, and celestial supernatural places.

Further clarification may come from the Early Classic Tomb 19 at Río Azul, where an avian Jester God head is painted on the north, east, and south wall of the burial chamber (fig. 2.12). All three show vegetation sprouting from their heads, and the foliation of at least the northern and southern beings is marked by the *te'* sign for tree. Only the eastern being, however, has the "*ajaw*-face" inset into its forehead. It may be that the eastern head, based on its specific location, was meant to suggest new growth. In such a context the "*ajaw*-face" may signify a seed from which the tree grows. Supporting this interpretation, the central figure is distinguished by a marking under its eyes that seems to be the **YAX** sign, emblematic of the color blue-green, unripeness, and newness.[55] Alternatively, the eastern head may signify centrality within the triad. As Taube observes, triads of Jester Gods appear throughout Preclassic and Classic period iconography and may relate to the three sacred hearthstones or more specifically the axis mundi.[56] Whether the "*ajaw*-face" in this context relates to a seed or centering (or both), it is notable that the head of the deceased was oriented toward the central Jester God and framed by the other two beings. Recall that heads have symbolic value as seeds and were regularly "centered" within Maya mortuary spaces to facilitate soul departure and renewal (see chapter 3).

Another important use of the "*ajaw*-face" sign comes from Classic period texts, where it appears as part of a glyph whose reading remains elusive but can be interpreted from context to mean "child of father" (fig. 2.13). The sign is similar to the **SAK** "*ajaw*-face" sign but can be distinguished by its distinctive curls that are emblematic of smoke or flame (*k'ahk'*). This may relate to the father's role in conferring the hot vitality necessary to create life or to the warm nature of Maya souls more broadly. A striking example of an "*ajaw*-face" sign that is similar to the "child of father" sign floats above the body of a sacrificed child on Yaxha Stela 13, surrounded by the smoke created by three burning stones or perhaps smoldering copal (fig. 2.14). The scene is quite similar to the sacrificed animals in the San Bartolo murals, whose abdomens are opened and filled with three burning stones.[57] Overall, the sign on the Yaxha stela seems to signify the transference or exchange of vitality or soul (an issue revisited in the discussion of sacrifice in chapter 3).

Considering the sign's use in such diverse contexts—parentage statements and scenes of sacrifice (burning "*ajaw*-face"), royal diadems (foliated "*ajaw*-face"), and death expressions (SAK "*ajaw*-face")—it seems that the sign must originally have had some very general quality as precious or sacred vitality and perhaps, as noted above, a seed. The complex significance of this sign is exemplified in the imagery of an unprovenanced Early Classic period incised vase curated at the Ethnologisches Museum in Berlin (fig. 2.15). This scene is perhaps the best illustration of the soul in all of Classic Maya art. One side of the vase shows the bundled corpse of what is likely the Maize God atop a funerary bier-throne. A solar cartouche rises above his body. Solar, lunar, and stellar cartouches are widely associated with ancestor spirits in Maya art, especially on monuments from the Yaxchilan kingdom.[58] The head of the Sun God is located at the center of the solar cartouche on the Berlin Vase, with his forehead shown as a burning "*ajaw*-face," similar to the vital essence shown on the Yaxha stela. The three curls may also connect the sign to Late Classic renderings of the SAK *ajaw*-face soul signs (compare with figs. 2.8 and 2.9). The back of the Sun God's head is marked by a moon sign for reasons that are not clear.

FIGURE 2.13. (*left*) "Child of father" from Tikal Stela 5, Late Classic period (drawing by author).

FIGURE 2.14. (*above*) Child with three spheres (stones or copal) burning in its chest cavity from Yaxha Stela 13 (drawing by author). Note the soul emblem floating above the child.

FIGURE 2.15. The funeral of a Maya king or the Maize God as shown on the Berlin Vase (drawing by N. Grube).

Perhaps the combined sun and moon reference some concept of male-female duality.[59] The Sun God head seems to float before the glyph *bih* (road) in a graphic representation of the departed spirit entering the celestial path of souls. The left leg of the funerary bier-throne shows the glyph for road-entering, *och bih*, recognizable as a serpent (*och*) entering the center hole of a disc marked by four dots (*bih*). In sum, the Berlin Vase shows the departure of the soul likened to solar (or lunar) ascent (a trope revisited throughout the course of this book). That we are indeed witnessing solar ascent is underscored by the location of the mourners; they are standing in the ocean, recalling the ascent of the sun from the eastern sea. Moreover, we are not looking at a scene merely of death but also of rebirth, as evidenced by what seems to be new maize growth that spills over the bundled corpse of the Maize God.

PRECIOUS BODIES

Aside from the iconography of Maya souls, we can also access Maya understandings of the soul through treatments of the body in life and death. Inescapably, we must begin with the data-rich tombs and burials of Maya royalty and nobility. By virtue of the volume of goods that they contain, elite mortuary contexts provide the greatest depth and variety of information for exploring the ideology of Maya souls in mortuary contexts. Nevertheless, we cannot ignore the graves of nonelites. Such deposits also reveal much about Maya concepts of the soul in life and death. Yet we must also acknowledge that royal bodies were distinct from those of the rest of the populace. As Stephen Houston, David Stuart, and Karl Taube have shown, royal bodies established hierarchy, personified beauty, generated moral authority, were metaphors for kingly consumption, and possessed *k'uh*, distinct from the rest of Maya society.[60] Such socially and spiritually charged bodies were subjected to intense rituals to mend the rupture created in their death.

Herein was a fundamental paradox that shaped much of Classic Maya royal mortuary ritual. Alive, royal bodies were the embodiment of the vitality of the kingdom. Dead, their decaying corpses signified a kingdom in transition and quite possibly turmoil. The royal funeral and subsequent veneration of the royal corpse was among the successor's most important obligations. Mistreatment of a respected royal body would have been disastrous, and the veneration of revered kings and queens was channeled into immense political capital. If the former ruler was unpopular (or, even more challenging, if the claimant to the throne had little familial or personal connection to the dead king), however, lavish mortuary rites and elaborate funerary monuments may have been avoided. Alternatively, new rulers with tenuous rights may have used sumptuous mortuary rites as a means to legitimize their claim to the throne. The complex political nuances of royal funerary rites and subsequent acts of veneration are explored more fully in chapter 4.[61]

In both life and death bodily adornments distinguished Maya kings and queens from the rest of the

FIGURE 2.16. Adorning the Maize God prior to his resurrection (drawing by author after K1566).

populace. As Bartolomé de Las Casas reports for the burial of a Pokom lord, "They dressed him in the best clothes and richest blankets that he had, adorning him with certain jewels of gold and stones that he had left and set aside, so that they buried these with him, because the rest, being already come near death, he had divided among his sons and wife, and brothers and relatives." Las Casas further explains that the other lords who came to witness the burial brought "clothes and some pieces of gold to adorn the body of the dead man, each according to his mind, consoling the living and showing the weight and grief that all his servants and friends had had at the death."[62] Francisco Fuentes y Guzmán provides a similar description of Pokom elite funerary rites: "They dress him afterwards in rich and figured clothes, in the style that he wore in life, with the same insignia which he wore reigning; and in this manner, they place him on a platform above figured tapestries of beautiful colors."[63]

Maya iconography shows that the costume of Maya kings and queens was often the same as that of the Maize God, embodying the principles of beauty, life, and vitality. An important station in the Maize God's mythic cycle as evidenced on Classic period polychromes is dressing him as king in his jade adornments (fig. 2.16; see also fig. 2.4). In this he is often shown assisted by maidens in a watery underworld place.[64] Once adorned, he is ready for resurrection. This mythic scene is apparently an allusion to the dressing of the royal body for burial in preparation for a similar rebirth. On some polychrome vessels that depict the Maize God's adornment, he sits upon or stands in front of a large platform (see fig. 2.4). This platform may be in reference to the biers upon which some Maya kings were interred, in parallel to the thrones on which they sat in life (see fig. 2.15 and the discussion in chapter 3).

Of all aspects of the royal costume, adornments of jade are the most distinctive. The significance of jade partly relates to its rarity; though not particularly uncommon at its source in the Motagua River valley, it had to be transported by foot or by canoe along the coast into the Maya lowlands. Moreover, the value of jade was in the labor required to work this hard stone. Jade also had symbolic association with life, fertility, rejuvenation, and preciousness—concepts broadly linked to Maya royal dynasties. One of the most impressive jade assemblages was recovered from Burial 116 of Tikal (fig. 2.17). Among other jade objects, a massive collar of fifty-six long, tubular jade beads covered the kings' shoulders and chest. Another series of strands with 120 spherical beads was draped across his stomach. Such collars are shown worn by both kings and queens. A similar device made of a composite of over 150 small jade and *Spondylus* discs was draped over the chest and shoulders of the Red Queen of Palenque.[65] Among the Tzotzil, the huipils worn by women, with their elaborate embroidered collars, are understood to situate them at the center of their worlds.[66] Maya kings and queens may have donned jade and shell collars for similar reasons, embodying themselves as the axis mundi. The interlaced beads of this regalia recall the bumpy surface of a turtle's carapace or crocodile's back, both of which were metaphors for the surface of the earth.

Despite the significance of jade as a royal adornment, not all kingdoms had access to the prized material in equal amounts. At Piedras Negras, Rulers 3 and 4 (Burials 5 and 13) were each interred wearing a necklace of clay balls painted blue-green to look like jade beads (fig. 2.18). Although this may reflect a stingy court holding back on funerary goods, the inclusion of a spectacular *Spondylus* cloak in the tomb of Ruler 3 suggests that the

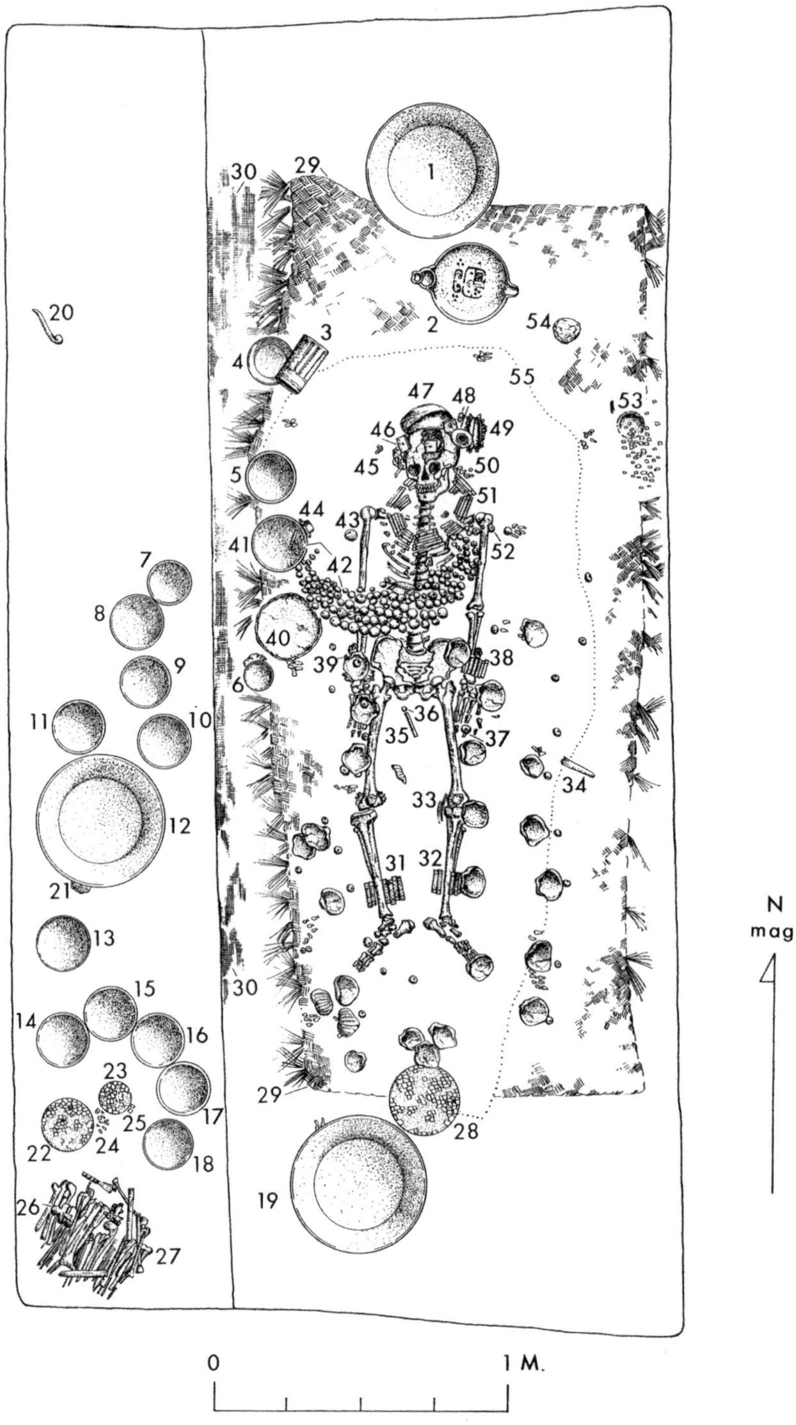

FIGURE 2.17. Tikal Burial 116 (drawing from Coe, *Excavations in the Great Plaza, North Terrace, and North Acropolis of Tikal*, fig. 260, courtesy of the University of Pennsylvania Museum of Archaeology and Anthropology).

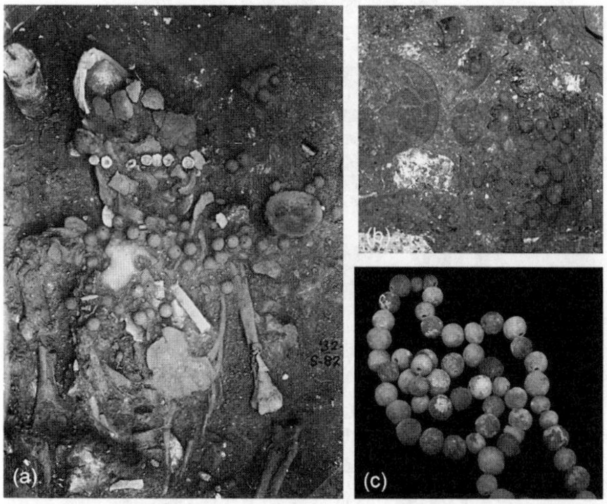

FIGURE 2.18. Ceramic beads painted to resemble jade beads: (a) Piedras Negras Burial 5 in situ (photograph courtesy of the University of Pennsylvania Museum of Archaeology and Anthropology); (b) Piedras Negras Burial 13 in situ (photograph courtesy of Stephen Houston); and (c) ceramic beads from Piedras Negras Burial 5 (photograph by author).

motivation was not frugality but simply that jade was in shorter supply at Piedras Negras in comparison to even neighboring kingdoms like Palenque.[67] Nevertheless, jade sometimes appears in surprising places. A relatively modest burial at Budsilha, a subsidiary community within the kingdom of Piedras Negras, produced a jade necklace (fig. 2.19). The necklace was a composite of relatively poor quality jade beads, brought together from a variety of different sources. Objects such as the Budsilha necklace indicate that prestige materials circulated outside the purview of the royal courts and that we should probably pay greater heed to the quality of craftsmanship, not simply the material of manufacture. As Brigitte Kovacevich notes for the distribution of jade in burials at Cancuen, "jade and greenstone may not have been restricted to elites, but the quality and type of artifact may have been the important indicator of status."[68]

As a point of contrast to the Piedras Negras tombs, Pakal, who reigned at Palenque decades earlier than Ruler 3 of Piedras Negras, was interred with a necklace of 118 jade beads of various shapes. Many were carved to represent fruit and other vegetal motifs, underscoring the king's fertility (fig. 2.20).[69] Over his shoulders and across his chest Pakal wore a massive jade drape of 189 tubular jade beads, similar to the one that K'inich Kan

FIGURE 2.19. Necklace of jade beads from Budsilha Burial 4 (photograph by author).

FIGURE 2.20. Jade mask, ear ornaments, and collar worn by Pakal in his sarcophagus (photograph by author).

Bahlam II is shown wearing on the western sanctuary jamb of the Temple of the Cross. Each of Pakal's fingers was adorned with a ring. His forearms were covered in bracers of approximately 200 jade beads, transforming his lower arms into cobs of verdant maize. In his right hand he held a jade cube and in his left a jade sphere—a provocative arrangement whose meaning has yet to be deciphered. Jade spheres were located near the toes of each foot and probably relate in some fashion to the sphere and cube in Pakal's hands, perhaps marking the four corners of his body.

Indeed the discrepancy in the amount of jade between the tombs at Piedras Negras and Palenque calls into question the very notion that the quantity of jade and perhaps other prestige goods should be understood to correlate directly with wealth, status, and especially power among the Maya, particularly in attempts to generalize such measures across polities. Despite being "poor" in jade, the lords of Piedras Negras were "rich" in other material resources (for example, sculpture of unparalleled quality, monumental architecture) and exercised significant influence over the populations of the Middle Usumacinta River Basin, as evidenced by the long-term success of the polity. Thus explorations of wealth and status among the Maya, particularly as evidenced by the royal body, must be sensitive to how geography, political fortune, and other factors influenced access to prestige goods among competing Maya kingdoms.

Keeping in mind the political dimension of royal funerary contexts, the tombs of queens are rarer than those of kings, presumably because of the diminished sociopolitical importance of women at Maya courts. Furthermore, claims to the throne via the matriline were rarer and presumably more tenuous than those made through the patriline.[70] Nevertheless, the tombs of Maya kings and queens are remarkably similar, with significant overlap in their adornments and funerary furniture. For example, Pakal and the Red Queen of Palenque were each buried with a greenstone-encrusted headband across their brows, large mosaic greenstone

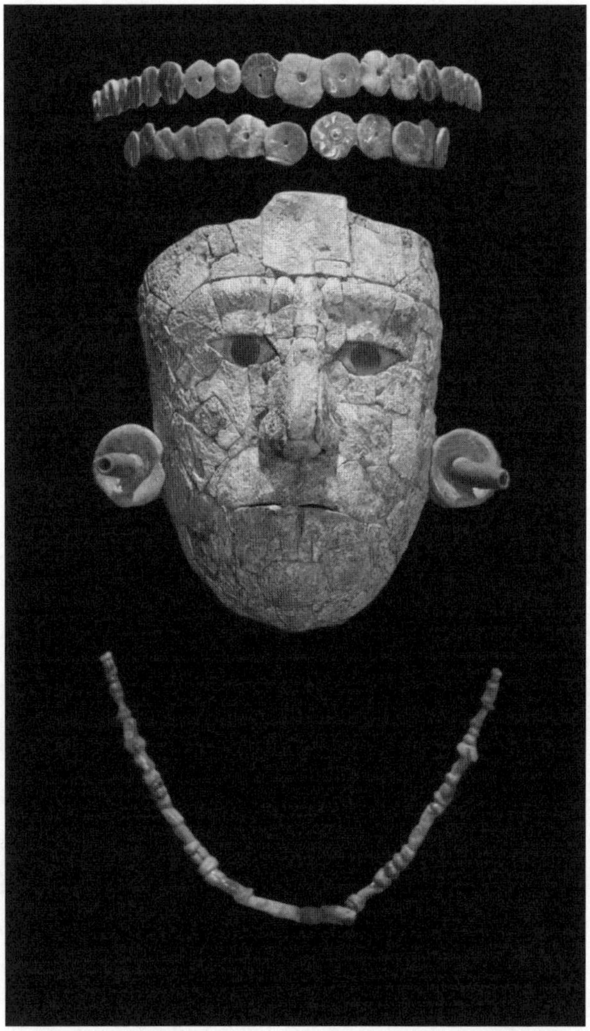

FIGURE 2.21. Malachite mask, ear ornaments, necklace, and diadem from the tomb of the Red Queen of Palenque (photograph by author).

FIGURE 2.22. Bald dancing supernatural on a vase from Altar de Sacrificios (drawing by Linda Schele, © David Schele, courtesy Foundation for the Advancement of Mesoamerican Studies, Inc.).

masks over their faces, and smaller greenstone jade mosaic masks (an ancestral portrait), with stone celts at their waist (figs. 2.20 and 2.21; see also fig. 2.51).[71]

Nevertheless, some objects placed within royal tombs point to the distinct duties and obligations of Maya kings and queens. A monolithic effigy axe (the handle and blade are of a single piece of obsidian) that was recovered from a tomb attributed to Shield Jaguar III of Yaxchilan alludes to his duties as war-maker and sacrificer of captives.[72] In contrast, weaving implements have been found in a number of queens' tombs, underscoring their importance as the head weavers of Classic Maya courts. For example, eighty-one bone needles were recovered from the tomb attributed to Shield Jaguar's wife, Lady K'abal Xook.[73] Although some of these needles may have been used to pin bundles, including her burial shroud, or were used in bloodletting, the sheer quantity suggests that some must have been used in the production and embellishment of cloth. Similarly, a bone needle and spindle whorl lay atop the sarcophagus lid of the Red Queen.[74] A similar assemblage of bone needles was recovered from the Margarita tomb at Copan.[75] As a contemporary analogue, Redfield and Villa Rojas note that women of the Yucatec village of Chan Kom are often buried with a needle and thread.[76]

ADORNMENTS OF THE HEAD

Heads are rarely depicted without hair and other adornments in Classic Maya art. In the few extant cases of unadorned heads, they look especially unusual and idiosyncratic (fig. 2.22). Much like the head modeling and dental modification discussed in chapter 1, the adornment of the head was essential to the construction and transformation of the Classic Maya person. Thus Maya royalty dressed their heads with rare and distinctive regalia that, among other things, operated as emblems of authority and served to distinguish them as unique, distinct from the masses. Remarkably, many of the same ornaments that are shown worn by Maya lords and ladies in Classic period images have been found by archaeologists with the dead and in other contexts.

Among the most important of these regalia were the royal headbands that bound Maya kings into the authority of office.[77] Excavations at the Kaqchikel capital of Iximche uncovered the seated body of an adult whose skull was encircled by a band of gold.[78] Based on reference to headbands in hieroglyphic inscriptions, some Classic period forms of this regalia were made of paper, mostly likely from the *amate* (fig) tree. Specifically, these are the headbands that seem to have been donned in accession rituals.[79] Paper headbands, however, have never been recovered by archaeologists.

Jade disc diadems have been found in Classic period tombs, including Tikal Burials 77 and 196, Piedras Negras Burial 5, and the burials of Pakal and the Red Queen of Palenque (see figs. 2.18 and 2.21).[80] Monuments at Palenque confirm that both the kings and queens donned these jade disc diadems (see fig. 1.26c). It is unclear, however, what (if any) connection existed between the jade disc diadems that are commonly shown in Maya art and the paper headbands mentioned in texts and occasionally shown, as in the scene on the Palenque Temple XIX platform.[81]

Classic period imagery shows a range of diadems depicting supernatural beings, affixed to royal headbands and other headgear. These diadems are frequently glossed as the Jester God, though David Stuart has recently shown that at least three distinct entities are represented by these diadems.[82] One of these entities, Ux Yop Huun, seems to be the spiritual embodiment of the paper headbands (fig. 2.23). As noted earlier, Ux Yop Huun is an avian creature possibly inspired by the Principal Bird Deity of Preclassic times. As Taube suggests, when Maya kings wore this avian creature atop their heads they likened themselves to the world tree as the axis mundi.[83] As an actual physical ornament worn on the forehead, Ux Yop Huun diadems have never been found. Depictions of this being on headbands may simply have been a referential motif that, as Stuart suggests, was intended to call attention to the specific nature of the headband as sacred paper.[84]

The other two diadems that are mistakenly lumped together as the Jester God are a trident blossom and a creature with an upturned nose and jagged, triangular teeth.[85] Although the blossom is occasionally anthropomorphic, in other contexts it appears simply as a flower, as depicted on a Late Preclassic mask from Tikal Burial 85 (fig. 2.24). The emblem is likely related to maize and associated concepts of fertility and vitality. Stuart speculates that it is a rendering of a maize tassel. More likely it is derived from earlier Olmec depictions of the Maize God, who was often shown with a cob of maize flanked

FIGURE 2.23. Ux Yop Huun as he emerges from the bicephalic serpent on the Palenque sarcophagus lid (modified from drawing by Linda Schele, © David Schele, courtesy Foundation for the Advancement of Mesoamerican Studies, Inc.).

by foliation emerging from his cleft head.[86] The diadem with the upturned snout and jagged teeth is generally recognized to be a shark (*xook*). Certainly its swirling pupils confirm that this is an underwater or underworld creature, in contrast to Ux Yop Huun, who is celestial. The *xook* adornment also appears in the imagery of Palenque and has been found archaeologically at Maya sites, including in the tomb of Pakal and in excavations at Aguateca (fig. 2.25; compare with the headdress held by Lady Sak K'uk' in fig. 1.26c). Although this shark diadem has been assumed to have served as a generic emblem of rulership, it must have had more specific connotations, perhaps related to warfare or bloodletting, in light of the aggressive nature of the creature depicted.

In addition to headbands and diadems, Maya lords and ladies are also shown wearing complex headdresses, especially in scenes of public dance and other ritual performances. Headdresses, like masks, were employed on some monuments as part of supernatural impersonation costumes, allowing the wearer to become the beings represented. The headdresses depicted on the Bonampak murals seem to have been made of diverse materials, probably a mix of carved wood, molded ceramic laminate, stone, feathers, and other animal products (fig. 2.26).[87] The few headdresses that are known archaeologically are fragments and residues. A rare example was identified in Tikal Burial 195, only because the tomb chamber flooded at some point in antiquity, burying the tomb objects in a layer of fine silt that preserved impressions of the perishable goods within (fig. 2.27). The headdress was located in the southeastern portion of the tomb and consisted of a frame (presumably of wood) mounted with feathers, some of which were 80 cm long, presumably from a quetzal.[88]

FIGURE 2.24. Late Preclassic mask from Tikal Burial 85. Note the trident blossom diadem (photograph courtesy of the University of Pennsylvania Museum of Archaeology and Anthropology).

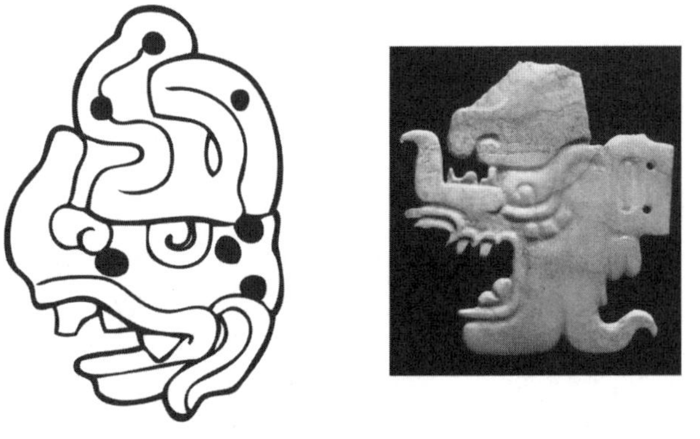

(a) (b)

FIGURE 2.25. *Xook* diadem: (a) tomb of Pakal (drawing by author after Ruz Lhuillier, *El Templo de las Inscripciones*, fig. 218); and (b) Aguateca (photograph courtesy of Takeshi Inomata).

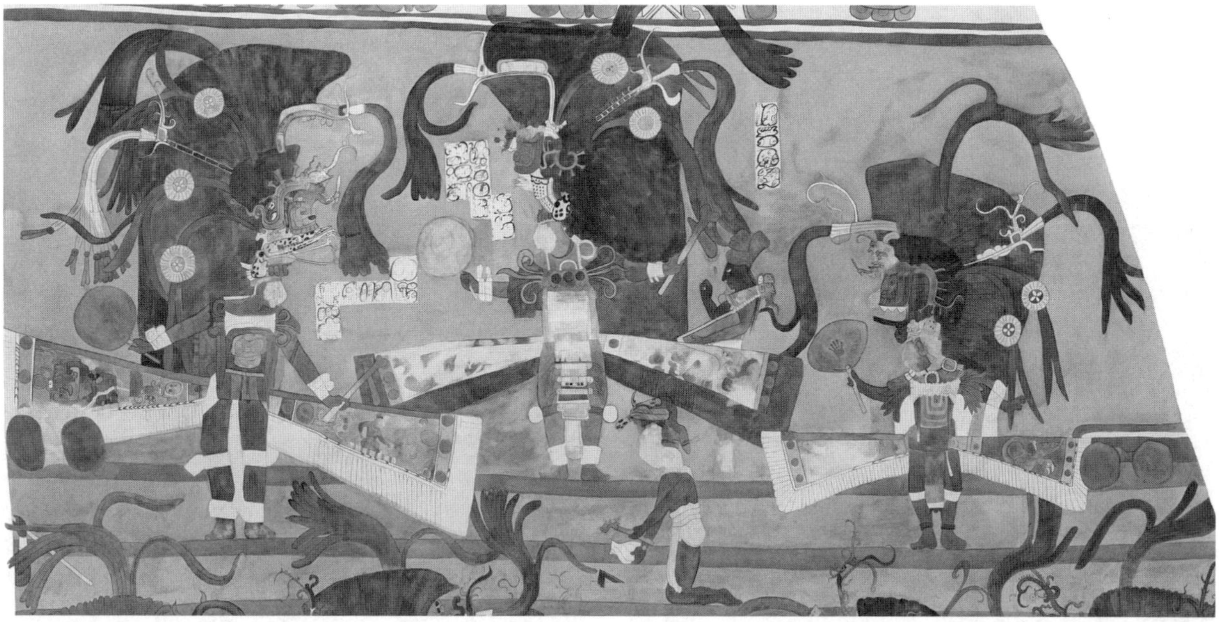

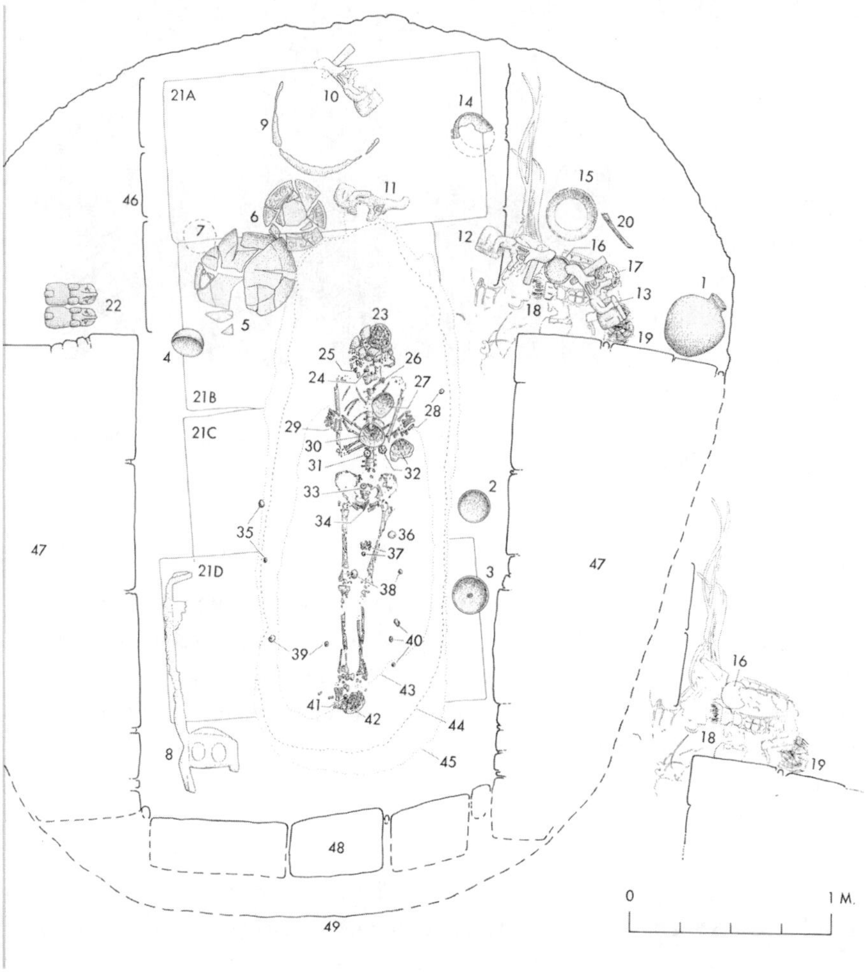

FIGURE 2.26. (*above*) Dancers from the mural of Room 3, Structure 1, Bonampak. Note the composite materials of the headdresses (reconstruction painting of Bonampak, Mexico, Yale University Art Gallery, Gift of Bonampak Documentation Project, illustrated by Heather Hurst and Leonard Ashby).

FIGURE 2.27. (*left*) Tikal Burial 195 (originally published as Coe, *Excavations in the Great Plaza, North Terrace, and North Acropolis of Tikal*, fig. 198, image courtesy of the University of Pennsylvania Museum of Archaeology and Anthropology). Note the remains of the headdress at sixteen and eighteen and the *Spondylus* shells at 23, 27, 30, 32, and 42.

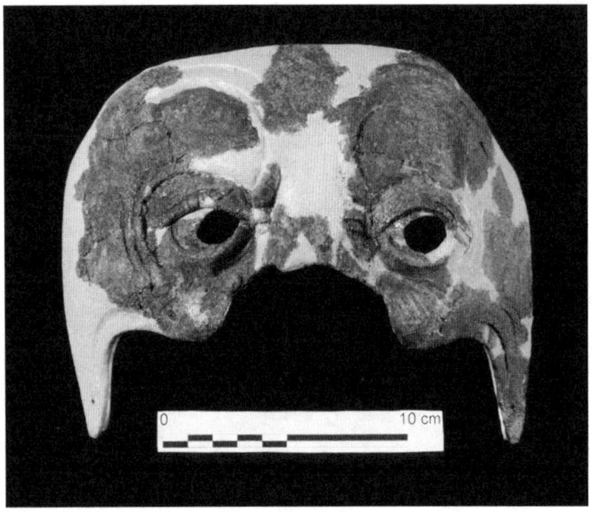

FIGURE 2.28. Ceramic laminate mask from Aguateca (photograph courtesy of Takeshi Inomata).

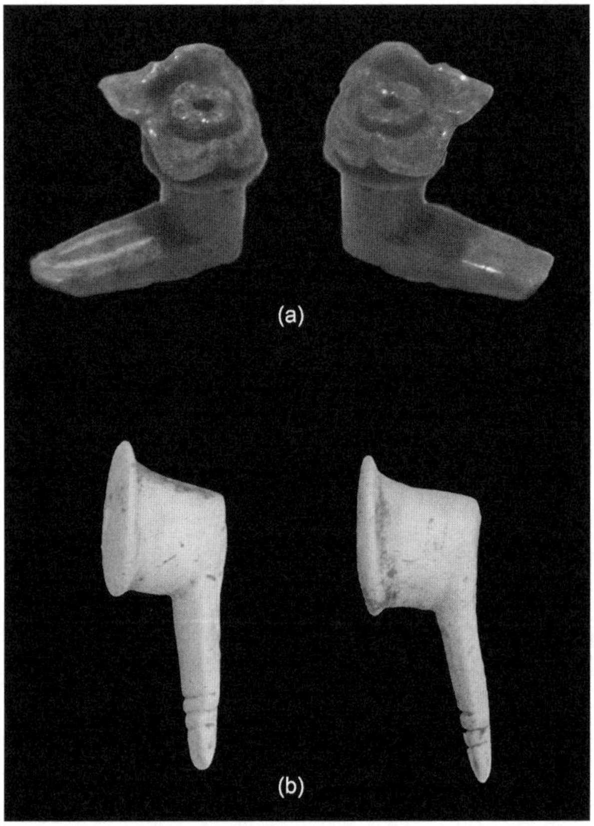

FIGURE 2.29. L-shaped stemmed ear ornaments: (a) floral design from the tomb of Pakal; and (b) Rancho Búfalo Burial 1 (photographs by author).

Río Azul Tombs 19 and 23 contained "decomposed, spongy, brown matter" around the heads of the primary occupants.[89] This distinctive material may be similar to the ceramic laminate that was identified by Rae Beaubien as the material used to construct a mask at Aguateca (fig. 2.28).[90] Because of its fragility, this material has been largely overlooked by archaeologists.

The Precolumbian Maya universally pierced their earlobes in order to insert a wide range of ornaments. Although such ornaments were certainly appreciated for aesthetic purposes, like dental and cranial modifications, these ornaments likely also had a deeper social significance relating to both the humanization of the body and conceptions of vitality and souls. In the early sixteenth century, refusing to return to Spanish society after his adoption by the Yucatecan Maya, the shipwrecked Gonzalo Guerrero explained: "I have my face tattooed and my ears pierced[,] what would the Spaniards say should they see me in this guise?"[91] Even in the rare depiction of nonelite members of Maya society, such as the recently uncovered murals at Calakmul, their earlobes are filled with large disc plugs (see fig. 1.19).

Diverse ear ornaments are evident in Classic period imagery and have been recovered archaeologically. Among the most common forms are small L-shaped earplugs that were either carved to resemble flowers or had floral disc inserts attached to the disc ends of the stem. For example, a pair of jade L-shaped stems was found near Pakal's skull with floral inserts (fig. 2.29a). A similar pair of L-shaped stems with floral jade flares was found in Tikal Burial 23.[92] These L-shaped stems date back to the Late Preclassic, as evidenced by a shell pair found in a Late Preclassic tomb at Rancho Búfalo (fig. 2.29b).

FIGURE 2.30. Squash flowers blooming in a milpa, Chiapas, Mexico (photograph by author).

In addition to the small L-shaped stems, Pakal was also buried with a pair of gigantic jade flower ear ornaments (see fig. 2.20). Why the Palenque king was buried with two pairs of ear ornaments is not clear. Perhaps Pakal wore the smaller pair in quotidian life, with the larger ornaments reserved for ceremonial affairs. Pakal's large jade ear assemblage consists of wide earflares that would have been attached to his earlobes. Each disc is incised with four leaves that surround a large opening in the center. A long jade bar was mounted in that space, replicating either a vine or a stem, and was capped by a small flower. Counterweights were hung behind the earlobe to keep the stem in place. A similar massive pair of floral ear ornaments was recovered in Tikal Burial 196.[93] These stems are relatively rare. More often royal tombs only contain a large jade or shell earspool. At Kaminaljuyu the Carnegie Institution excavated a pair of ear ornaments with stems made of perishable material, suggesting that we are likely missing important parts of other ear ornament assemblages.[94]

Recall the common Maya trope of metaphorically linking heads with gourds. It may be that some of the Classic period ear ornaments are specifically referencing flowers of the family Cucurbitaceae, which includes calabash, squash, and other gourds (fig. 2.30). This parallel is especially evident in the assemblages that include long stems capped with tiny flowers, such as those from Pakal's tomb. The tiny bell flowers that cap the stems resemble the bell-shaped blossoms of gourd flowers shortly before they bloom. Alternatively (or additionally), the floral symbolism of Maya ear ornaments may refer to the flowery nature of Maya souls and to the beauty of the speech and sounds that reached the ears of lords and ladies.[95]

Ears, along with mouths and noses, were understood to be portals that emitted and received winds into the body.[96] The glyph for road or passage (*bih*) is a stylized earspool that is often worn by *witz* and other supernatural beings in Maya iconography. On a *witz* mask from Tikal Temple 5D-33 serpents pass through the earspool, effectively the glyph for *och bih* (road entering), a euphemism for death (fig. 2.31a). The meaning of the Tikal *witz* mask is complex, perhaps referencing the passage of supernatural beings into and out of sacred mountains but perhaps also identifying Temple 5D-33 as a funerary monument. A similar element appears as a diadem in the murals of San Bartolo, in a scene of what seems to be birth at the beginning of creation (fig. 2.31b). Notably, the tongue of the serpent on the San Bartolo murals ends with the "*ajaw*-face" sign. Taube shows that the motif of serpents exiting discs is pervasive in Mesoamerica, as evident, for example, on the Temple of the Feathered Serpent at Teotihuacan, which he likens to a Flower Mountain.[97] Thus the passage of snakes through portals may be a metaphor for the movement of souls and spirits at birth, death, and perhaps other times.

CAPTURING LIFE, MASKING DEATH

For the contemporary Maya, death is not so much an event as a process. As Redfield and Villa Rojas explain for the Yucatec of Chon Kom, "death is a time of struggle,

(a) (b)

FIGURE 2.31. *Och bih* ornaments: (a) ear ornaments on *witz* mask from Temple 5D-33 of Tikal (originally published as Coe, *Excavations in the Great Plaza, North Terrace, and North Acropolis of Tikal*, fig. 182, courtesy of the University of Pennsylvania Museum of Archaeology and Anthropology); and (b) worn by a figure on the San Bartolo murals (illustration by Heather Hurst).

FIGURE 2.32. Jade bead (*arrow*) that was placed in the mouth of the primary occupant of El Zotz Burial 9 (photographs courtesy of Stephen Houston, El Zotz Archaeological Project). *Inset*: laboratory photograph of the same bead. In the field photograph a shell bead necklace overlays the assemblage.

the soul seeking to leave and yet to remain, the forces of evil contending with Heaven for his soul."[98] The struggle ends with the body's final exhalation, marking the departure of the spiritual essences from the body. Numerous contemporary Maya communities perceive souls as entering and exiting through the mouth. June Nash notes that the Tzeltal believe the soul exits through the tongue.[99] Gossen similarly observes that the *chulel* of the Tzotzil are located at the tip of the tongue; they are the first type of soul to enter the child at birth and the last to leave after death.[100] Such a perspective seems a meditation on the actual physiological process of death. Final breaths are often labored and may be preceded by the so-called death rattle, caused by the accumulation of saliva in the throat of the dying when they can no longer swallow properly. Breathing becomes shortened and irregular until the final exhalation.

Writing shortly after the conquest, Las Casas notes for the Pokom that "when it appears then that some lord is dying, they had ready a precious stone which they placed at his mouth when he appeared to expire, in which they believe that they took the spirit, and on expiring, they very lightly rubbed his face with it."[101] Subsequent Pokom rulers supposedly kept this jade bead "because the stone was esteemed as a divine thing," imbued with the sacred essence of the now dead ruler. Based on this passage, Michael Coe proposes that the isolated jade beads commonly found near the mouths of the dead in Classic Maya burials relate to this concept of capturing the last breath (fig. 2.32).[102] These breath-beads tend to be larger than most other burial beads (if there are any). They are also perforated, indicating that they had been worn prior to their use in the mortuary rites. This practice of placing beads at or in the mouth dates as far back as the Late Preclassic period, as evidenced in a recently excavated tomb at Chiapa de Corzo.[103]

The Classic Maya understood jade to be the embodiment of life and vitality not only because of its preciousness but specifically because of its blue coloration, which hinted at fertility.[104] As Taube shows, it was a material that was also closely related to "wind, carrier of rain, and the essence of the life spirit."[105] Among the Aztec, jade was thought to respire and attract moisture.[106] According to Diego de Landa's *Relación de las Cosas de Yucatan*, the mouths of the Yucatec dead were filled with ground maize during the sixteenth century.[107] Considering the close association of maize, jade, life, and fertility, this may have been either a local variant of the practice of placing jade in the mouth or an alternative practice employed by people without access to jade. We can only wonder if the Classic Maya nonelite stuffed the mouths of their dead with maize.

Houston and Taube propose that greenstone funerary masks placed over the faces of the dead were, like jade beads, used to capture the spiritual essence of the dead.[108] The tradition is especially prominent in the central and northern lowlands at Oxkintok (Tomb 5, Structure CA-3), Dzibanche (Tomb 1, Building 1; Tomb 1, Temple II; Chamber 203, Temple of the Cormorants), Calakmul (Tomb 1, Structure II; Tomb 4, Structure II; Tomb 1, Structure III; Tomb 1, Structure VII), Tikal (Burials 10, 48, 160), and Palenque (Pakal and Red Queen) (see figs. 2.20 and 2.21).[109] At Calakmul (and perhaps also these other sites) funerary masks were placed over the faces of corpses after they had been wrapped in a funerary shroud.[110] In this sense the mask was a projection of the persona bundled within, an image that would not be affected by the rapid putrefaction of the face beneath.

One of the Calakmul masks is especially evocative of Houston and Taube's suggestion (fig. 2.33). Breath volutes and shell ornaments emerge from the nostrils and mouths of the mask. Similarly, *ik'*-shaped teeth are visible on Pakal's mask, a rare glimpse of human teeth in Maya art and, like the Calakmul mask, a rendering of his vital essence in stone (see fig. 2.20). The anthropomorphic masks from Palenque and Calakmul capture a youthful visage, despite the advanced age of some of the decedents, and were likely intended to equate the deceased with the Maize God. Funerary masks were like the ancestral heads worn by Maya kings in that both embodied the spiritual essence of the revered dead.

SOLAR VITALITY

The Maya have painted their dead in red pigment since Preclassic times.[111] This funerary treatment was reserved primarily for the bodies of Maya royals. In part this is because the red pigments used to paint the bodies (usually cinnabar or red hematite) were difficult to procure or produce and thus had intrinsic value. Cinnabar is red mercury sulfide and occurs within volcanic geological strata, especially those associated with hot springs. Thus it has to be imported from the highlands. Extensive exposure to cinnabar can cause mercury poisoning, including

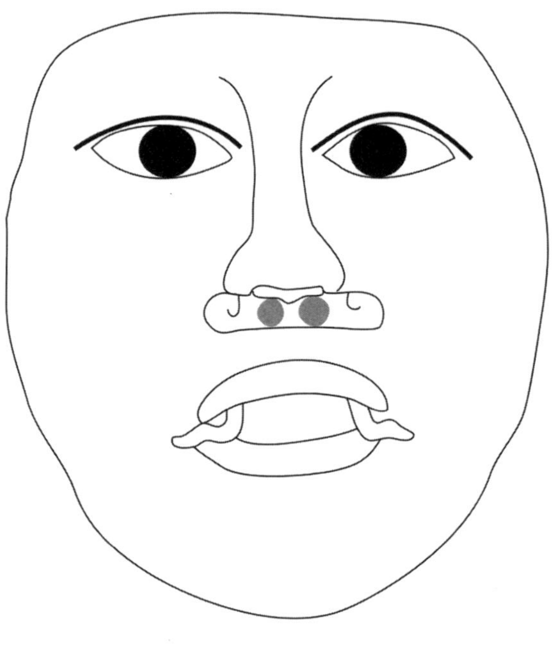

FIGURE 2.33. Jade mask from Tomb 1, Structure II-D, Calakmul (drawing by author).

FIGURE 2.34. Hematite clay bricks and *Conus spurius* shell tinklers from Burial 9, El Zotz (photograph courtesy of Stephen Houston, El Zotz Archaeological Project).

shaking, loss of senses, and even death. Although there has been no systematic survey of ancient and modern sources of cinnabar in the region, reports of cinnabar locations in Mesoamerica include the Sierra Gorda of Querétaro, Mexico; central San Luis Potosí, Mexico; and western Honduras.[112] Red hematite (iron oxide) is far more common and occurs not only in banded iron formations but also as a precipitate in weathered tropical soils. For example, my colleagues and I noted an abundance of hematite nodules on the surface of La Mar, Mexico, a subsidiary site in the Piedras Negras kingdom. Through a labor-intensive process, both cinnabar and hematite were ground into powders that could then be dissolved in water and painted on a variety of media, including human bodies. Excavations at El Zotz in 2010 uncovered an Early Classic tomb, Burial 9, that contained fifteen blocks of specular hematite (fig. 2.34). Presumably the hematite clay was packed into fist-sized bricks for transport and trade.

For the Maya, red is the color of sunrise in the east. The Tzotzil of Chamula see life as an accumulation of heat. Mature people, especially shamans and other powerful individuals, are understood to possess significant quantities of heat.[113] Moreover, heat is associated with reproduction, which may help explain why the "child of father" glyph involves the *k'ahk'* (flame) sign. Fire and combustion are understood to be regenerative in the Maya worldview; incense is used to feed ancestors, and spirits and fields are burned in preparation for planting. All heat and fire was ultimately derived from the sun, whose winds were depicted as square-nosed serpents marked by flame volutes, the so-called zip-monsters (fig. 2.35).[114] Heated breath emanates from solar beings and royalty, both of whom were textually identified as *k'inich Ajaw* (great sun lords). Solar objects and beings infixed with the *k'in* (sun) sign are closely linked to the fertility of the sun and the four cardinal directions.[115]

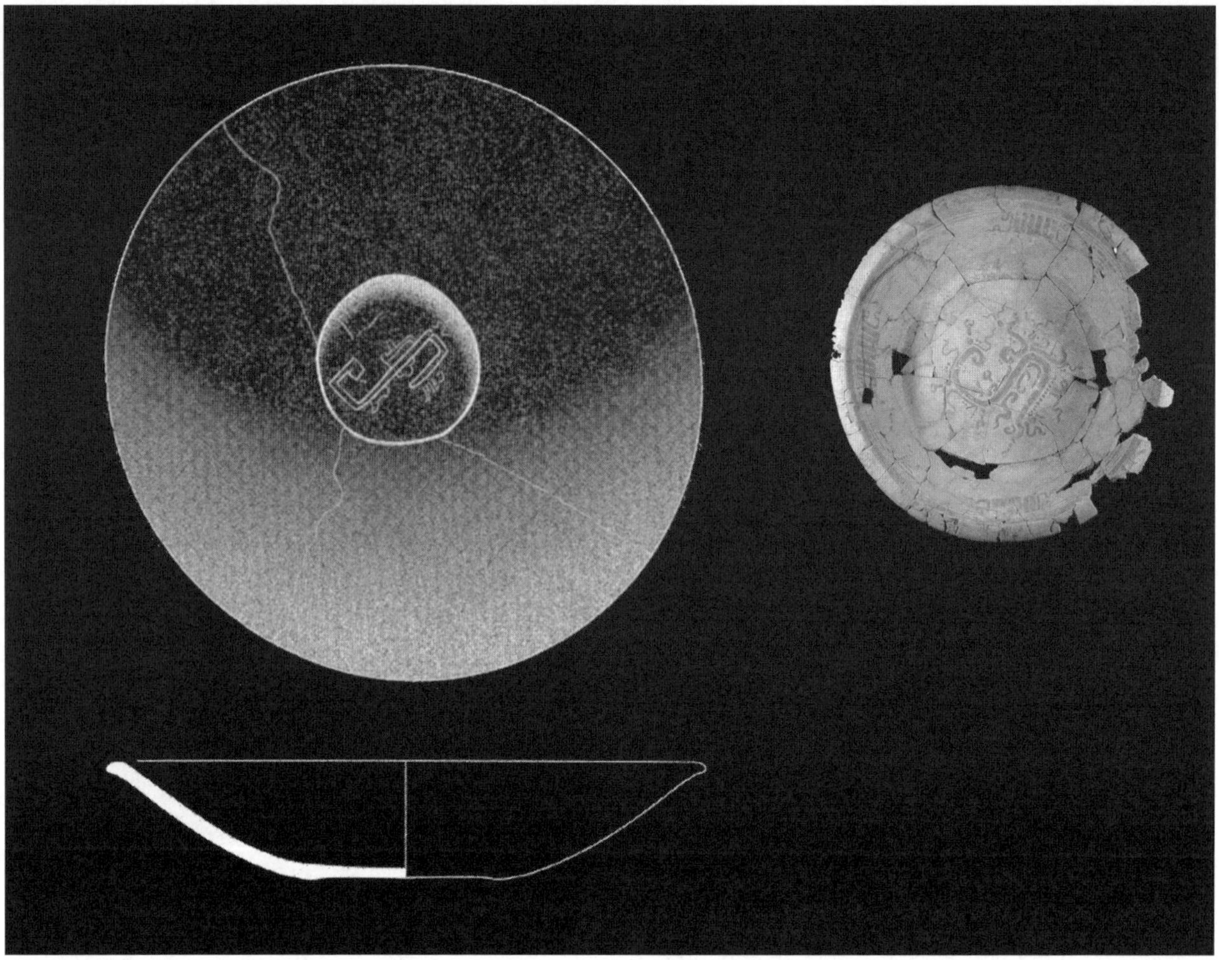

FIGURE 2.35. Late Classic period vessels from Piedras Negras showing the fiery square-nosed serpent (images by A. René Muñoz). The vessel at left is from Burial 82.

By painting the royal corpses red, the Maya symbolically imbued them with solar heat, combating the cold earthiness of decay. The act may have been meant to replicate the travels of the sun, which descends into the underworld in the west (*och k'in*, "entering sun"), passes through the underworld, and is reborn in the fiery sea to the east (*el k'in*, "exiting sun").[116] The symbolism is most overt in the tomb of Pakal. The sarcophagus lid shows the dead king rising as the new sun from the underworld atop a *k'in*-marked object (see fig. 2.05). Beneath the lid the king's body was painted in blazing red cinnabar that covered his body and the contents of the sarcophagus.[117] In the adjacent temple the inside of the sarcophagus of the possible wife of Pakal, the so-called Red Queen, was saturated with cinnabar in preparation for her fiery rebirth.[118]

The remains of the primary occupant of El Zotz Burial 9 were painted first in a layer of red hematite, followed by a layer of red cinnabar (fig. 2.36). Cinnabar is the more vibrant, brighter of the two pigments. Specular hematite is darker, has a brownish hue due to its iron content, and appears to sparkle as a result of mineral inclusions. Some of the objects from the El Zotz tomb were also intentionally painted with red cinnabar. Other materials—such as the remains of the sacrificed youths—only demonstrated trace amounts of red pigments, however, reflecting the unintentional spread of pigment during corpse decomposition and partial tomb collapse. It is possible that the different shades of red produced by hematite and cinnabar had subtly different meanings for the Maya. Cinnabar obviously suggests the sun, as evidenced by

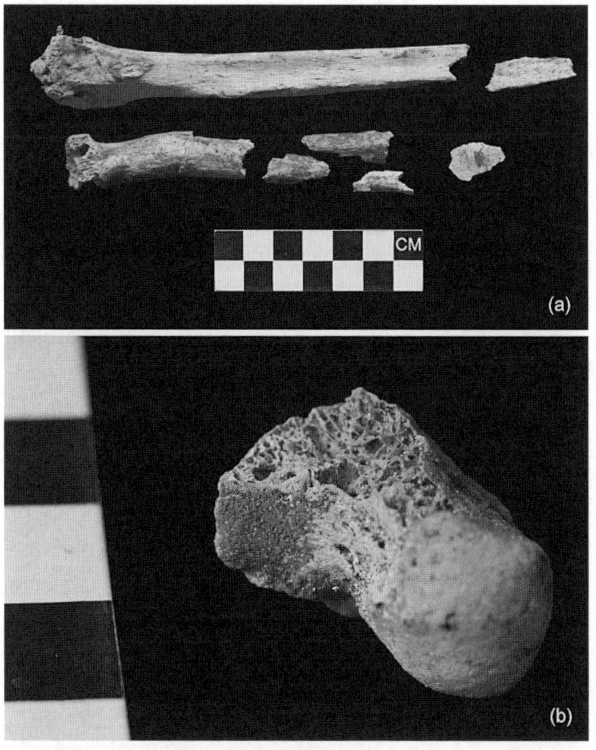

FIGURE 2.36. Painted bones of the primary occupant from El Zotz Burial 9: (a) specular hematite (brick red) on the bones of the right forearm; and (b) a layer of specular hematite overlain by cinnabar (vibrant red) on a right wrist bone (capitate) (photographs by author).

the Palenque tombs. Yet it may be that hematite, with its ruddy hue, was meant to replicate the flesh and blood of the living and to hide the discoloration of *livor mortis*. In that sense the painting of Maya corpses was much like the red paint applied to many Maya buildings, to show them as coursing with animate spirit.[119]

Red pigment was applied directly to the royal corpses and funerary shrouds throughout the Maya lowlands, including Río Azul Tombs 19 and 23, Calakmul Structure III Tomb 1 and Structure II Tomb 4, and the royal tombs of Tikal.[120] Some scholars have erroneously assumed that the presence of red pigment on skeletal remains means the bone was directly painted. In most cases of purported postinterment painting, however, the evidence is ambiguous, as the pigmentation could just as easily have transferred during corpse decomposition.[121] One of the best examples of direct bone painting was noted by Lori Wright in her study of the remains from Burial PNT-019 of Tikal.[122] She noted drips of red pigment running down the sides of the horizontally placed bones and that some of the paint had entered the medullary cavities of long bones that had already fallen apart. Such concentrated drip marks do not occur in contexts where the pigment transfers to the bone in a dilute solution of human putrefaction, and rarely do endosteal surfaces receive pigment in such instances of painted corpse decomposition.

The painting of Maya bodies may also have had an underlying functional purpose. Decomposition sets in rapidly in the hot and humid lowlands, triggered first by the invasion of microorganisms and macroorganisms. Mercury, the element that constitutes cinnabar, is an effective pesticide and may inadvertently have prevented infestation of the corpse, especially by flies that lay eggs on bodies hours after death. Vera Tiesler notes that the cinnabar on Pakal's remains was laid down in multiple layers interlaced with deposits of an unknown organic substance.[123] It is possible that substance was a resin, similar to the material used to coat funerary bundles at Calakmul (see the section on bundles later in this chapter), in an attempt to protect the body against attack from insects and other organisms.

ARRANGING THE BODY

The modern Maya are deeply concerned with the expression of self through action of the body. As Pitarch observes among the Tzeltal, "attention to bodily movements and the economy of gestures is extraordinary. The way of sitting, leaning, standing; hand movements; facial expressions—everything should transmit the impression of serenity and self-control. There is a constant concern with maintaining a vertical position and corporal symmetry."[124] Such control obscures the inner emotions of the heart, the location of the souls. The concern with body posture is quite evident in Classic period iconography. The passive, emotionless faces of Maya kings and queens hide all feeling, whereas the expressed agony of tortured captives exposes their inner animal nature. Rather than using facial expression, meaning is conveyed in Classic period imagery through movement of the limbs, especially the arms and hands, in dance and otherwise (fig. 2.37).[125] After the head, the hand is the second most common body part used in Maya hieroglyphic writing (for example, see the glyphs in fig. 2.37b).

Considering the importance of bodily comportment and the conveyance of meaning through hand gestures, the manner by which the Maya arranged the bodies of the dead undoubtedly was imbued with symbolic significance. Across the social strata, the most common way the Maya placed the bodies of their dead within the grave was in the extended supine position—a body on its back, legs extended, with the face usually oriented skyward (see figs. 2.17 and 2.27). This is especially true in the western Maya lowlands. At Piedras Negras 87 of the 93 (93.5 percent) undisturbed bodies were encountered in an extended supine position. Of the 29 bodies at Palenque for which information regarding body position was available, 26 (89.7 percent) were in an extended supine position.

FIGURE 2.37. Expression through hands in Maya art and text: (a) Maize God statue from Copan (drawing by Linda Schele, © David Schele, courtesy Foundation for the Advancement of Mesoamerican Studies, Inc.); and (b) Stela 1 from Piedras Negras. Note the hand glyphs pertaining to the passage of the lunar cycle (drawing by John Montgomery © Foundation for the Advancement of Mesoamerican Studies, Inc.).

Although the extended supine position is a relatively straightforward way to arrange the body, it is not necessarily the most convenient. Extended bodies require more space than flexed corpses and involve digging more earth or laying more stone in the construction of a crypt or tomb. Contrary to common perception, rigor mortis (the stiffening of the corpse that occurs when calcium ions concentrate in the muscles following cessation of adenosine triphosphate production) is not permanent. Rigor mortis peaks about twelve hours after death, after which the rigidity dissipates as early muscle decay sets in. Thus rigor mortis was not a dictating factor in body position in the past just as it is not in modern funerary contexts. In Euro-American tradition bodies are placed horizontally in coffins, with hands over the abdomen, in an arrangement that is meant to suggest the body at "eternal rest." It is likely the Classic Maya were referencing a similar trope.

The contemporary Maya associate verticality with the awakened body and rational self. Maya parents frequently chastise their children for slouching. Horizontality is submission of the body to the inner soul. Pitarch suggests that this concern with verticality among the Tzeltal relates in part to a preoccupation with protecting spiritual co-essences: "The loss of verticality may be total, as when lying down on one's back, which is when the *ch'ulel* or the *lab* gain complete autonomy from the body and are able to separate themselves from it." Moreover, the Tzeltal note that corpses cast no shadow: the shadow itself is "not so much a projection of the body as of the *ch'ulel* lodged in the heart." Thus "while one is on one's feet, in the daylight hours, one's shadow, projected by the *ch'ulel*, keeps itself horizontal, clinging to the ground, but when the body is horizontal, the *ch'ulel* raises itself up."[126] Horizontality is thus considered appropriate for sleep, as it allows the soul to come to the fore.

The stumbling and frequent collapse associated with

extreme drunkenness is understood to occur because the rational self is no longer in control of the body: the inner soul has taken control. Horizontality is needed in healing ceremonies, as it makes it easier for souls to be coaxed back into the body. The supine body of Classic period inhumations likely was not just a convenient arrangement that suggested sleep but may have also been important for untethering immortal souls from bodies in decay.

Certainly Classic period artistic conventions emphasize the importance of vertical control of the rational self. Protagonists are invariably depicted rigidly upright, even when seated.[127] This position contrasts with the depiction of captives and drunkards, who are shown crumpled, cowering, prone, and fallen (fig. 2.38; see also fig. 1.8). Early Mediterranean art, such as that of ancient Egypt and archaic Greece, seems to share this preoccupation with verticality of the body. But this concern was lost in Western artistic traditions long ago, giving way to bodies in a variety of both horizontal and vertical positions in Classical and later artistic traditions. A rare exception of a horizontal protagonist in Classic period Maya art is the Berlin burial vase, which depicts a horizontal dead body contrasted with vertical living mourners (see fig. 2.15). The composition is striking and unorthodox in the Classic period canon. In another rare depiction of a deceased royal corpse, Stela 40 at Piedras Negras shows the barely visible bundled body of a royal woman, laid out horizontally atop a bier-throne (fig. 2.39). Yet the artisans were ambivalent about showing the queen supine: her shoulders and head are prominently vertical, as she sits atop her bier-throne. As an entombed dead body, this royal woman was shown horizontal. Yet as a royal protagonist the sculptors were obliged to position her head and shoulders vertically, as is typical of depictions of active ancestors (see fig. 2.2).

Of the other six undisturbed, nonsupine bodies at Piedras Negras, five were in a flexed supine position—lying on their back with their arms and legs bent over the thorax or to one side (fig. 2.40). Although rare in the

FIGURE 2.38. A vertical Early Classic period lord crushes a horizontal captive on Tikal Stela 39 (drawing by Linda Schele, © David Schele, courtesy Foundation for the Advancement of Mesoamerican Studies, Inc.).

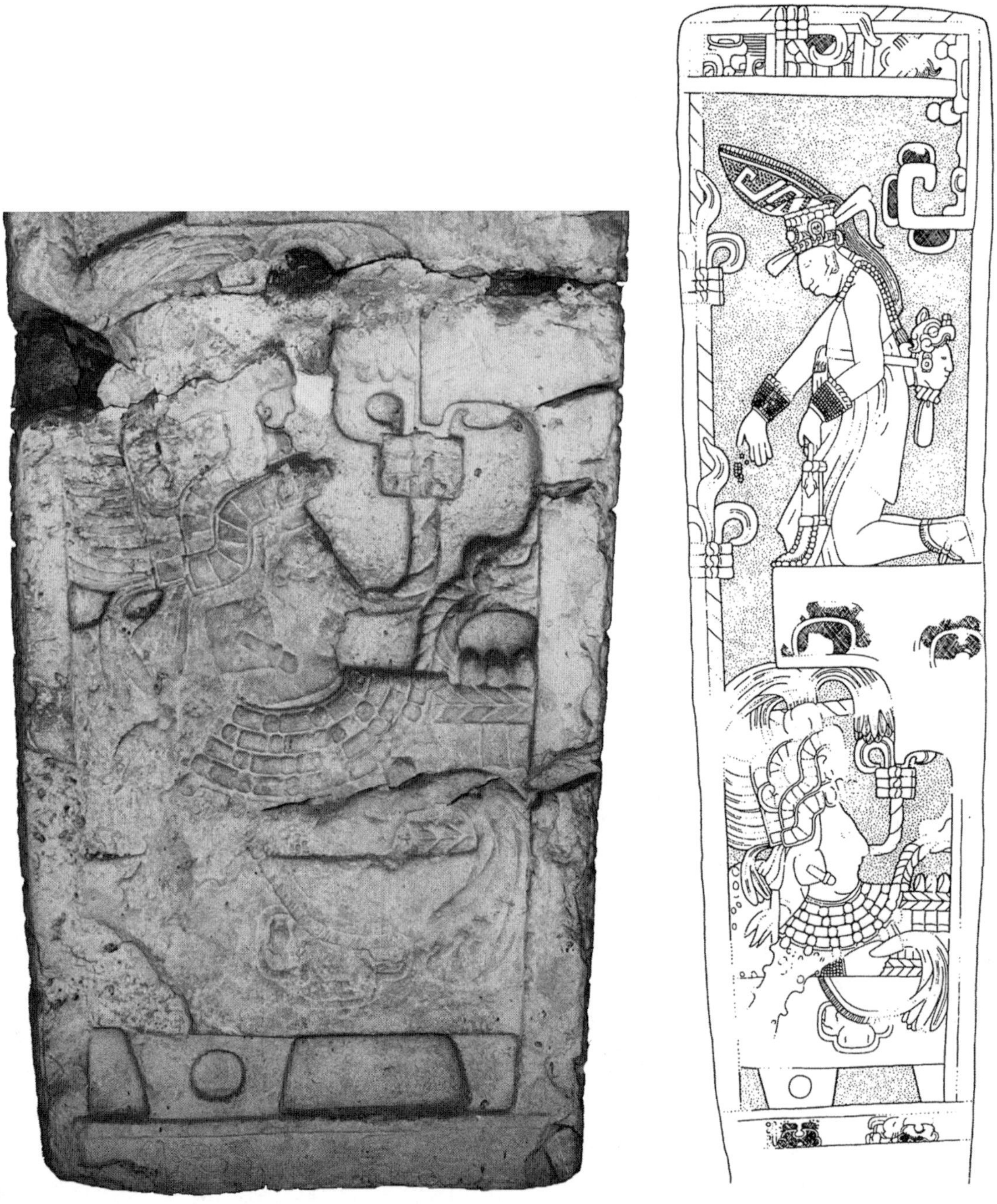

FIGURE 2.39. Supine yet upright posture of a royal entombed woman on Piedras Negras Stela 40 (drawing by John Montgomery, © Foundation for the Advancement of Mesoamerican Studies, Inc.; photograph by Charles Golden). Her bundled body is visible to the side of her left shoulder and right hand (compare with the bundled body in fig. 2.15), yet her head and arms are shown upright and active, as is typical of depictions of ancestors.

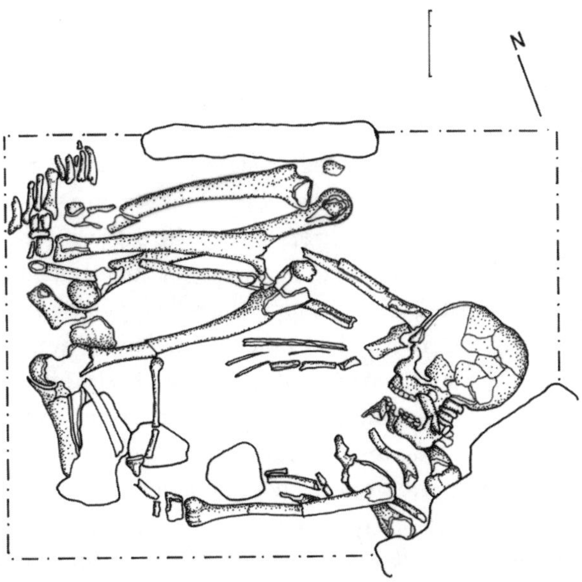

FIGURE 2.40. Bundled, tightly flexed body from Piedras Negras Burial 26 (drawing by Zachary Hruby).

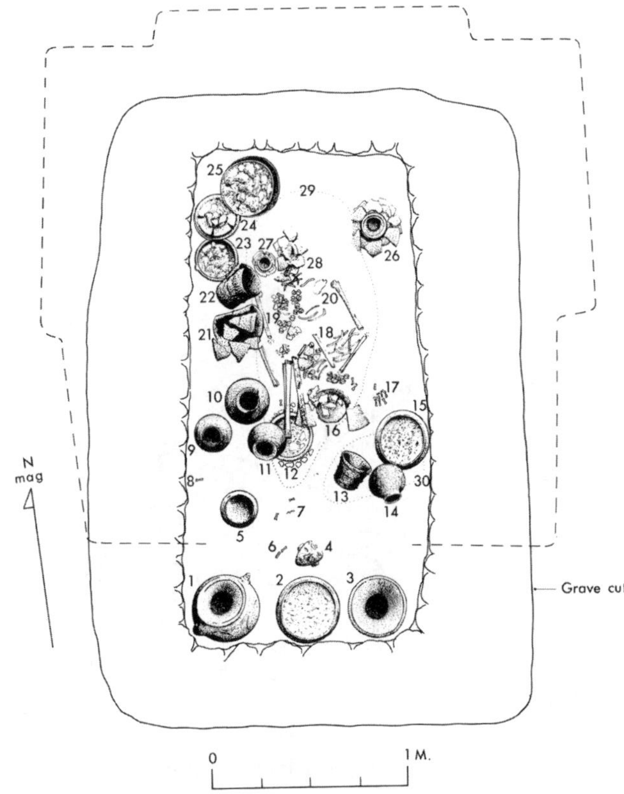

FIGURE 2.41. Late Preclassic Tikal Burial 85 (originally published as Coe, *Excavations in the Great Plaza, North Terrace, and North Acropolis of Tikal*, fig. 28, image courtesy of the University of Pennsylvania Museum of Archaeology and Anthropology).

west, flexed burials are more common in other parts of the Maya lowlands. In his survey of mortuary patterns in the southern Maya lowlands, W. Bruce Welsh reports 298 flexed burials out of a total of 788 (37.8 percent); the remaining 490 burials were extended.[128] Of the 130 complete bodies with observable body position at Tikal, 70 were flexed (53.8 percent). Of those, the majority ($n = 61$) lay on either their right or left side. Another 9 burials were flexed in a supine position, similar to the 5 examples noted at Piedras Negras. One is a probable Late Preclassic ruler interred in Burial 85.

Tight body flexure was accomplished by wrapping the corpse in a funerary shroud or with bindings, similar to "mummy bundles" of Precolumbian South America. Among other things, this arrangement facilitated the transport and display of corpses. Flexed corpse bundling is especially common in Mesoamerica at Early Classic period Teotihuacan and Kaminaljuyu, where such bundled corpses were interred in an upright position.[129] Four of the five flexed supine burials at Piedras Negras date to the Early Classic period, perhaps reflecting the influence of Teotihuacan (or Teotihuacan-inspired ideas and practices) during this time. Yet these bodies were not set upright (as at Teotihuacan), so it may be more appropriate to consider other explanations for the few flexed burials at Piedras Negras.

In contrast to Teotihuacan and Kaminaljuyu, most bundled bodies in the Maya lowlands were not placed in a seated position but instead on their backs, as at Piedras Negras (fig. 2.40). In fact a careful review of the literature reveals that most supposed seated burials in the Maya area (such as Tikal Burials 85 and 121) are in fact supine flexed burials, as noted above (fig. 2.41). True seated burials (bundled bodies set upright) have a different

FIGURE 2.42. Rebirth of the Maize God on a vase from Calakmul (drawing by author). Note the fetal position atop the capped "*ajaw*-face."

pattern of bone distribution. In such positions, as the body decomposes, the funerary wrap prevents lateral displacement of the bones of the skull and thorax; they then collapse internally and will be surrounded by the bones of the appendages, as is typical of the Teotihuacan seated bundles.[130] Perhaps the paucity of seated burials among the Maya relates to the dichotomy of verticality for the waking and living, horizontality for the sleeping, dreaming, and dead.

The occurrence of flexed supine burials in some parts of the Maya area, especially in the Central Petén, warrants further explanation. Iconographically, this posture is closest to the birth position in Maya iconography. This posture is most famously shown on the Palenque sarcophagus lid, where Pakal is reborn from the underworld (see fig. 2.5). As is widely recognized, Pakal's posture is in emulation of the birth of the Maize God (fig. 2.42). Perhaps these supine flexed bodies were also meant to replicate this mythic event. If so, the birth posture, which we more typically refer to as the fetal position, points to the regenerative nature of Maya souls, which are reborn to new bodies in later generations.

Piedras Negras had only one example of a flexed body placed on its side (fig. 2.43). This body was a sacrificial victim placed within an Early Classic tomb (Burial 110). The tomb's primary occupant, though poorly preserved, was arranged in an extended supine position. The second individual, a youth, was crammed at his feet in a position that remarkably mirrors that of the captive from Tikal Stela 39 (compare to fig. 2.38). In both instances the subordinated body had both knees flexed, one knee in front of the groin and the other extending below the body, with both hands brought up to the chest. Unfortunately, we do not know the identity of the youth from the Piedras Negras tomb; the arrangement may suggest that he is a captive from an enemy kingdom. The precedent for youthful captives certainly exists: centuries later the lords of Yaxchilan celebrated the capture of a *ch'ok ajaw* (unripe lord) from Piedras Negras on a lintel from La Pasadita, undeniably a vital blow against the livelihood of the dynasty.[131]

WRAPPED AND BUNDLED

According to Landa's *Relación*, at the time of the conquest, the Yucatec wrapped the bodies of the dead in burial shrouds.[132] The wrapping of corpses is attested on both the Early Classic period Berlin Vase and Stela 40 from Piedras Negras (see figs. 2.15 and 2.39). Archaeologically, burial wraps may be detectable when fibers of the surrounding cloth or ligatures are preserved. Bundling is more commonly discerned, however, based on the distribution of the skeletal remains, as noted in the discussion above. Even in the case of extended bodies, corpse wrapping can be inferred with careful taphonomic consideration. When the majority of joint surfaces are still in articulation, there is a good likelihood that the body was constrained, usually by clothing or by some other material that wrapped the body. As a basic principle, human flesh decomposes faster than garments and wrappings made of cotton, leather, animal hide, or just about any other material. When the corpse is unconstrained, decomposition causes displacement of

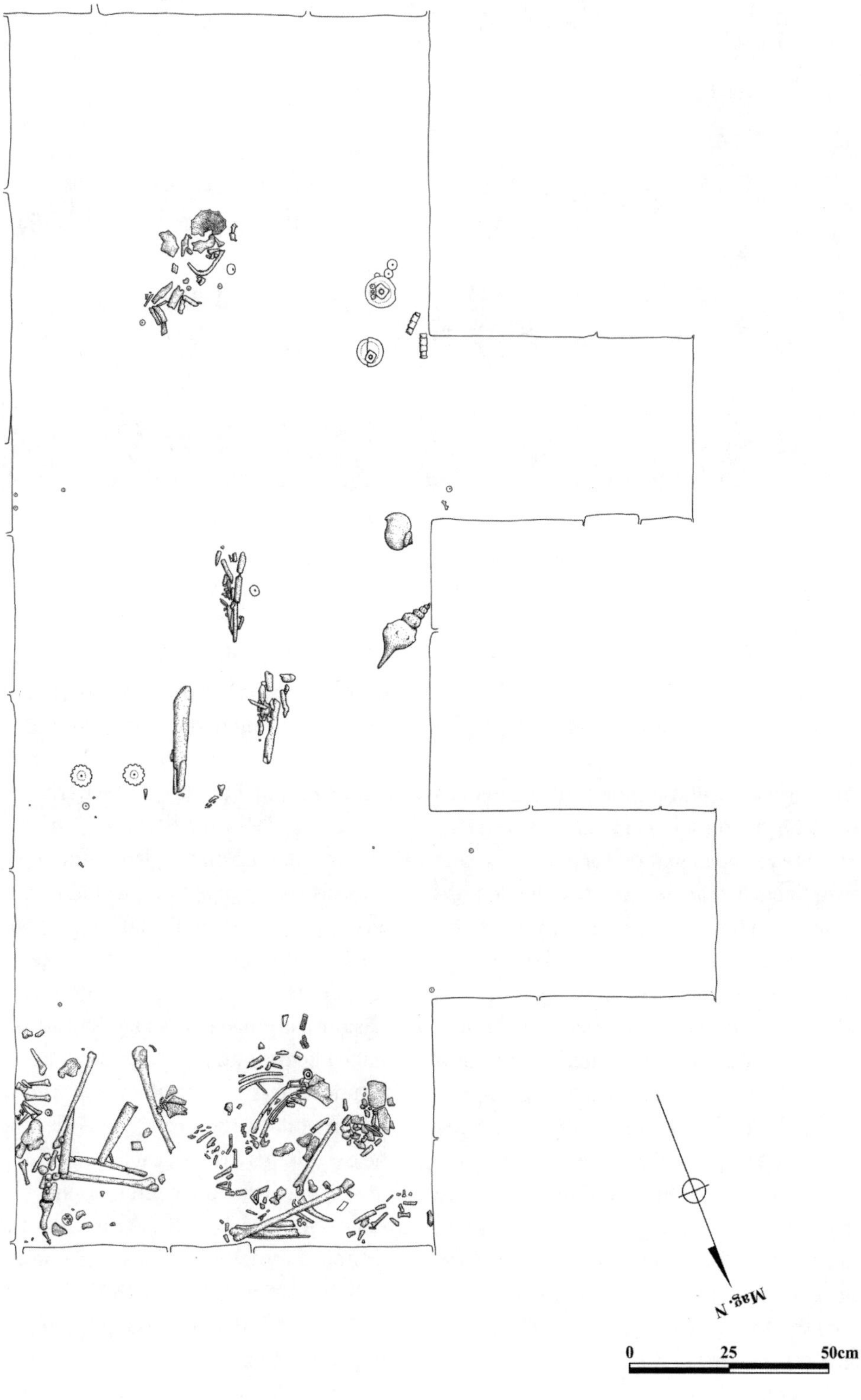

FIGURE 2.43. Piedras Negras Burial 110 (drawing by Heather Hurst).

FIGURE 2.44. Bone pin from a burial shroud in Rancho Búfalo Burial 1 (photograph by author).

the bones as the body first expands during bloat and then liquefies in the process of putrefaction.[133] Dirt can also serve to keep the skeleton articulated during the process of decomposition. In crypt and tomb chambers, however, we should expect significant postmortem shifting of bones unless the body was otherwise constrained. The loincloths, skirts, and huipils worn by the Maya helped to keep the hips (os coxae and femora) articulated, as Tiesler observed for the remains of Pakal.[134]

Bodies that were wrapped in mortuary shrouds were especially constrained during the process of decomposition. Such shrouds can be inferred indirectly, based on the presence of isolated bone pins recovered from Maya graves (fig. 2.44). The tradition of corpse wrapping seems to date as far back as at least the Late Preclassic period. As noted above, Tikal Burial 85 contained the remains of a probable early king, placed in a flexed supine position (see fig. 2.41).[135] Textile fragments were recovered from around the skeleton, presumably the remains of the wrap. The skeleton was missing the cranium and both femora, perhaps removed during reentry or retained prior to inhumation (see the discussion below regarding the collecting of bone from tombs). A greenstone mask recovered from the tomb was too small to have been worn (see fig. 2.24). The mask was likely bound to the outside of the funerary bundle.

Quite a few other royal tombs at Tikal have produced textile fragments suggestive of body wrapping, including Burials 10, 23, 48, 116, 164, 195, and 196. Burial 48 also included elements of a mosaic mask. In the Early Classic Río Azul Tomb 19, remnants of the textile wrapping were preserved along with knotted chords that were used to bind the funerary bundle, recalling the knots surrounding the bundled bodies on the Berlin Vase and Stela 40 of Piedras Negras (see figs. 2.15 and 2.39).[136] Tombs at Lamanai and Altun Ha also contained bundled bodies, and corpses wrapped or covered in matting have been excavated at Copan (Burials 92–3, 95–1).[137]

Elite and nonelite shared a wide spectrum of arm arrangements in Classic Maya graves. The repetition of these arm arrangements suggests that they had meaning, much like the hand gestures noted earlier, though their significance is largely elusive. In the case of wrapped supine bodies, the arms were often bound closely to the decedent's side or placed over the pelvis (fig. 2.45). Pitarch notes that the contemporary Tzeltal arrange the bodies of the sick with their arms at their sides during healing ceremonies.[138] Less commonly Classic period bodies were wrapped with their arms crossed over their chest, with their hands gripping the opposing arm (fig. 2.46). Captives and noble visitors are frequently shown in Classic period iconography with their arms crossed during acts of submission or deference. It is unlikely that the crossed arms of the deceased could suggest submission, however, as some Maya royalty

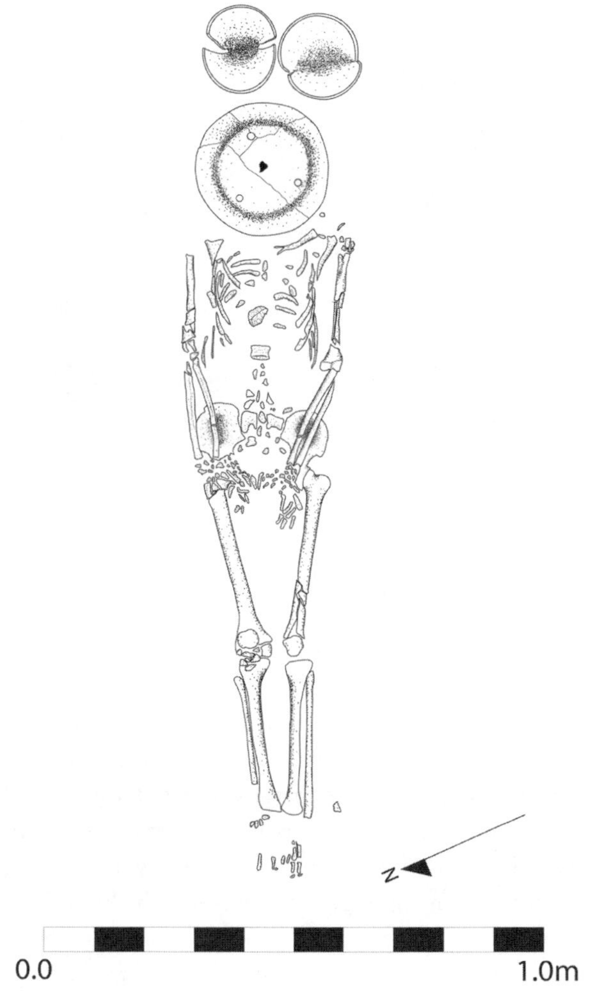

FIGURE 2.45. Nonelite corpse wrapping inferred from the tight articulation of the body in Late Classic El Kinel Burial 5 (drawing by author).

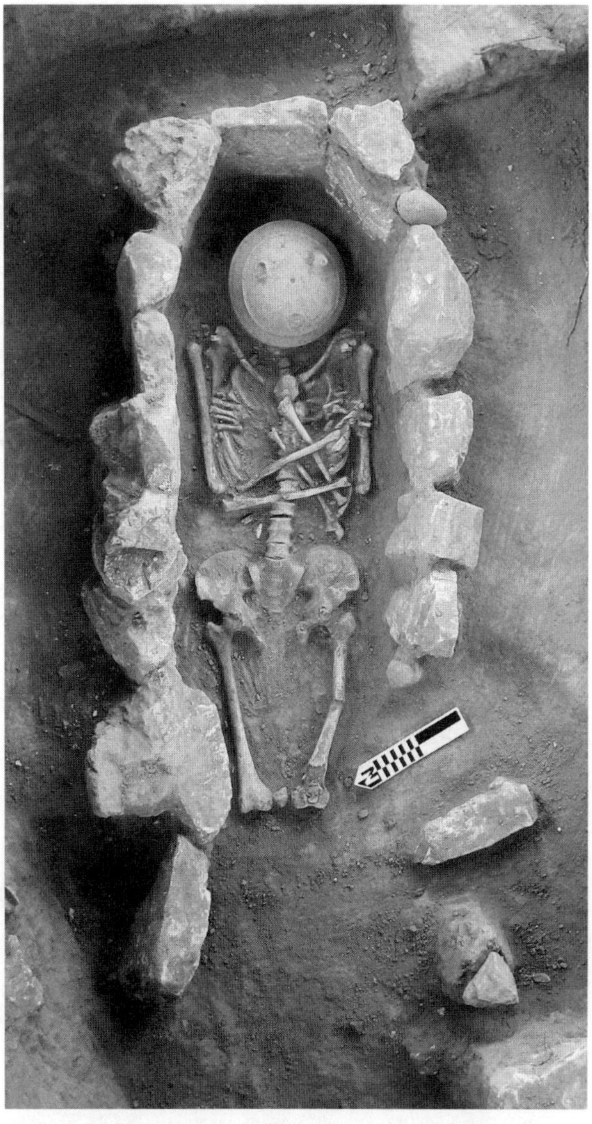

FIGURE 2.46. Nonelite corpse wrapping with arms crossed over the chest in Late Classic El Kinel Burial 10 (photograph by author).

were buried in this position. It may simply be that such bodies were understood to be inactive or stopped. This dichotomy is certainly apparent in Classic period imagery: the protagonists (such as kings and queens) are the most active, with arms gesticulating and pointing, in contrast to the crossed arms of their subordinates.

Aside from its ritual implications, the wrapping of corpses (particularly royal bodies that were subject to protracted public funerals) may have been a strategy to mitigate and manage the rapid (and messy) bodily decomposition that ensued in the hot and humid Maya lowlands. The primary occupant of El Zotz Burial 9 was bundled in a textile that had been impregnated with clay in order to harden the wrap (fig. 2.47). Similar hardened funerary wraps have been found in Early and Late Classic period tombs at Calakmul.[139] For example, the Late Classic period Tomb 1 of Structure 15 contained a woman who had been wrapped in three layers of textile that was hardened by resin, perhaps rubber or chicle (sap from the sapodilla tree). Considering the association between

FIGURE 2.47. Textile fragment from El Zotz Burial 9 (photograph courtesy of Stephen Houston, El Zotz Archaeological Project).

nobility and fragrant winds, the Maya may have felt compelled to obscure any signs of decomposition and decay associated with the royal body. These wrapped corpses were likely repeatedly censed and covered with flowers to combat the stench of putrefaction. In Tomb 19 at Río Azul the leaves of what was apparently allspice (*Pimenta dioica*) were placed between the body and the funerary wrap to soak up fluids, mask the stench, or perhaps both.[140] In at least the cases from El Zotz and Calakmul, the wraps were hardened and reinforced with resins to prevent disintegration as bodily fluids seeped from the corpses. When hardened these wrapped bodies would also have been easier to manage throughout the ritual process, much as coffins and caskets are used to contain and transport bodies in modern Euro-American funerals.

Texts from Piedras Negras, Copan, Dos Pilas, and Quirigua suggest that between three to ten days elapsed between death and burial.[141] Las Casas describes the period between death and inhumation for a Pokom lord at the time of the conquest: "When he [the lord] stops breathing they immediately dispatched messengers to all the towns subject to him and to those other lords that they had as friends making known to them the death and that they waited to bury him until such and such a day."[142] There seems to have been a particular need for keeping the royal body intact for at least a few days if not a week. In the hot and humid lowlands this was ample time for decomposition to proceed. Measures were necessary to manage the rotting corpse and obscure the obvious signs of decay, especially as no evidence indicates that the Maya embalmed their dead. As Francisco Fuentes y Guzmán writes of the Pokom of the sixteenth century, "as soon as he was dead, the first thing that keepers and gentlemen of his chamber did was to hand over the body . . . and they bathed it and purified it with decoctions of aromatic herbs and flowers, without using other ceremony or unguent to preserve the body from decay."[143] Although tomb chambers may have been prepared in advance for old and ill lords, many royal deaths nevertheless would have required some time for the construction of the tomb chamber. For example, Houston suggests that the Early Classic period tomb at El Zotz was left open for weeks or more to allow protracted funerary rituals even after the king was interred.[144]

With the body completely obscured by the funerary wrap, masks provided an image of the dead with which the mourners could interact. Conceptually, the act of masking wrapped bodies was similar to decorating buildings with stucco masks: both embodied the supernatural essences and persona within. Functionally, even in instances of unwrapped bodies, funerary masks may also have served to retard or hide decomposition by covering and protecting the orifices of the face that are targeted by flies and other pests almost immediately after death.[145] It is probably no coincidence that the sites

FIGURE 2.48. The decomposition of a Maya king or the Maize God as shown on the Berlin Vase (drawing by Nikolai Grube).

of greatest political influence, such as Calakmul and Tikal, also have the most complex traditions of wrapping and masking of the dead. At these especially significant political centers the display of the royal body prior to inhumation may have been more prolonged than at less significant political centers. As a result, even greater efforts were needed in these kingdoms to conceal the decomposing corpse in order to provide sufficient time for distant dignitaries to reach the political capitals prior to entombment.

Ideologically, body wrapping relates to a long and complex history of bundling spiritually potent objects. Kathryn Reese-Taylor and colleagues suggest that corpse wrapping may have been done to tether the soul to the body.[146] This seems unlikely, however, considering the preponderance of ethnographic evidence that suggests a greater concern with encouraging Maya souls to depart after death. For the Tzeltal of Cancuc, the *ch'ulel* have already departed the body prior to death yet will linger until urged to move on to the underworld via the cross or altar in the church.[147] If contemporary ethnography is any clue, the bundling of the Classic period dead likely relates more to concerns of the body than of the soul. Nevertheless, even isolated bones of the dead may have remained spiritually charged, so bodies in general needed to be cared for and protected. In that sense the wrapping and interring of a body—particularly a royal body—was akin to the careful disposal of used ritually potent objects, such as stingray spines or lithic eccentrics, which were also bundled or sealed in containers prior to placing them in the ground.[148] The wrapping of Mesoamerican bodies may also have related to concepts of mythic rebirth. One side of the Berlin Vase shows a bundled corpse (see fig. 2.15), while the other side shows a decomposing skeleton feeding the roots of the maize-being (or ancestor) sacred trees (fig. 2.48). In a similar fashion, the bundled bodies at Teotihuacan may suggest the cocoon from which Teotihuacano soul-butterflies emerged.

DANSE MACABRE

Even unwrapped bodies seem to have been arranged in positions that were suggestive of a deeper meaning. One such example is the dance posture that was first noted by my colleague Charles Golden during our excavations at the Yaxchilan secondary center of El Kinel.[149] Burials 2 and 4 were interred with one leg extended, the other leg slightly flexed (fig. 2.49; see also fig. 1.29). The single flexed leg is iconic of dance in Classic Maya art.[150] The right hand of the child in Burial 2 rested on the hip, while the left hand was brought to the chest. This one-hand-up, one-hand-down posture has many parallels in Classic period images of dance (compare fig. 1.29 to fig. 2.49b).[151] The tight articulation of the bones of the young man's feet in Burial 4 indicates that he was likely wearing footwear when he was interred. One of his hands rested on his pelvis, while the other was at his side. Considering the particular arrangement of his left hand, he might have been holding a perishable staff or some other object typically used in Maya dance.

These two El Kinel burials (located within the greater

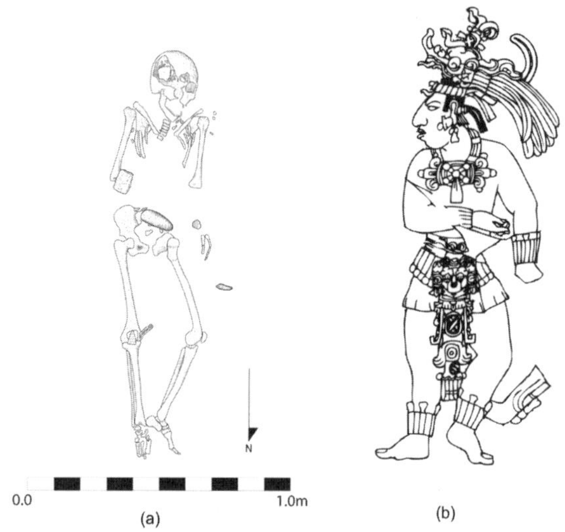

FIGURE 2.49. Classic Maya dancers: (a) body from El Kinel Burial 4 (drawing by author); and (b) K'inich Kan Bahlam from Palenque Temple XIV Tablet (drawing by Linda Schele, © David Schele, courtesy Foundation for the Advancement of Mesoamerican Studies, Inc.).

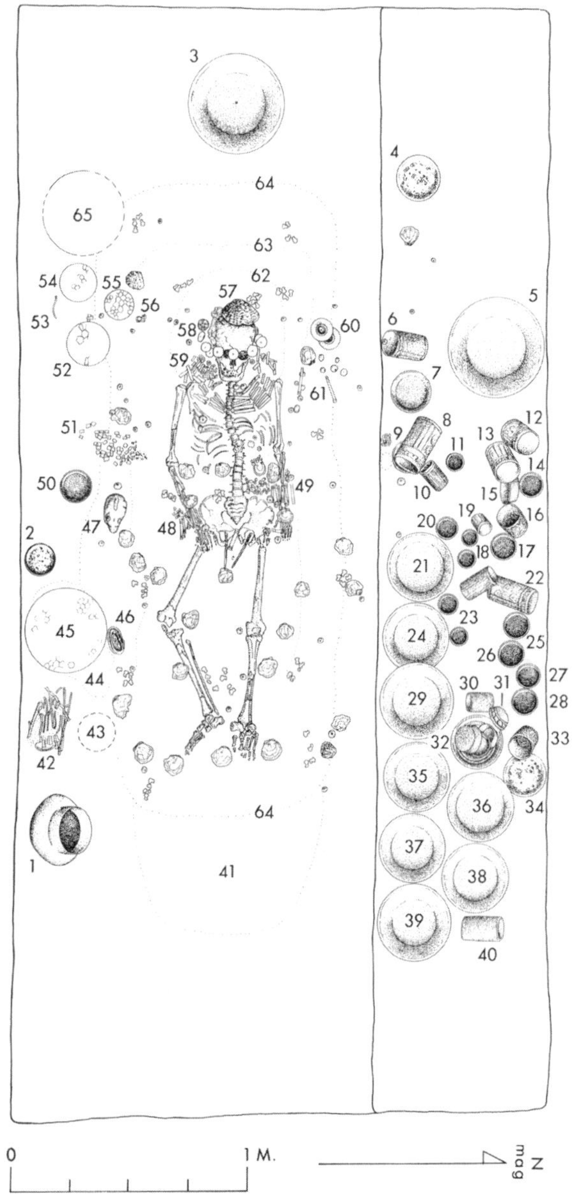

FIGURE 2.50. Tikal Burial 196 (originally published as Coe, *Excavations in the Great Plaza, North Terrace, and North Acropolis of Tikal*, fig. 282, image courtesy of the University of Pennsylvania Museum of Archaeology and Anthropology).

Yaxchilan polity) may reference the importance of dance in the ritual life of the kingdom, as evidenced by numerous depictions of dance on its monuments at both the polity capital and its many subordinate centers. But dancing dead are not restricted to this kingdom. A royal tomb at Tikal excavated by Nicholas Hellmuth (Burial 196) contained the remains of a Maya king (possibly Ruler B, Yik'in Chan K'awil) also interred in a dancer's position (fig. 2.50). As with so much of the symbolism in Classic period graves, the meaning of this distinctive body position relates to myths of maize, life, and death and especially beliefs pertaining to the quintessential dancer of Maya lore: the Maize God (see figs. 1.5, 1.12, 2.37a). As Mary Miller suggests, "in nature, maize plants sway to and fro, their crisp, green leaves moving like limbs of the human body; the Maize God, too, is in motion, often seeming to dance and sway."[152] The Maize God's role as a principal dancer can be traced at least as far back as the Late Preclassic with the San Bartolo murals. According to Taube, "dancing maize gods closely related to courtly life, and reflected the beauty, health and abundant wealth of the ruler. However, in dance, the maize god could also symbolize the resurrection of the deceased ruler as maize emerging from the earth."[153]

Beyond body posture, dance symbolism is evidenced by adornments worn by the dead, especially in royal funerary contexts. Dance was intimately linked to the performance of kingship and was one of the dominant subjects on Classic period stelae. Belts of ancestral heads were essential royal dance regalia, and through dance Maya lords accessed their forebears (fig. 2.51; see also fig. 1.7b).[154] The heads are smaller than life-size and are typically mosaics composed of jade, shell, pyrite, and other materials. Monuments invariably depict three celts suspended beneath each ancestral head; archaeologically they are also found in triads, most often crafted of jade, though slate, chert, and other materials are also known. As the kings danced, the celts would have produced a loud metallic clanking sound accompanied by the rattle of shell tinklers (often *Oliva* sp. or *Conus* sp.), which are often shown worn around the waist and legs of Maya dancers (fig. 2.52). El Zotz Burial 9 contained just such a dance costume, including a mosaic belt ancestor mask ornament, three stone celts, and an impressive assemblage of 98 *Conus spurius* shells strung with dog canine clappers (see fig. 2.34). These objects were found near the waist and legs of the dead king, suggesting that he may have been dressed in his dance regalia. Ancestor head–celt assemblages have been found in Tikal Burial PNT-19 (Early Classic) and Calakmul Structure III Tomb 1 (Early Classic) and are especially prevalent at Palenque, where at least eight sets have been recovered.[155] Rarer are mortuary assemblages that include the masks, celts, and shell tinklers. In addition to the El Zotz tomb, one other example is the Early Classic period Hunal tomb from Copan. Shell tinklers (but without the full dance assemblage) have been recovered in a variety of contexts, including Tikal Burials 123 and 164.[156]

The ancestral heads worn by dancing Maya kings

FIGURE 2.51. Belt assemblage from the tomb of Pakal (photograph by author).

FIGURE 2.52. Waxaklajuun Ubaah K'awiil as the world tree–axis mundi on Copan Stela B (modified from drawing by Annie Hunter in Maudslay, *Biologia Centrali Americana*, plate 34). His three ancestral head belt ornaments and attached celts are highlighted in gray.

likened them to fruit trees, akin to Classic period depictions of heads hanging from trees and reminiscent of the Hun Hunahpu's calabash-head that was hung in a tree in the K'iche' *Popol Vuh* (see fig. 1.11).[157] Recall that sacred trees represent the axis mundi. On Copan Stela B, Waxaklajuun Ubaah K'awiil wears three ancestor head-celt assemblages that may refer to the three hearthstones, another symbol for the axis mundi. Maya kings-as-trees were the spiritual center of their polities: beings that bore the fruit of their ancestors, descendants, and the kingdom itself. The concept is reinforced on quite a few Maya stelae. On Stela B the Copan king clutches a bicephalic serpent bar to his chest, mirroring the bicephalic serpent that winds through the branches of the tree on the Palenque sarcophagus lid. The concept of the lord-as-tree is ubiquitous throughout Maya history. In the *Crónica de Oxkutzcab*, the founder of Uxmal is shown with the Xiu lineage springing as a tree from his backside.[158] In the *Rabinal Achi* the king is likened to a tree when the protagonist expresses his desire to slay the lord of Rabinal and all of his people: "Come on out, lord who's been pierced, lord who's been fitted with gems, however that may be, let me take the lead, since I'm not finished, *chopping through the root, the trunk*, of that Lord of Walkers, Lord of Workers" (emphasis added).[159] Today the Maya continue to liken their spiritual and political leaders to trees.[160] The notion of lords as trees and centers is more fully explored in chapter 3.

COMPLICATING FACTORS

To be able to reconstruct the original position of ancient corpses, the body must have been relatively undisturbed after its initial inhumation and skeletal preservation must be reasonably good. In the hot, humid tropical forest of the Maya lowlands, with its very active biota, burial disturbance and poor skeletal preservation are more common than not. Archaeologists often blame the presumed acidity of lowland soils as the key factor in poor skeletal preservation. In reality there is far more variability in soil pH than is acknowledged. Soils in the

Usumacinta River region generally range from neutral in well-drained areas to strongly acidic in poorly drained *bajo* (swamp) regions.[161] In transects in and immediately adjacent to Piedras Negras, however, soils ranged from slight to moderately alkaline.[162] This soil alkalinity is likely a product of the Maya-built environment, the result of so much limestone reduced to plaster, rubble, and other construction materials. Alkalinity, like acidity, also adversely affects skeletal preservation. Though we do not have any definitive evidence for the practice, the Maya may even have added lime to their burials to hasten decomposition and mask the smell of decay emanating from the graves.

Although a detailed study of Maya burial taphonomy has yet to be conducted, open-aired crypts and tombs generally tend to have more poorly preserved skeletons than simple burials dug into the earth. (For an example of skeletal preservation in an earthen mound, see fig. 2.46; for preservation in a tomb, see fig. 2.43.)[163] Alkalinity resulting from the heavy use of plaster inside crypts and tombs is likely one factor contributing to their poor preservation. Another is fluctuating levels of humidity within graves, especially between the wet and dry seasons. Hydrological cycling results in expansion and contraction of bone, causing fissuring and fragmentation of the periosteal (outer) bone surface, which dries faster than the endosteal (inner) surface. The result is the extensive fragmentation characteristic of Maya burials. Disintegration and scattering of the bone is further hastened by the movement of water through the burial chamber. This was especially evident in Burial 4 at the site of Budsilha, where a drain was converted into a small crypt (fig. 2.53).[164] The drain was originally built to run water off the patio into the *bajo* below. The drain seems to have continued to function even after the body was placed within. As a result we found quite a few of the bones washed to the foot of the burial. It is unclear whether the Maya placed bodies in drains for convenience or whether the practice had a deeper meaning, perhaps likening the burial spaces to the watery underworld.

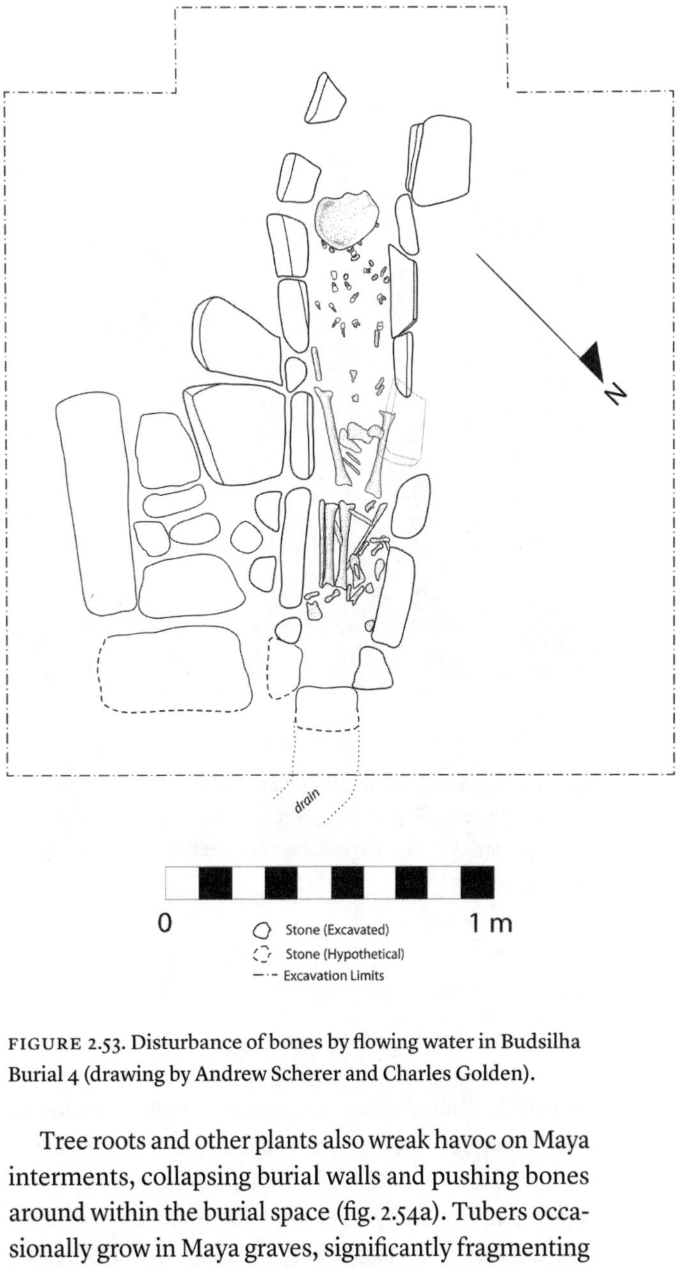

FIGURE 2.53. Disturbance of bones by flowing water in Budsilha Burial 4 (drawing by Andrew Scherer and Charles Golden).

Tree roots and other plants also wreak havoc on Maya interments, collapsing burial walls and pushing bones around within the burial space (fig. 2.54a). Tubers occasionally grow in Maya graves, significantly fragmenting both the skeletal remains and any associated burial goods (fig. 2.54b). As frustrating as it is to uproot a tuber or cut through a root with a machete during burial excavation, it can be equally problematic if the damage was done decades or centuries ago, so that the nature of the disturbance must be inferred from the disposition of the bone and not confused with anthropogenic disturbance.

FIGURE 2.54. Destruction of burials by: (a) invasive tree roots in Tecolote Burial 8 (photograph by Charles Golden); and (b) a tuber growing through the middle of El Kinel Burial 9 (photograph by author).

BONE

As the most enduring aspect of the human body, skeletal remains attained a rich polyvalence in the Maya worldview. Depending on context, bones indexed the whole body, the inevitability of death, yet also referenced fertility and the potential for new life. Contemporary ethnography provides some clues for understanding the meaning of Classic Maya bones. As Christenson points out, many highland communities view skulls and other skeletal elements as seeds and thus as the source of new life.[165] The Classic Maya word for bone was *baak*, as it is in most modern Maya languages. In K'iche' the word for bone (*baq*) is a near homophone for their word for seed (*baq'*).[166] The Tz'utujil see inhumation as an act of sowing; the remains of the dead are the seeds from which future generations sprout.[167] The Yucatec also recognize the fertile nature of bone, which is understood as the principal ingredient in semen. As one informant explained to Miguel Astor-Aguilera upon being served a maize gruel, "*tun taal le sa'il keep u bakal*, that is, 'here comes the corn-bone semen.'"[168] In this quotation the Yucatec informant is employing *bakal* as a double entendre. Ripe kernels of corn that squirt a milklike substance when they are scraped off a stiff cob are an easy metaphor for semen. The fertile nature of Classic Maya bones is also evident in depictions of the rebirth of the Maize God, which often show him emerging, or literally germinating, from a skull (see fig. 1.12), much as Pakal is shown reborn above an animate skull on the sarcophagus lid (see fig. 2.5).

Classic period depictions of bone are rich in anatomical detail. Skulls are depicted with sutures that separate the individual bones of the cranial vault and are typically marked with dots near the eyes, recalling supraorbital, infraorbital, and mental foramina (openings in bone for the innervation of blood vessels and nerves). Long bones are also marked with dots, likely recalling the nutrient foramina that allows blood vessels to pass to the medullary cavity. The long wavy lines that run along the length of long bones are likely meant to show the medullary cavity and underscore the biological and metaphysical understanding that bone is alive and impregnated with vital essence. Yet bones are also notably stylized in Maya art. Long bones are rarely identifiable to a particular bone in the human body. Skulls are never shown to be modified, even though the majority of Classic Maya skulls were artificially shaped. It seems that Maya artists honed in on certain physiological aspects of bone that interested them and ignored other details that were irrelevant. Instead accurate depictions of human anatomy were reserved for the living body, perhaps suggesting an indifference or even concern about rendering the dead too realistically in Maya art.

Crossbones are often partnered in the iconography with eyeballs that suggest illness, death, and the underworld, as noted in chapter 1. As crosses, these bones may allude to the quadripartite order of the world and movement between the living world and the underworld. They are also shown on the wings of bats, mythological decapitators, and emissaries of underworld lords (fig. 2.55).[169] As noted, crossbones are also linked to the supernatural midwife, Goddess O (see fig. 1.35). Associated with this being of healing, crossbones may represent ailments to be cured or again suggest the underworld. Crossbones are merged with a flower on the Temple of the Foliated Cross at Palenque, where they specifically seem to underscore the notion of vitality embodied in bone (fig. 2.56).[170] Bones are also body parts of supernatural beings associated with illness, death, and the underworld. They are most evident in the Death Gods, beings with skulls for heads and skeletal limbs and thoraxes of bone yet with fleshed hands and feet, bloated bellies, and distended eyeball body ornaments (see fig. 1.20d). These beings are embodiments of dread and decomposition, the conceptual inverse of the Maize God.

FIGURE 2.55. Crossbones and eyeballs on the wings of a bat (drawing by author after K5367).

FIGURE 2.56. Crossed bones and flower motif from the Palenque Tablet of the Foliated Cross (drawing by author after Robertson, *The Sculpture of Palenque*, vol. 4, fig. 153).

FIGURE 2.57. Crypt containing the remains of an exhumed skeleton in a Yucatec community in Campeche (photograph by author).

AFTERLIFE OF BODIES

In modern Tzeltal there is a linguistic linkage between bone (*bak*), skeleton (*bakelal*, literally "all of the bones"), flesh (*bak'et*), and body (*bak'etalil*, literally "all of the flesh").[171] In this regard the Tzeltal and other contemporary Maya groups name specific body parts as a means to reference the whole. For example, *ni'il chikinil* (literally "nose ear") is a reference to the face.[172] As discussed in chapter 1, the same ideology is evident in Classic period conventions that treat the head as a referent to the rest of the body and as the locus of personhood and identity. The naming of parts to indicate the whole is evident in the dialogue of the *Rabinal Achi*.[173] Throughout the drama the speakers use the coupled expression "in my teeth, in my face" to refer to themselves.

In this fashion Classic Maya bones may have been used to index the bodies and persons from whom they originated, were subject to acts of veneration and desecration, and served as conduits by which the living communicated with the dead. For both the ancient and modern Maya, the significance of postmortem bodies and body parts is reinforced by the importance of the dead in the lives of the living.[174] The souls of the dead are powerful advocates for the living, yet disgruntled ancestral souls can otherwise be the source of much hardship. In some contemporary Yucatec and K'iche' communities, skeletons are exhumed, bundled, and housed within shrines where they can be cared for and communicated with (fig. 2.57).[175] According to Astor-Aguilera, the focus is primarily the long bones and skulls, where "*pixano'ob*, soul-like person(s) tend to concentrate," whereas vertebrae and other small bones may be discarded.[176] As described in Landa's *Relación*, the skulls of Cocom lords were given new faces, retained in shrines, and provided with offerings of food on festival days.[177] Modern bundled bones are wrapped in white cloth, placed in boxes, and housed in the home of the living or placed in the family crypt at the community cemetery. The placement of bones in familial crypts may have a Classic period analogue in the mausoleums of Caracol.[178]

A rich epigraphic and iconographic corpus reveals a diverse range of activities involving the interaction and manipulation of human remains.[179] Tikal Altar 5 illustrates the exhumation of a skeleton by two priests or lords (fig. 2.58). The remains are those of a noble woman who had been buried for eight years and seems to have been moved from an unknown place (Maasal) to Tikal itself. Something is being done to her remains that may have involved scraping (possibly with the stone jaguar paw and knife that are wielded by the two protagonists) and also fire drilling, as suggested by the long staffs that

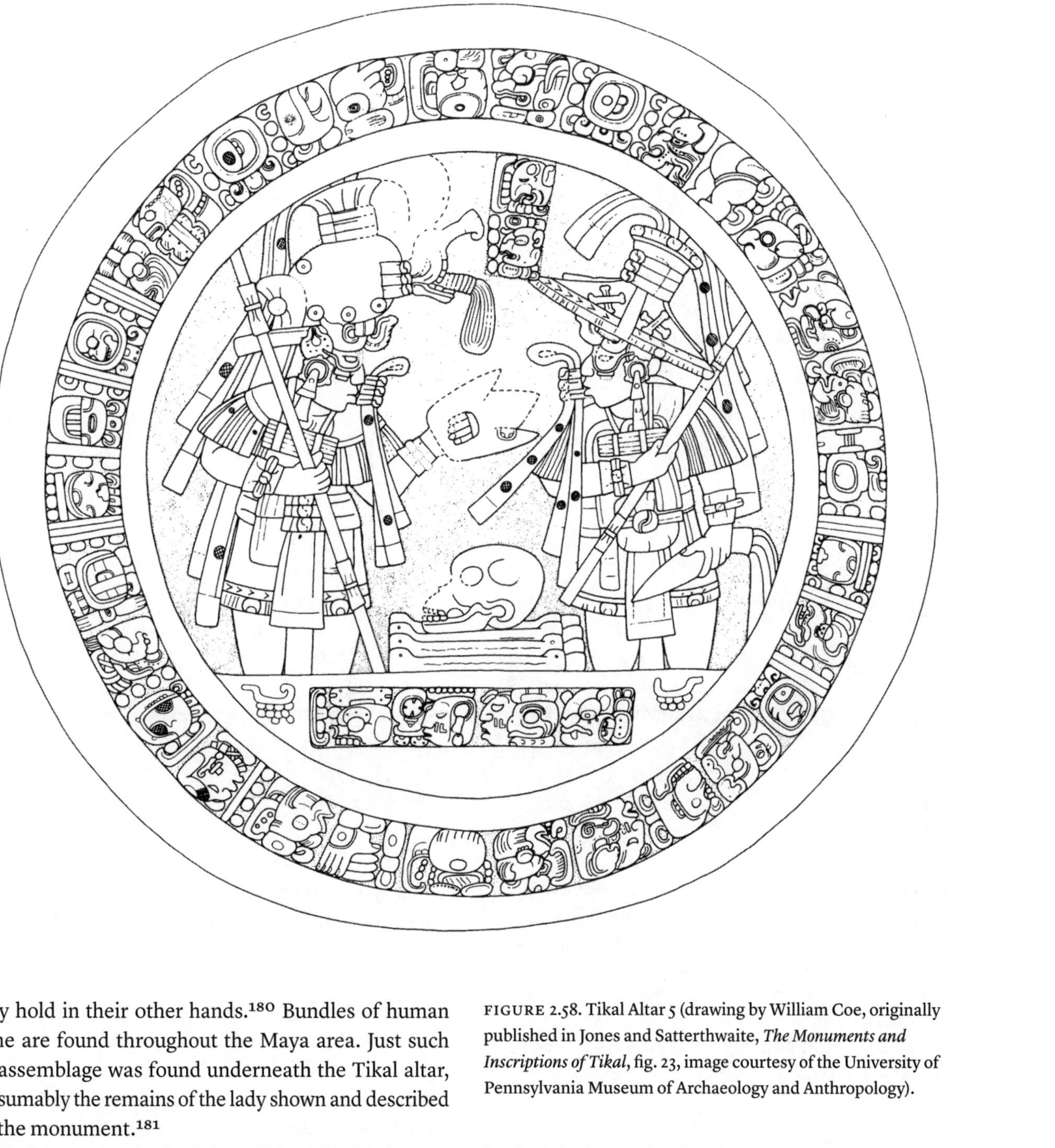

they hold in their other hands.[180] Bundles of human bone are found throughout the Maya area. Just such an assemblage was found underneath the Tikal altar, presumably the remains of the lady shown and described on the monument.[181]

FIGURE 2.58. Tikal Altar 5 (drawing by William Coe, originally published in Jones and Satterthwaite, *The Monuments and Inscriptions of Tikal*, fig. 23, image courtesy of the University of Pennsylvania Museum of Archaeology and Anthropology).

Piedras Negras had a rich tradition of using human remains and mortuary spaces in rituals to access the souls of the dead. Although he died before his son took the throne, Ruler 2 remained an active agent throughout the reign of his son (Ruler 3) and even "danced" on the three *k'atun* anniversary of his birth, almost forty years after he died.[182] Although the tomb of Ruler 2 is yet to be found, the tomb of Ruler 4 was excavated in 1997 and demonstrates significant evidence for ancient disturbance that can be linked in part to an *el naah umukil* (house burning at the burial) in AD 782.[183]

Within the kingdom of Yaxchilan we have explicit reference to isolated bones, especially long bones and skulls, used in ancestor and spirit conjuring.[184] Yaxchilan

(a)

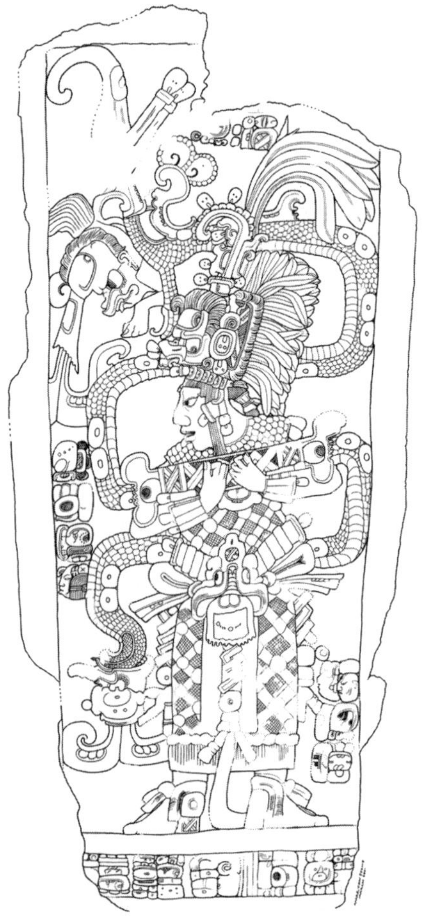

(b)

FIGURE 2.59. Classic Maya noble ladies conjuring with bone: (a) Yaxchilan Lintel 25 (drawing of Lintel 25, front edge, by Ian Graham, from Graham and von Euw, *Corpus of Maya Hieroglyphic Inscriptions*, vol. 3, part 1, *Yaxchilan*, 3:55 © 1977 by the President and Fellows of Harvard College, Peabody Museum of Archaeology and Ethnology, Harvard University, PM# 2004.15.6.5.22 [digital file# 101240034]); and (b) unprovenanced monument in the de Young Museum, Fine Arts Museums of San Francisco (drawing by Mark Van Stone).

Lintel 25 (ca. AD 723) provides an important royal precedent for such practice. Lady K'abal Xook uses an unidentified skull to conjure a Yaxchilan supernatural, Aj K'ahk' O' Chahk, an important supernatural patron of the city (fig. 2.59a).[185] Notably, the monuments at Yaxchilan and elsewhere show that women were bone-conjurors, apparently an important ritual role of Classic Maya queens.[186] An unprovenanced monument that stylistically links it to the Usumacinta region shows a woman conjuring with an oversized femur that she holds much as kings clutch the bicephalic serpent bar (fig. 2.59b). The role of women in conjuring from bone may relate to the fertile (perhaps even masculine nature of bone), so that women were the necessary gender counterbalance, able to coax from the bone the serpents from which supernatural beings emerged. Recall the Yucatec informant who likened semen to bone paste discussed above. The reproductive potential of body parts is also suggested in the K'iche' *Popol Vuh* when the maiden is impregnated by the severed head of Hun Hunahpu.

The use of bone in conjuring may explain the absence of select skeletal elements in graves throughout the Yaxchilan kingdom. A recently excavated crypt at Bonampak was reported to have been missing the cranium, yet a mandible was present, suggesting that the cranium was removed sometime after the original interment of a fleshed body. Other examples come from the Yaxchilan subordinate center of El Kinel, where two well-preserved and perfectly articulated burials are missing entire long bones. Burial 4 had both radii and ulnae removed from the grave, even though the humeri and the bones of the hands were untouched (see fig. 2.49a). Similarly, the lower legs of El Kinel Burial 10 were completely removed (see fig. 2.46). Scattered blocks at the "foot" of the crypt suggest that the burial chamber was originally longer than it was when I encountered it during excavation. The absence of cut marks on the well-preserved distal femora indicates that the bones of the lower legs were once there and were removed at some point after decomposition of the flesh by the ancient Maya. The bones may have been removed for conjuring or for the manufacture of bone objects (two possibilities that are not mutually exclusive). Budsilha Burial 6 in the kingdom of Piedras Negras was missing its entire skull, yet six maxillary teeth were present, indicating that it had once been inside the chamber (fig. 2.60). Nearby Burial 5 was also missing much of its skeleton, though poor preservation may also have been a factor in that interment. Both burials were located less than a meter below the surface of the final floor of the site's main plaza and would have been easily accessible to occupants of the community.

In the Yucatec community photographed in figures 2.1 and 2.57, bodies are put into a crypt for three years before they are exhumed and moved to smaller crypts that house the bundled remains of fellow family members. Similarly, Redfield and Villa Rojas indicate that burials

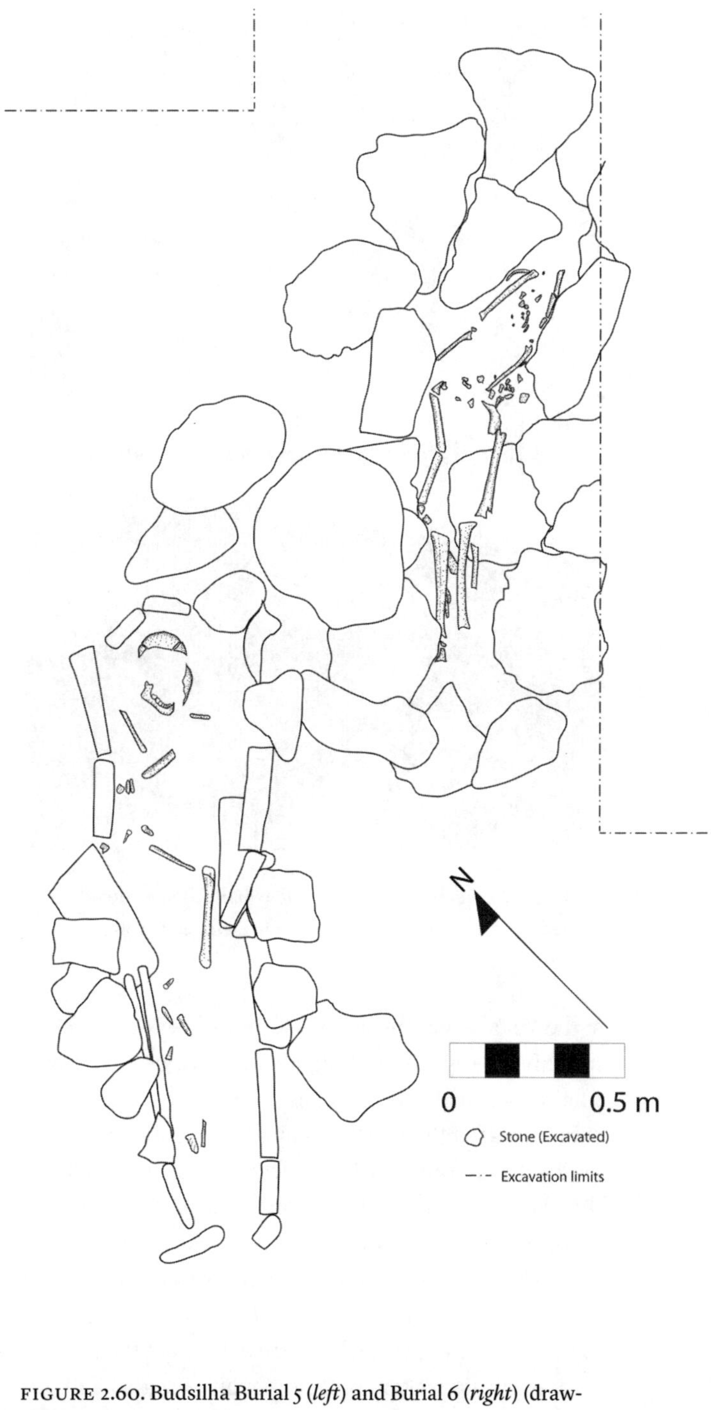

FIGURE 2.60. Budsilha Burial 5 (*left*) and Burial 6 (*right*) (drawing by author).

FIGURE 2.61. Reopening of the *pib* (earth oven) to remove cooked food in a Yucatec community in Campeche (photograph by author).

at the Yucatec village of Chan Kom were exhumed after roughly three years, and Miguel Astor-Aguilera reports that skeletons in the villages where he worked were exhumed after three to five years.[187] According to one of his Yucatec informants, the skeletons are exhumed when they have "cooked long enough in the *pib* [earth oven] and are now ready to serve."[188] A Maya *pib* is essentially a hole in the ground, lined with heated rock and covered by leaves, sheet metal, and dirt (fig. 2.61). In part this analogy captures the parallel between exhuming the cooked meal (such as chicken, pork, and maize breads) and the finished body, now reduced to a skeleton. However, the metaphor also makes a specific link between the state of cooked and decomposed flesh, the point at which the meat is falling from the bone. The reentry of burials at Bonampak, El Kinel, Budsilha, and elsewhere demonstrates a deep history of Maya concerns for monitoring the status of buried bodies, reentering the graves, and recovering select (or complete) skeletal elements after the decomposition process was completed.

The Maya also retained body parts of their enemies, fashioning their corporal elements into flayed skin masks on shields, stuffed heads, skull pectorals, and various other objects to be worn and displayed (fig. 2.62; see also fig. 1.7a).[189] When the remains themselves were unobtainable or no longer preserved, the Maya crafted effigy body parts out of shell and other materials.[190] Invariably these objects emphasize decay, decrepitude, and the incompleteness of the enemy's body, contrasted with the beauty and vitality of the owners of such trophies. The violence directed against enemy bodies was quite evident in a mandible recovered from the Acropolis at Piedras Negras (fig. 2.63). The mandible demonstrates extensive and deep cutting along the mylohyoid line of the lingual aspect of the body. The cuts were made on fresh bone. The density and depth of the marks indicate that these cuts were not made by careful defleshing or cleaning of the bone but instead were the result of forceful and vigorous slashing of the throat, either to kill the person or to cut through the muscles of the floor of the mouth in an attempt to separate the mandible (or head) from the rest of the body. A powerful blow fractured the mandible at the midline of the body (mental symphysis; visible at right in fig. 2.63).

As indices to a complete body, captured and kept body parts of enemies can be understood as part of a broader continuum of captive display, one of the dominant themes of Classic period monumental art (see figs. 1.8 and 2.38). The depiction of captives and the display of their body parts operated as perpetual claims of dominion of the captors over the corporal existence of their enemy. For this reason the Classic Maya, almost

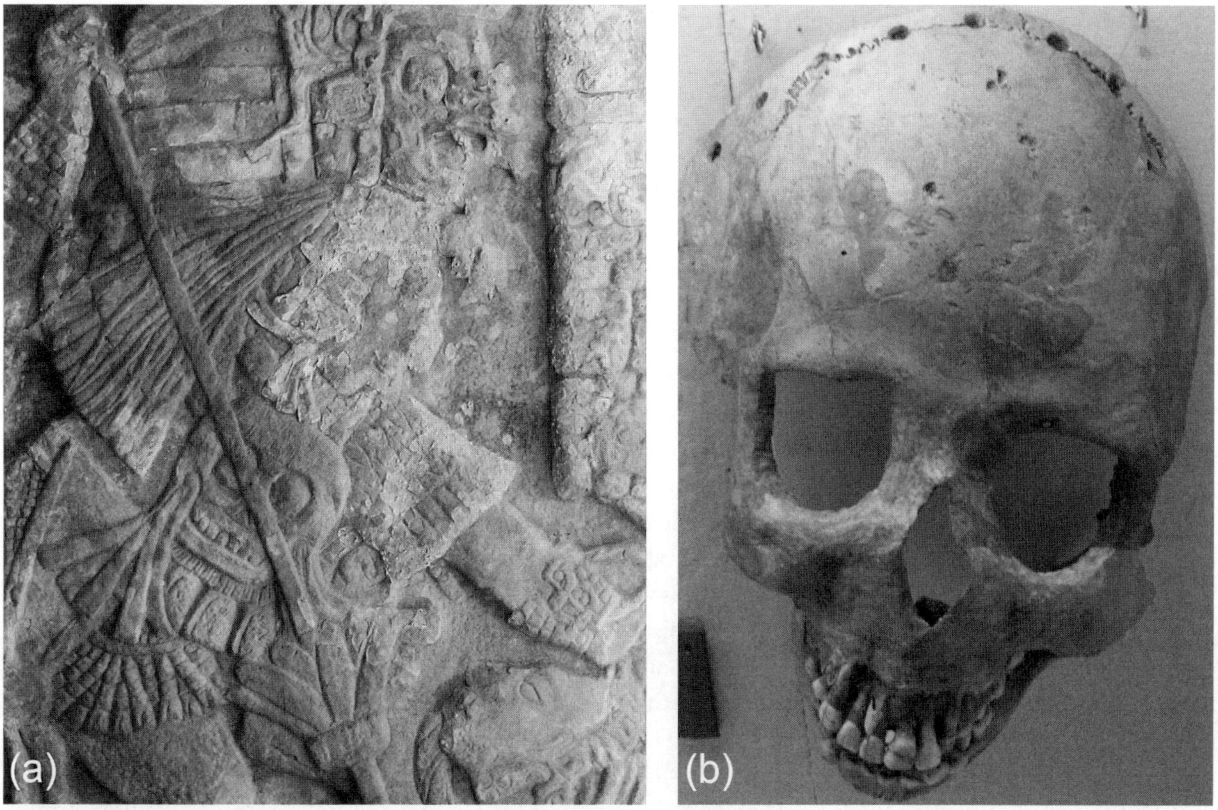

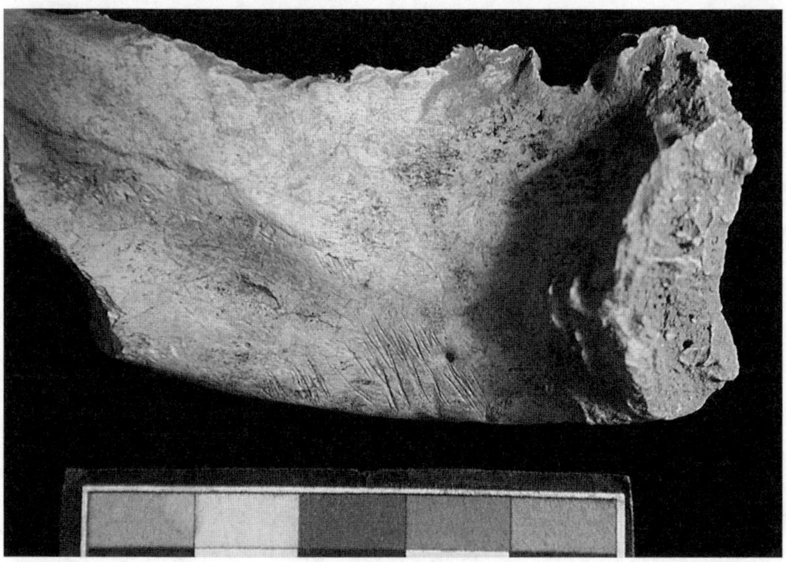

FIGURE 2.62. (*above*) Classic period skull pectorals: (a) worn by a lord on Bonampak Lintel 3; and (b) from the site of Copan (photographs by author).

FIGURE 2.63. (*left*) Lingual view of left mandible recovered from the Acropolis (Operation 46F) of Piedras Negras, demonstrating perimortem fracture of the mental symphysis and cuts along the lingual aspect of the bone (photograph by author).

without fail, included the names and place of origins of their captives on their monuments.[191] The enemies and their body parts operated as metonyms for the places and polities from which they were taken. This is especially true in the polity of Yaxchilan, where the kings of the eighth century all used epithets that named their favorite captives. Shield Jaguar III is named at least thirty-two times in the hieroglyphic record as the "Guardian of Aj 'Nik,'" Bird Jaguar IV is known as the "Guardian of Aj Uk," and Shield Jaguar IV is proclaimed to be the "Guardian of Torch Macaw" (Tah Mo').[192]

"Captive" is one of the many homophonies of *baak* (bone). One of Bird Jaguar IV's other epithets was *aj k'al baak* ("he of twenty captives," with a double meaning of "he of twenty bones"). The lords of Yaxchilan and its subsidiaries are frequently depicted wearing the body parts of their enemies, as on the murals and lintels at Bonampak (see figs. 1.7 and 2.62a). Unlike protagonists that bear their names near their heads, captives are named on their body parts, as noted in chapter 1. In the sixteenth century anxiety about becoming the skeletal possession of the enemy is exemplified in the *Rabinal Achi*, where the K'iche' captive ponders a drinking gourd and his imminent fate: "Could this be the skull of my grandfather? Could this be the skull of my father? Is this what I'm looking at, what I see before me? Then won't this also become a work of some kind, an artifact, this bone of my crown, bone of my head, carved in back, and carved in front?"[193]

As a possible Classic period precursor to the skull drinking vessel of the *Rabinal Achi*, bowls fashioned from human frontal and parietal bones have been recovered at Tikal, Palenque, and elsewhere in Mesoamerica (fig. 2.64).[194] At the main pyramid at Postclassic period Mayapan, headless skeletal bodies are capped by open niches, presumably to receive the heads of enemy captives. Archaeologists have recovered cranial fragments within one of the niches.[195]

But not all isolated human bone need be understood as either the venerated remains of ancestors or the violated bodies of enemies. The modern Maya treat some human remains with indifference, depending on the particular meanings ascribed to those parts. The Yucatec Maya discard or ignore small bones in favor of skulls and long bones during reburial, a practice that may seem unfathomable to modern readers—especially to Americans, who have a preoccupation with recovering every tiny fragment of victims of mass disaster and war.[196] Yet consider the disregard that modern people pay to other parts of our body, such as hair and nail clippings or the five liters of blood flushed into the sewer system as part of the embalming process. For the Maya, the significance of certain body parts likely reflected the degree to which a given bone was identifiable as human and thus emblematic of the individual from whom it was taken.

In the crowded mortuary landscape of the Maya, it was inevitable that the bones of long-dead individuals were encountered when new graves were dug. At Piedras Negras, where house mounds were densely packed with burials, it was not uncommon to see later graves dug into earlier inhumations. The older remains were moved aside to make way for the new burial. Tombs at Caracol operated as mausoleums: earlier skeletons were swept aside to make room for later bodies.[197] In a similar fashion, Vogt notes that at Zinacantán skeletons were regularly disturbed during later inhumation and were carefully set aside and reinterred above the coffin of the later burial.[198]

Quite a few objects crafted from human bone have been recovered by archaeologists, in contexts where we simply do not know whether the bones were collected from an ancestor, an enemy, or an anonymous person (see figs. 2.64 and 4.33). Regardless of its source, bone (with connotations of both life and death) is a liminal material that bridges the world of the living and that of the supernatural. Bones are fragments of a person that no longer exists. In manufacturing those objects, the Maya selected bone that was most emblematic of individuals—skulls and long bones, especially the femur. More than

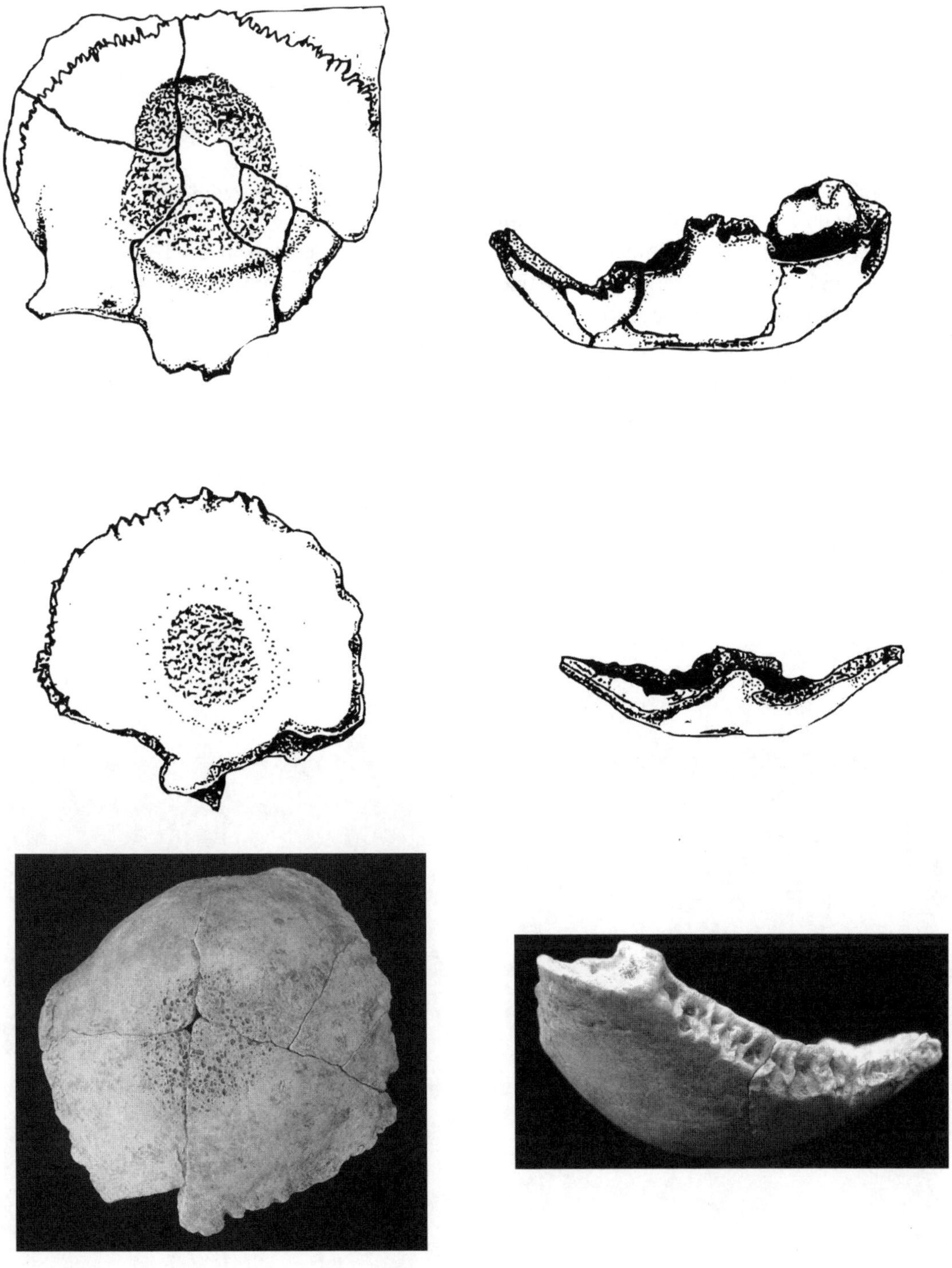

FIGURE 2.64. Bowls made from human crania showing base (*left*) and profile (*right*) of three different vessels. The top two vessels are from Tikal; the bottom vessel is from Palenque (drawings by author after Laporte, "Contexto y función de los artefactos de hueso en Tikal, Guatemala," fig. 1; photographs by author).

any other long bone, the human femur is easiest to distinguish from the equivalent bone in animals. It is the largest long bone in the human body, much larger than any animal long bone in the Maya area, and has a distinct morphology that relates to human bipedality. Dancers on the Bonampak murals wield axes whose heads are mounted into femora (or perhaps wooden hafts carved to look like femora) (see fig. 2.26).

THE DEAD BODY

Historically the study of Maya skeletons has focused on issues of biology: age and sex, diet and health, ancestry and migration. Yet for the Maya no material was more imbued with complex, sometimes contradictory, significance than human bone. Bone hinted at the enduring potential for new life, and rich agricultural metaphors were applied to human remains. But bone was also emblematic of death, associated with horrific underworld beings. As indices of the people that they once were, bodies and bones were the subject of intense rites of veneration and desecration as the living sought to manage the souls of the dead. The final two chapters of this book examine those mortuary rituals and their implications for understanding Classic Maya society.

CHAPTER 3

RITUAL, LIMINALITY, AND THE MORTUARY SPACE

For many decades now archaeologists have used tabulations of burial objects as a means for measuring wealth and status in the past.[1] Although it is certainly true that the graves of Maya royalty are easily distinguishable from those of the hoi polloi, reducing burial goods to measures of wealth ignores the objects' ritual function and social significance and neglects a valuable opportunity to explore Classic Maya worldview. Maya graves were liminal spaces: the sites of intense ritual activity relating to the separation of the dead from the lived community.[2] Burial sites were points of departure for Maya souls on their journeys to the otherworld. For the living, graves were loci where contact was maintained with the spirits of the dead. Ritual action provided the framework for maintaining that contact, and material culture was employed to create meaning within the burial landscape.

The decipherment of the social and ritual significance of objects and even other people within the burial space is no simple task, however. We must look for patterns in practices evident in Classic Maya mortuary contexts: repetition across space, through time, and even among the graves of individuals of different social positions. As this chapter shows, the symbolism, meaning, and material culture of Maya graves relate to concepts of life, death, and the status of souls. Central to Classic Maya mortuary rites was the establishment of the grave and body as an axis mundi, an act that relates to the departure of the soul and the rebirth and resurrection of the dead (see chapter 2). These practices are most evident in the graves of the Maya elite. Nevertheless, the archaeology of nonelite graves shows that such beliefs were also enacted in the interment of Maya commoners. To unravel meaning in the mortuary contexts we must consider that a certain spatial syntax was at work in the placement and arrangement of funerary objects. This is of course the great tragedy of looting in the Maya area. Mortuary goods that are divorced from their original context are mute in regard to their original role within the grave. Sacked graves offer only hints of what they once contained.

Death is a source of societal rupture, calling for intense rituals to navigate the disruption that follows the loss of life.[3] This is especially evident in the mortuary rites of Classic Maya royalty. The delicate balance maintained among the living, their ancestors, and the supernatural world was disrupted with the demise of especially important social personas: the kings and queens. The Maya navigated these stormy waters with human sacrifice and other forms of corporal offerings (see the discussion near the end of this chapter).

CONCEPTUALIZING THE MAYA WORLD

Maya cosmology strongly influenced the organization of the mortuary space.[4] Ancient and contemporary Maya spatial ordering follows basic principles that are pervasive in the worldview of the indigenous peoples of the Americas. Most important is the perception of the earth as a quadripartite space with four corners linked to four directions. Each direction has its own symbolic attributes and meanings.[5] These directions only vaguely correspond to modern cardinal directions. The principal axis in ancient and modern Maya cosmology is the east-west path of the sun. East is the direction of the rising sun, which is reborn each day from the sea. East is represented by the color red and is associated with new life.[6] West is the direction of the setting sun; it is associated with darkness and its color is black. Less significant is the north-south axis. In contemporary Maya geography this axis is blurred with spatial understandings of verticality (up-down) as it relates to the travel of the sun. As Gossen explains for the Tzotzil: "Informants consistently said that east is the sun's position at *slok' htotik* ('the sun appears' or 'dawn'); north is the horizontal equivalent to the sun's vertical position at *olol k'ak'al* ('half heat,' 'half day,' or 'noon'); west is *sbat htotik* ('the sun departs' or 'sundown'); and south is the horizontal equivalent to the sun's vertical position at *olol ak'obal* ('half-night' or 'midnight')."[7] The Tzotzil conceptually rotate the daily celestial path of the sun so that it has an equivalent circuit on the horizontal plane of the earth. By doing so the Tzotzil and other Maya are able to replicate the movement of the sun in their ritual performances by walking or dancing a counterclockwise circuit, starting at east and moving first north then west, south, and back to the east.[8] North and south are also used to define the boundaries of the sun's journey. As Evon Vogt explains in regard to Tzotzil cosmology, "north and south are called 'sides of the sky,' that is, the sides of the path of the sun."[9]

Western concepts of directionality hinge on the ordering of space as a 360° circle, anchored at true north. We conceptualize north as the direction toward the North Pole on an earth that we understand to be spherical in shape. The other three cardinal directions represent 90° intervals along the circle. The primacy of north in the Western worldview is demonstrated by mapping conventions; north is placed at the top of most maps. We abbreviate the compass rose with a north arrow, implying the remaining directions relative to the position of north. In contrast, Maya maps from the colonial period consistently have east at the top.[10] Moreover, Western understandings of directionality treat north as a fixed direction, always centered on the North Pole of the earth. That direction does not change based on our location or the time of the year.[11]

The ancient Maya had no concept of the terrestrial poles, much less a round earth; nor did they have compasses to ascertain its close equivalent, magnetic north. For contemporary and contact period Maya, east-west was likely the primary axis, based on the path of the sun. This axis subtly changes over the course of the year, as sunrise and sunset migrate north in the summer and south in the winter. This principle is highlighted in contemporary Cho'rti' concepts of directionality and color.[12] The four corners of the earth are marked as the locations of the sunrise and sunsets of the winter and summer solstice. Thus red is the color of the rising sun at the June solstice, the height of the first growing season. Yellow, associated with maturity (as in the color of dried maize husks and kernels at harvest time), is the color of the rising sun at the December solstice. Black and white are the colors associated with the setting sun at the winter and summer solstices, respectively. Thus, rather than being distinctly different directions, north and south are different aspects of the principal east-west axis for the Cho'rti'. These concepts are more fully discussed in chapter 4 as they pertain to the orientation of Classic Maya graves.

The Maya did not conceive of directions as perpendicular axes with right angles anchored at true north, so Precolumbian urban centers were never oriented to true north and generally lacked a gridded settlement pattern, even at densely populated centers like Palenque and Mayapan. Despite the lack of gridded settlement, the ancient Maya were very much concerned with the construction of quadrangular space, as evident in the form of structures, platforms, and plazas, all of which can possess corners of 90°. Nevertheless, the Maya never united these quadrangular constructions into a greater gridded system. There is no evidence that Maya rulers controlled the settlement design of entire communities or had any reason to seek such a unified civic plan beyond the civic ceremonial precinct. For the ancient and modern Maya, the concern with quadrangular space has nothing to do with spatial efficiency but instead reflects a fundamental logic as to how civilized space should be constituted. As Vogt notes for the Tzotzil of Zinacantán, "by making the universe square instead of round, as it is perceived from a mountaintop, Zinacantecos associate it with the sane, systematic, well-ordered world of culture, as opposed to nature. And by scaling, they can run the gamut in their thinking about houses, tables, fields—and the universe."[13] Karl Taube has shown that concern with an ordered, quadrangular organization of cultured space, in opposition to the disorderliness of wilderness, is demonstrable for the Classic period.[14] For both the ancient and modern Maya, squared spaces were essential for religious performance, facilitating the clockwise and counterclockwise circuits that are central to dance, processions, and other ritual performances.

The ancient Maya understood the earth to be a massive quadrilateral plane, supported by trees or pillars.[15] The corner supports of the earth and sky were anthropomorphized in the Postclassic period as the *bacab* or *pauahtun*.[16] In some conceptions the quadrilateral earth was likened to the back of a giant cosmic crocodile, each leg marking one of its corners.[17] The Maya also conceived of the earth as a great turtle floating in a cosmic sea.[18] Here the earth was understood to be circular, surrounded by water. The circularity both references human perceptions of the horizon and accords with the circular route of the sun on its daily travels.

THE AXIS MUNDI

For the ancient and modern Maya, spaces of any shape and size, from the interior of a house to the entire surface of the earth, could be partitioned into four parts and marked by a central point. The natal community typically is understood as the center of the earth. As Gossen notes for the Tzotzil: "They believe that their centrality on the square earth-island . . . gives them a special relationship with the sun, the principal deity, which no other Indian or mestizo community can hope to match." The Tzotzil perceive the center as civilized, whereas the margins of the earth are occupied by "demons, strange human beings, and animals."[19] The ancient and modern Maya landscape is, paradoxically, a place of many world centers. Not only is each Maya town conceptualized as the center of the world, but smaller segments of each community, including people, can be the axis mundi.[20]

Evidence from mortuary archaeology establishes that Classic Maya graves were also conceived as quadrangular spaces with ritual centers. Although the basic shape of the majority of Classic period tombs and crypts is rectangular (for example, figs. 2.17, 2.27, 2.41, 2.43, 2.50), the intended shape of simple crypts and cists can be more ambiguous. Nevertheless, conceptual clues to the spatial ordering of Maya mortuary space are evident in deposits such as the painted tombs of Río Azul.[21] Recall Tomb 19, where the east, north, and south walls

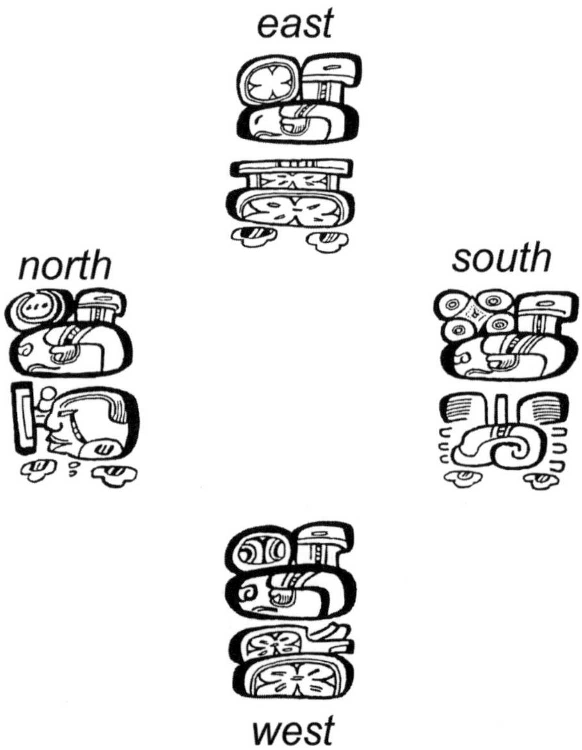

FIGURE 3.1. Direction glyphs on the walls of Río Azul Tomb 12 (drawings by David Stuart).

were painted with variants of Ux Yop Huun, effectively centering the deceased (see fig. 2.12). Directional symbolism and the centering of the dead is more overt in Tomb 12 of Río Azul: directional glyphs decorate each of the walls, each sign preceded by an eagle that seems to have the phonetic value of *kan* (sky) (fig. 3.1). Textually, east was shown in Classic Maya script by the *k'in* flower marking a solar receptacle. Recall from chapter 2 that the phonetic value of the sign is *el k'in* ("to rise, or exit the sun"; for more on solar receptacles, see chapter 4).[22] In Tomb 12 the eastern sky-eagle is also prefixed by *k'in*, the rising sun. West is written as *och k'in* ("the entering sun": the sunset when this solar being enters the underworld). As an inverse to the eastern sky-eagle, the western sky-eagle from Tomb 12 is prefixed by *ak'ab*, the sign for darkness. The significance of north and south in Tomb 12 is more opaque. The northern sky-eagle is marked by the sign for the moon and the southern sky-eagle is marked by *ek'*, the star. Combined, the north and southern sky-eagles mark the night sky and could refer to a nocturnal celestial path, perhaps the Milky Way (see chapter 2). Centered within the four directions of Tomb 12 was the body of a dead lord, situated as an axis mundi.

Centers and axes mundi are associated with the color green, indicative of new growth. Throughout Mesoamerica they are conceptually linked to crosses, trees, and poles and more generally to the concepts of strength and vitality (see chapter 2).[23] According to Robert Carlsen, at Santiago Atitlán, Guatemala, "the leader of the *cofradía*, the *cabecera*, is the tree's trunk . . . he is sometimes called just that, 'the trunk.'"[24] During the Classic period centers-as-trees were anthropomorphized as the Maize God and also as the so-called Patron of Pax.[25] Both of these beings were rendered as jade beads and placed within the graves of Maya elites, presumably to symbolize the axis mundi or to center the burial chambers. For example, Pakal was buried with four large jade ornaments placed at the end of each of his appendages (fig. 3.2). A fifth ornament was placed at his groin, carved as a hybrid of the Maize God and a descending crocodile, a motif that Taube shows to be emblematic of the tree-as–axis mundi, centering Pakal and his burial chamber.[26]

For the contemporary and contact period Maya, the axis mundi as world tree is conceptualized as one of the paths along which the souls of the dead travel between the world of the living and the celestial realm of afterlife. Alfred Tozzer described the beliefs of early twentieth century Yucatec Maya:

> There are seven heavens above the earth, each of which has a hole in the center, one directly above the other. According to one idea, a giant *Ceiba* growing in the exact center of the earth rears its branches through

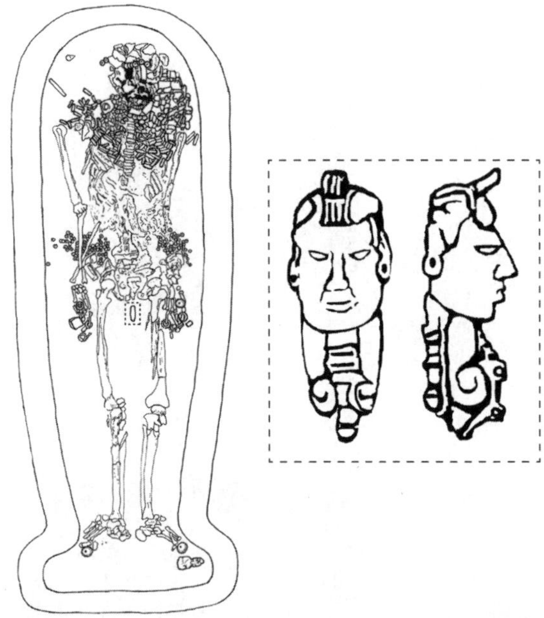

FIGURE 3.2. Location of hybrid Maize God–crocodile within the sarcophagus of Pakal (drawing of the burial plan by Linda Schele, © David Schele, courtesy Foundation for the Advancement of Mesoamerican Studies, Inc.; drawing of Maize God ornament by the author after Ruz Lhuillier, *El Templo de las Inscripciones*, figs. 242 and 243).

> the successive holes in the heavens until it reaches the seventh, where "El Gran Dios" of the Spaniards lives. It is by means of this tree that the dead spirits ascend from one world to the other until they reach the topmost one, where they finally remain. Another explanation is that there is a ladder made of vines running from the earth up through the holes in the heavens to the seventh, and it is by this vine that the souls ascend.[27]

A passage from the K'iche' *Popol Vuh* is further suggestive of the importance of centering in Maya concepts of death, renewal, and resurrection. The grandmother of the Hero Twins, mother of Hun Hunahpu, witnesses the death and rebirth of maize, which was left behind by the Twins as a sign of their journey into the underworld to challenge the Lords of Xibalba:

> The Grandmother was weeping, crying out before the ears of unripe maize that had been left planted. They had sprouted, but then they dried up when they [the Twins] were burned in the pit oven [by the Lords of Xibalba]. Then the ears of maize had sprouted once again, and the Grandmother had burned copal incense before them as a memorial. The heart of their grandmother rejoiced when the maize sprouted a second time. Thus they were deified by their grandmother. She named it Center House, Center Ancestral Plot, Revitalized Maize, and Leveled Earth. She named it Center House and Center Harvest for it was in the very center of the interior of their home where they had planted the ears of maize. She named it Leveled Earth and Revitalized Maize for it was upon level ground that the ears of maize had been planted. She named it Revitalized Maize because the maize had sprouted again.[28]

A tomb from Structure 10J-45 of Copan contained a large jade pectoral bar, placed at the center of the funerary slab, carved with the likeness of the Patron of Pax, a being emblematic of sacred trees (see fig. 2.11a).[29] Here the Patron of Pax is depicted with his arms curled over his chest, with the back of his hands touching one another.[30] Maya kings were also frequently shown with this posture, clutching bicephalic serpent bars to their chests (see fig. 2.52). The serpents that emerge from these bars are widely recognized as conduits to otherworldly places and are often depicted with supernatural beings emerging from their mouths. No such bar has ever been found archaeologically, so it is not clear whether these were real summoning devices or simply an artistic convention.

In either case the curled arm pose of the Classic Maya seems to imply an act of (or potential for) conjuring or facilitating otherworldly movement. In most

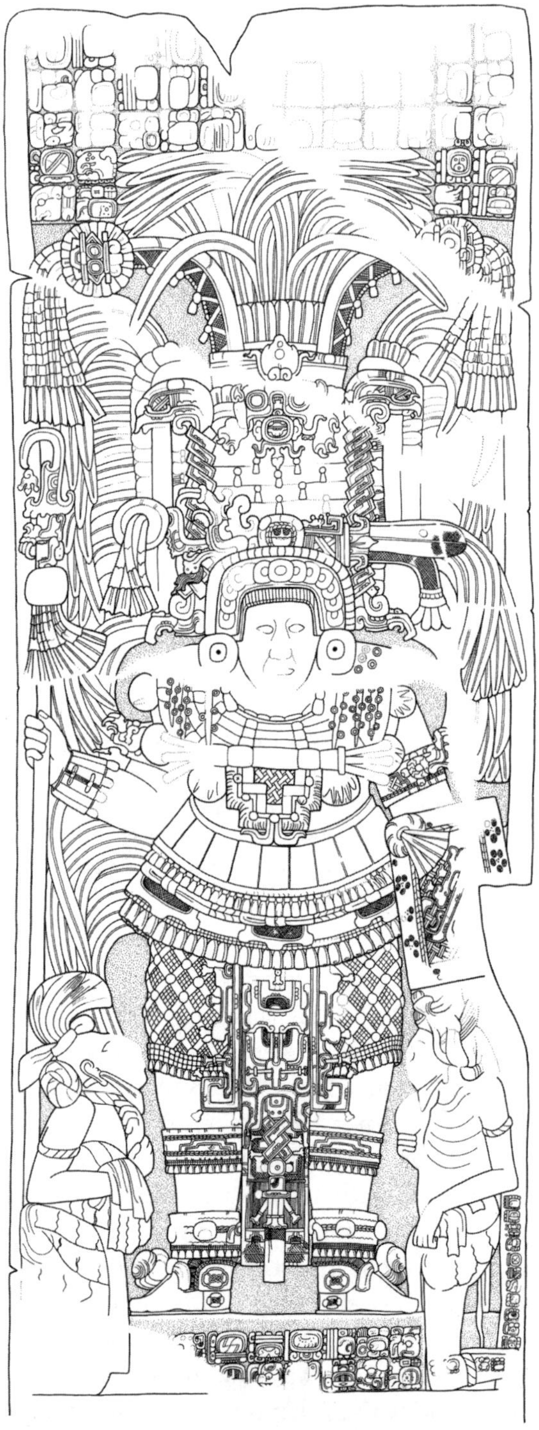

FIGURE 3.3. (*left*) Piedras Negras Stela 8, depicting Ruler 3 wearing a "shiner" loincloth framed by two square-nosed serpents (drawing of Stela 8, front, by David Stuart, from Stuart and Graham, *Corpus of Maya Hieroglyphic Inscriptions*, vol. 9, part 1, *Piedras Negras*, 9:44 © 2003 by the President and Fellows of Harvard College, Peabody Museum of Archaeology and Ethnology, Harvard University, PM# 2004.15.6.19.23 [digital file# 98880046]).

FIGURE 3.4. (*above*) Animate tree on an unprovenanced polychrome vase (modified from original drawing by Linda Schele after K1226 © David Schele, courtesy Foundation for the Advancement of Mesoamerican Studies, Inc.).

instances the conjuring device is shown, while in others it is implied by the pose. At Palenque bicephalic serpents wind through the branches of the trees on both the sarcophagus lid and the Temple of the Cross panel (see figs. 2.5 and 2.6). Recalling earlier discussions of humans-as-trees, I would suggest that Maya kings (or supernaturals, such as the Patron of Pax) embodied trees both as representing the axis mundi and as conduits to otherworldly places when they are shown with the distinctive curled arm pose.

A possible clue comes from the aprons that Maya lords are frequently depicted as wearing, especially in (but not limited to) scenes where they also hold the serpent bar (fig. 3.3; see also fig. 2.52). The cloths depict a cross-eyed being with a bar nose ornament. The apron being has an open mouth from which hang twisted cloth and other ornaments, attached to what may be a stylized *ik'* incisor (on fig. 3.3 it is shown as an inverted "*ajaw*-face" sign). The apron's face is framed by a pair of

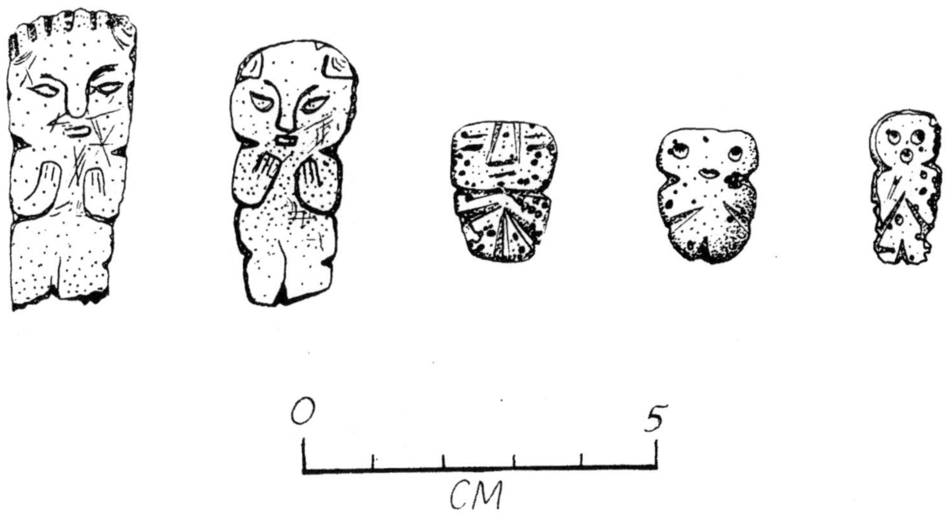

FIGURE 3.5. Shell "Charlie Chaplin" figurines from the caches of Piedras Negras (drawings by William Coe, originally published as Coe, *Piedras Negras Archaeology*, fig. 51d-h, image courtesy of the University of Pennsylvania Museum of Archaeology and Anthropology).

square-nosed serpents, and he is usually shown with the sign for shininess or specifically a jade celt inset into his forehead. The celt-in-forehead is also characteristic of God D and his various avatars or subordinates.[31] Linda Schele and Mary Miller recognized that the apron face was the same as the face on the Palenque sarcophagus and interpreted it as emblematic of the world tree (see fig. 2.5).[32] David Stuart has suggested, however, that the face in the apron is more simply an anthropomorphized celt representing the general concept of shininess and has nicknamed this being the "shiner."[33] Stuart suggests that this being is distinct from other animate trees embodied by the Patron of Pax, as depicted, for example, on an unprovenanced polychrome vase (K1226) (fig. 3.4). As he points out, the animate tree on K1226 has a **YAX** sign affixed to it, which identifies the tree as *yaxte'* (green tree, *ceiba*). Moreover, the face of the tree on K1226 has a jaguar paw above its left ear, symbolism typically associated with the Patron of Pax.

I suspect, however, that it may be premature to suggest that the "shiner" being from the aprons and shown on the Palenque trees is unconnected to other animate depictions of trees, such as the one on K1226. The K1226 tree (like many other depictions of animate trees) has a jade celt over its forehead, similar to the one inset into the foreheads of the "shiner" beings and shown to be an enlarged version of similar fruit that hangs from its branches. Andrea Stone and Marc Zender see a specific connection between jade celts and fruit and suggest a possible reading of **HUT** (fruit, seed, or face) for the celt sign.[34] Actual *ceiba* trees have fruit that resembles jade celts in size and shape and is green in color. Thus, when Maya lords wore the aprons with the animate "shiner" face and suspended jade celts around their waists, it may be that they became like-in-kind to the great *ceiba* tree and embodied the axis mundi. In a similar fashion the cradling of conjuring bars to the chest suggests both the tree and more generally conjuring or movement between otherworldly places.

Tiny figurines made of shell, stone, or bone that all bear the same distinctive curled arm pose have been recovered in caches at Tikal, Caracol, Piedras Negras, and numerous other sites (fig. 3.5).[35] These "Charlie Chaplin" figures may embody the concept of otherworldly movement, conjuring, or the axis mundi as anthropomorphic tree. These objects date primarily to the Early Classic period and are generally found within cache vessels and rarely in mortuary contexts.[36] Tikal Burial 10 is one notable exception and contained eight of

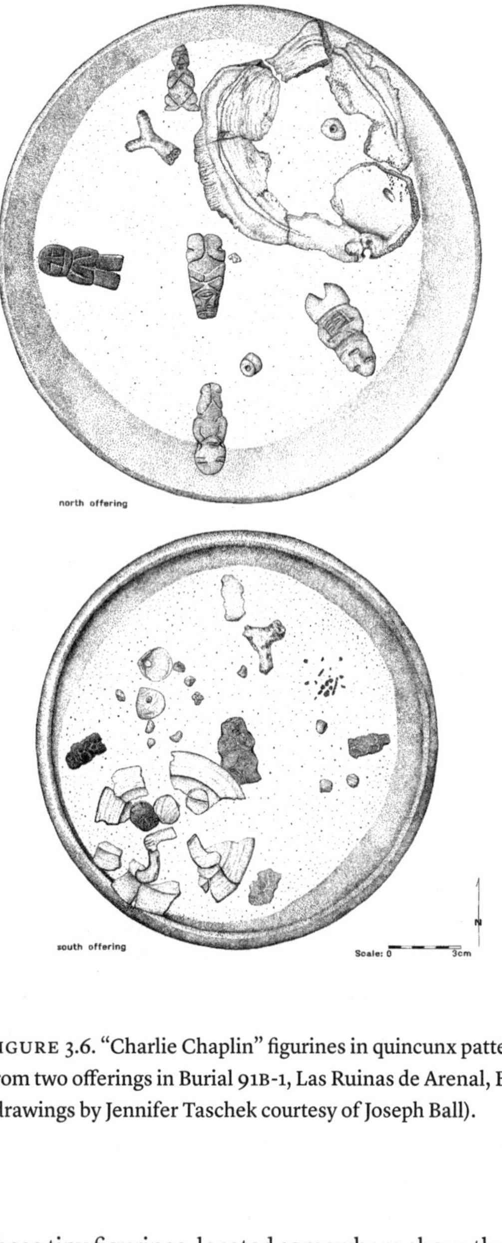

FIGURE 3.6. "Charlie Chaplin" figurines in quincunx patterns from two offerings in Burial 91B-1, Las Ruinas de Arenal, Belize (drawings by Jennifer Taschek courtesy of Joseph Ball).

these tiny figurines, located somewhere above the head of the deceased.[37] Unfortunately, many of the contexts that have produced Charlie Chaplin figurines are poorly documented, so their original depositional arrangement is unknown. A notable exception comes from a burial at Las Ruinas de Arenal, Belize (fig. 3.6).[38] Lidded cache vessels from the foot end (north) and head end (south) of the burial each contained five Charlie Chaplin figurines arranged in a quincunx. Remarkably, the figures in each deposit, carved of shell and obsidian, were respectively colored red, white, black, and yellow, corresponding to each of the primary directions in Maya cosmology. The fifth, central figure was made of greenstone, again establishing green, the color of new growth and semantically linked to the *ceiba*, as the ritual center.

Numerous royal burials across the Maya area have produced jade objects similar to the Charlie Chaplin figurines, depicting anthropomorphic figures carved with their arms curled over their chests (fig. 3.7). In Tikal Burial 23 a jade figurine with arms flexed over its chest was placed above the right shoulder of the tomb occupant (fig. 3.7a).[39] The corpse's arms were flexed over the body. The deceased wore a necklace containing a second humanoid with its arms curled over its chest (fig. 3.7b).[40] A jade ornament depicting a being with curled arms was also recovered near the mandible of Ruler 3 of Piedras Negras (Burial 5; fig. 3.7c). A similar pendant was recently found placed near the mandible of the noble woman in El Perú–Waka' Burial 39 (fig. 3.7d).[41] This unusual object also had a removable jade "eye piece" over the left eye, perhaps to mask a defect in the pendant or somehow related to a condition in the woman herself. Such pendants were also found at Palenque, in Tomb 2 of Temple XVIII, the substructure tomb in the Temple of the Skull, and within the sarcophagus of the Red Queen (see below).[42]

An especially stunning example of a flexed-arm being was found in an Early Classic lidded cache associated with the Structure 10J-45 tomb from Copan (fig. 3.8).[43] An apparent Maize God is shown as the axis mundi, situated in a double quincunx, recalling the tableau from the cache vessels from the burial at Las Ruinas de Arenal (see fig. 3.6). The first quincunx of the Copan cache was composed of four jade figurines. The second consisted of four *Spondylus* shells that each contained a jade bead or two. As a depiction of the axis mundi, the Copan tableau evokes the Maize God's resurrection from the back of a shell, usually shown as the turtle carapace–earth (see fig. 1.12).

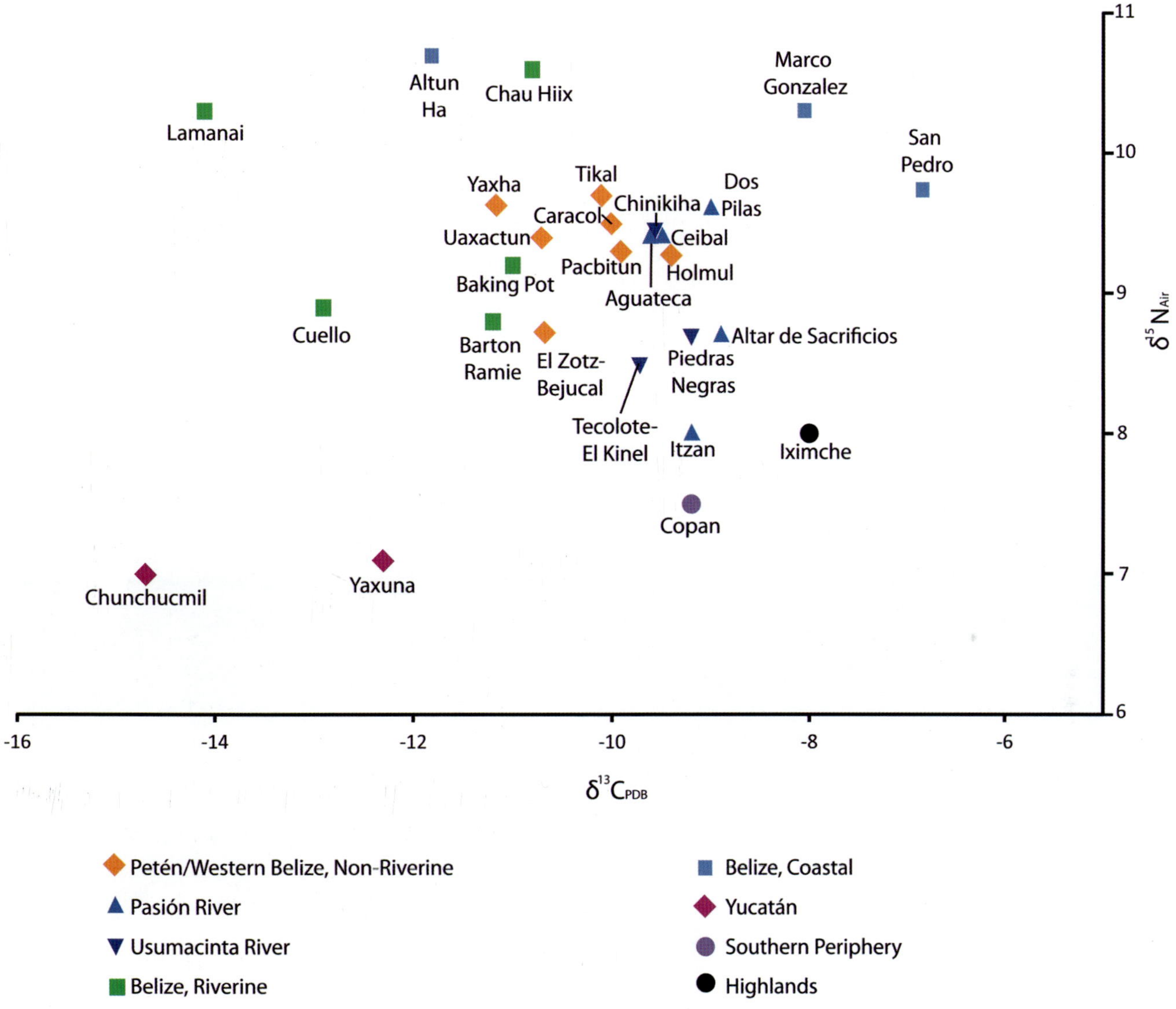

PLATE 1. Mean stable carbon and nitrogen isotope ratios from across the Maya area (figure by author based on published data).

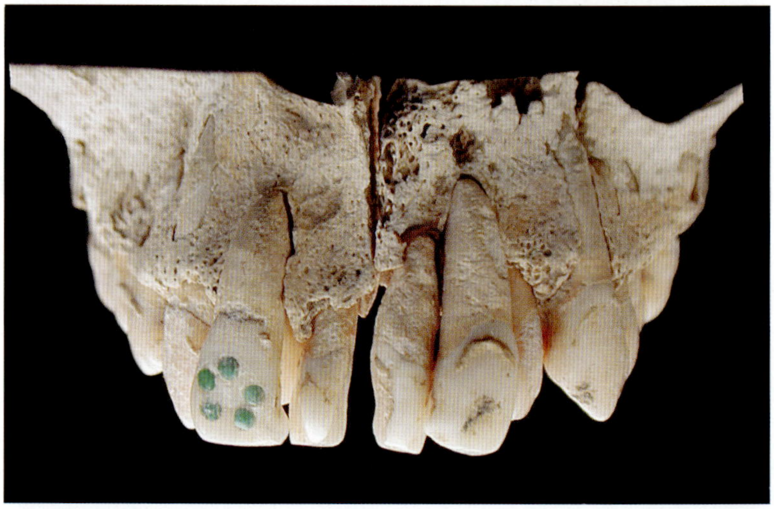

PLATE 2. (*above*) The modified teeth of the primary individual from Burial 5 of Piedras Negras, identified as the remains of Ruler 3 (photograph by author).

PLATE 3. (*right*) Jade inlays and supernumerary incisors of a young adult male from Piedras Negras Burial 45 (photograph by author).

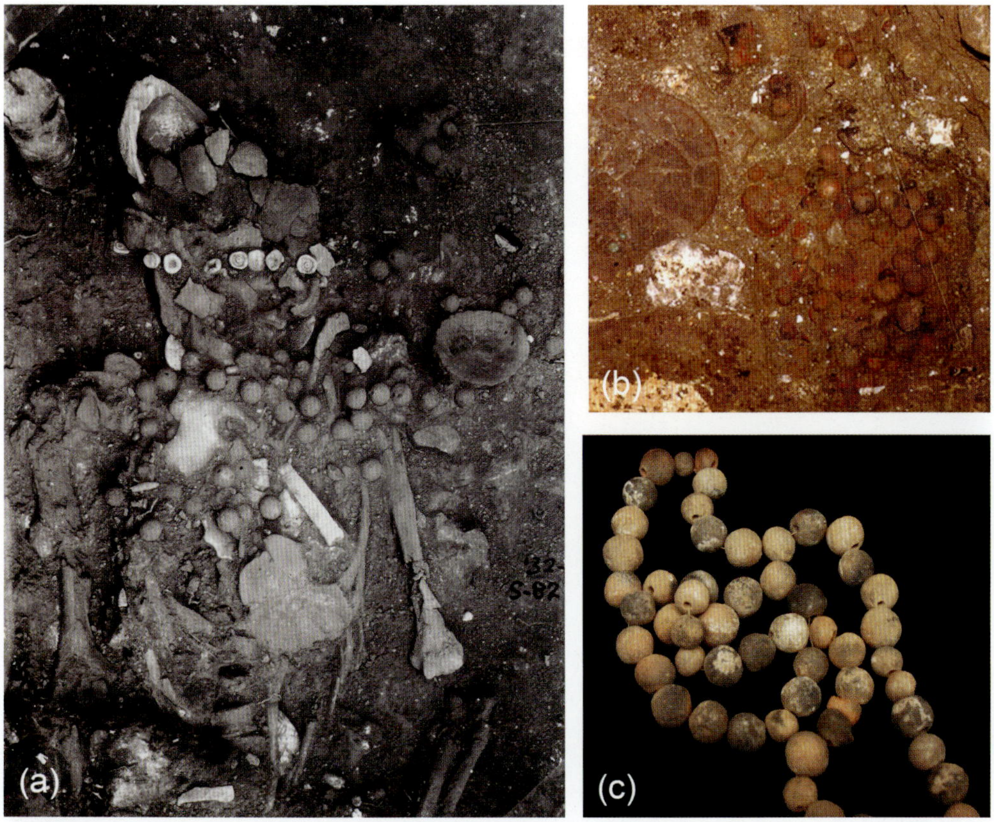

PLATE 4. Ceramic beads painted to resemble jade beads: (a) Piedras Negras Burial 5 in situ (photograph courtesy of the University of Pennsylvania Museum of Archaeology and Anthropology); (b) Piedras Negras Burial 13 in situ (photograph courtesy of Stephen Houston); and (c) ceramic beads from Piedras Negras Burial 5 (photograph by author).

PLATE 5. Necklace of jade beads from Budsilha Burial 4 (photograph by author).

PLATE 6. Dancers from the mural of Room 3, Structure 1, Bonampak. Note the composite materials of the headdresses (reconstruction painting of Bonampak, Mexico, Yale University Art Gallery, Gift of Bonampak Documentation Project, illustrated by Heather Hurst and Leonard Ashby).

PLATE 7. Jade bead that was placed in the mouth of the primary occupant of El Zotz Burial 9 (photographs courtesy of Stephen Houston, El Zotz Archaeological Project). *Inset*: laboratory photograph of the same bead. In the field photograph a shell bead necklace overlays the assemblage.

PLATE 8. Hematite clay bricks and *Conus spurius* shell tinklers from Burial 9, El Zotz (photograph courtesy of Stephen Houston, El Zotz Archaeological Project).

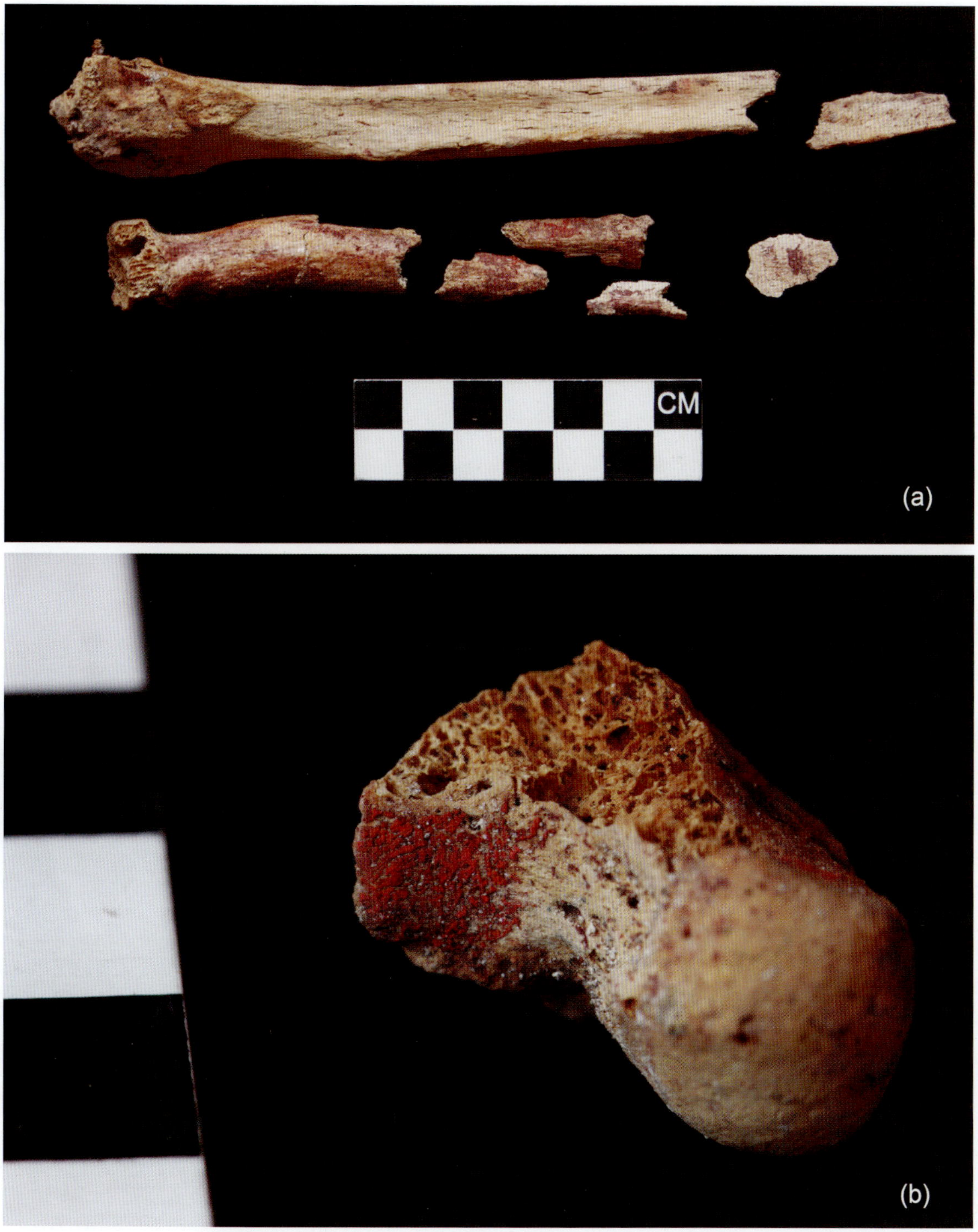

PLATE 9. Painted bones of the primary occupant from El Zotz Burial 9: (a) specular hematite (brick red) on the bones of the right forearm; and (b) a layer of specular hematite overlain by cinnabar (vibrant red) on a right wrist bone (capitate) (photographs by author)

PLATE 10. Early Classic cache from Structure 10J-45 tomb at Copan (photograph © Jorge Pérez de Lara).

PLATE 11. Internal and external surfaces of an unworked *Spondylus* shell from Piedras Negras Burial 5 (photographs by author).

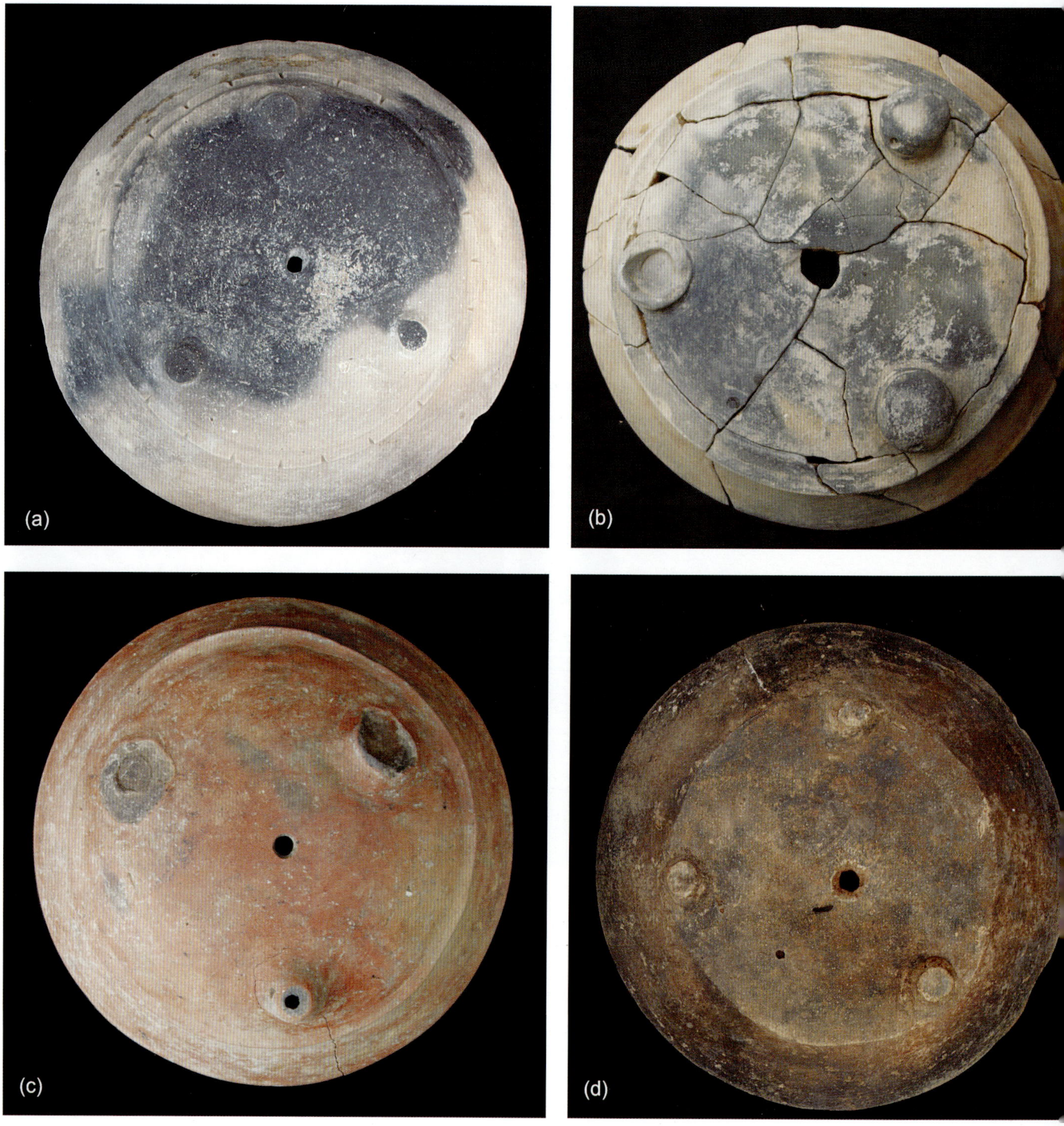

PLATE 12. Perforated dishes from burials within the kingdom of Yaxchilan: (a) El Kinel Burial 11; (b) El Kinel Burial 12; (c) El Kinel Burial 10; (d) Tecolote Burial 6 (photographs by Charles Golden). Note the postperforation burning in (a) and (b).

PLATE 13. Unprovenanced vessel recovered near Piedras Negras with its center painted with red hematite (photograph by author).

PLATE 14. Unprovenanced vessel from the Museum of Fine Arts, Boston, showing an enthroned lord, a sacrifice, and a dancer grasping the rope or tether of an entombed lord (K1377 © Justin Kerr).

PLATE 15. Burned subadult human remains from Piedras Negras Burial 13: (a) cranial vault showing differential burning to the frontal (*left*) and parietal bones that articulate along the coronal suture (*arrows*); and (b) anterior maxillary teeth showing burning of roots (photographs by author).

PLATE 16. The urban character of a modern Yucatec cemetery in Campeche (photograph by author).

PLATE 17. Solar masks on the Kohunlich Temple of the Masks (photograph by Holly Scherer)

PLATE 18. *Witz* cave entrance in Classic period iconography: (a) on the façade of the Labna arch building; and (b) marked with jaguar spots on a cache vessel from Tecolote (compare to the bloodletting vessel on Yaxchilan Lintel 24) (photographs by author).

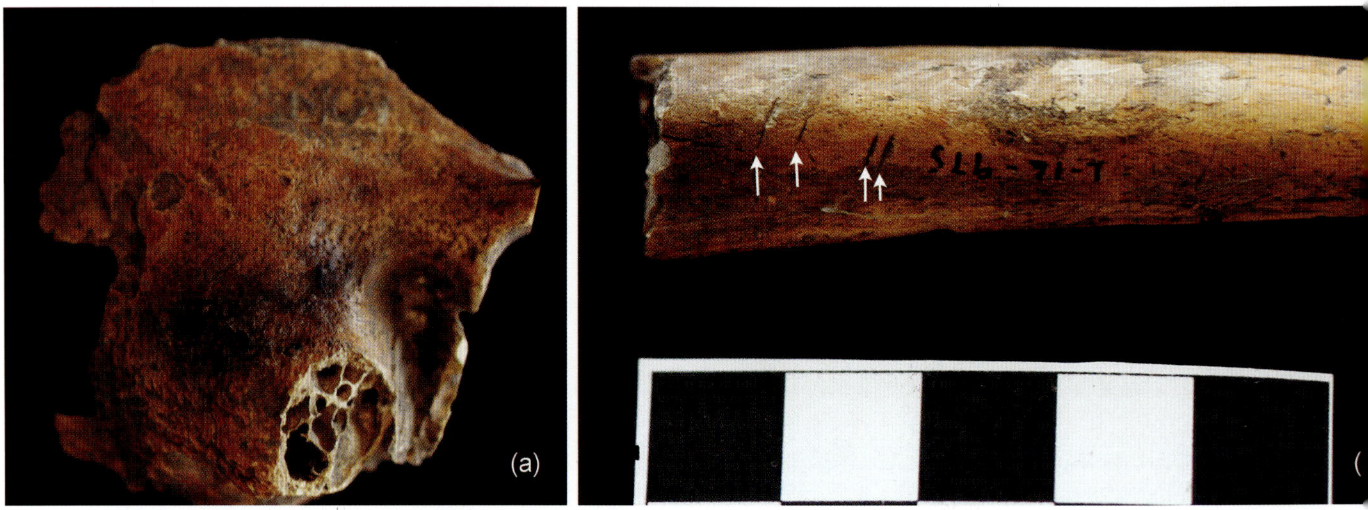

PLATE 19. Human remains from Piedras Negras Burial 4: (a) thermal blackening of the right temporal bone (damage to the mastoid process from excavation); and (b) cut marks on a radius fragment (photographs by author).

PLATE 20. Burned scatter of human bone and ceramic sherds from Cueva de las Tres Entradas (photograph by author).

HEART OF THE EARTH

Shell played a central role in the ritual life of the Maya and was an essential component of Classic period iconography. The Maya valued *Spondylus* shell for its vibrant red color, suggestive of life, heat, blood, and the sun, as well as its association with the distant sea, the place of solar rebirth (fig. 3.9). In some instances the solar quality of *Spondylus* shells was enhanced by painting their insides red with cinnabar. *Spondylus* and other types of shells seem to have been understood as surfaces that contained, protected, or transported animate spirit.[44] The Maya frequently placed jade beads inside *Spondylus*, as, for example, in the offerings left on the internal staircase of Temple of the Inscriptions at Palenque, in the royal tombs of Piedras Negras, and in the Copan cache tableau just noted (see fig. 3.8). As Taube demonstrates, both *Spondylus* and maize were symbolically linked as core Maya symbols for life and vitality. Jade beads were understood as seeds of new life, *Spondylus* as the shell or husk that contained it.[45] The placement of jade beads within *Spondylus* may have been inspired by the Maya's observation of pearls occasionally found within mollusk shells. Like jade, pearls were employed by the Maya as symbols of animate spirit in body adornments as, for example, in the ear ornaments worn by Pakal in his tomb (see. fig. 2.20).[46] In this particular framework *Spondylus*, with its deep red coloration, may have been a metaphoric womb for the germination of new life. Because of the

FIGURE 3.7. (*above*) Jade ornaments of anthropomorphic beings with arms curled over the chest: (a) and (b) Burial 23 at Tikal (originally published in Moholy-Nagy, *The Artifacts of Tikal*, fig. 108, image courtesy of the University of Pennsylvania Museum of Archaeology and Anthropology); (c) Piedras Negras Burial 5 (drawing by author after Coe, *Piedras Negras Archaeology*, fig. 47e); (d) El Perú-Waka' Burial 39 (drawing by author after Rich, "Ritual, Royalty, and Classic Period Politics," fig. 7.86). All figures are close to relative scale: 3.7c measures 7 cm in height.

FIGURE 3.8. (*below*) Early Classic cache from Structure 10J-45 tomb at Copan (photograph © Jorge Pérez de Lara).

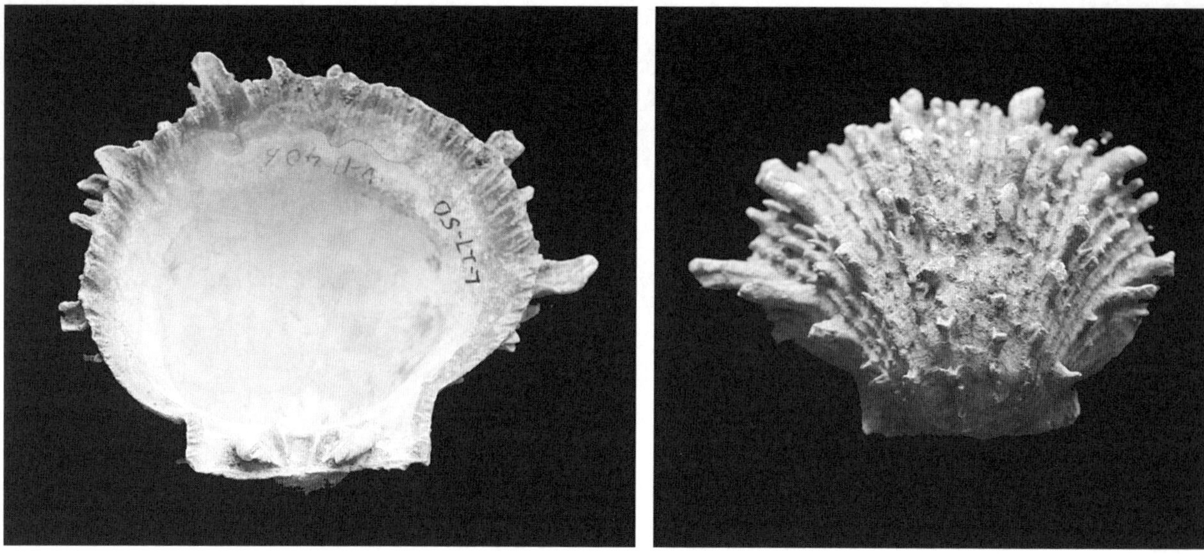

FIGURE 3.9. Internal and external surfaces of an unworked *Spondylus* shell from Piedras Negras Burial 5 (photographs by author).

many symbolic attributes that link *Spondylus* to concepts of vitality, renewal, and rebirth, this shell was frequently placed in elite Maya graves.

As containers for the seeds of life or surfaces from which those seeds sprout, *Spondylus* and other shells were widely employed as symbols of fertility and reproduction and were associated with both female and male genitalia. Landa's *Relación* reports that Yucatec maidens wore shell over their loins.[47] In a similar fashion a youthful woman, perhaps the wife of the Maize God, wears a *Spondylus* shell suspended from a belt over her genitalia in the San Bartolo murals, as do the maidens on the Berlin Vase (see fig. 2.15).[48] Depictions of Maya noblewomen dressed in the netted skirt of the maize beings are frequently shown with *Spondylus* over their groins, like the woman in figure 2.59b. The *Spondylus* shells on these sculptures are framed within the mouths of reptilian or aquatic creatures that Michel Quenon and Geneviève Le Fort connect to the underwater serpent from which the Maize God is shown to emerge on a number of ceramic vases (see fig. 2.4).[49] It may be that that the *Spondylus* is also an allusion to the maw of this being, a symbolic womb that facilitates supernatural birth. The reproductive symbolism of the *Spondylus* does not seem restricted to women, however. For example, the body attributed to the Calakmul king Yuknoom Yich'aak K'ahk' was buried with a pair of *Spondylus* shells over his pelvic area.[50]

As a variant of the bead-in-shell motif, a *Spondylus* shell was placed above the head of the Red Queen of Palenque. Inside was a diminutive limestone figurine (fig. 3.10).[51] The figurine is female (as is evident from her huipil). From her neck dangles a pectoral that supports three celts, recalling the axis mundi symbolism discussed earlier. The figurine's large eyes and unusually shaped mouth recall the "*ajaw*-face" sign, suggesting that the little object may be meant to represent animate spirit, perhaps even that of the deceased. Placed near the Palenque queen's head, it is likely that this tableau was meant to suggest the spiritual rebirth of this royal woman. In similar fashion a descending Maize God ornament was tucked underneath a *Spondylus* shell that was placed above the head of the deceased in Burial 196 of Tikal (fig. 3.11; see also fig. 2.50).[52] The practice of placing *Spondylus* shells at the heads of the deceased is very widespread in the Maya area during the Classic period, including most royal tombs at Tikal, Piedras Negras, and Río Azul (see fig. 2.18a).[53] In some cases the shells were tucked underneath the heads of the deceased. In others they are found atop the skull, situated directly above the parietal bones. The particular arrangement seems to suggest the body was below or within the shell, as further evidenced by a series of royal burials from Tikal. In Tikal Burial 195 one shell was placed at the

FIGURE 3.10. Anthropomorphic figurine placed in a *Spondylus* shell near the head of the Red Queen of Palenque (drawing by author after González Cruz, *La reina roja*: 175).

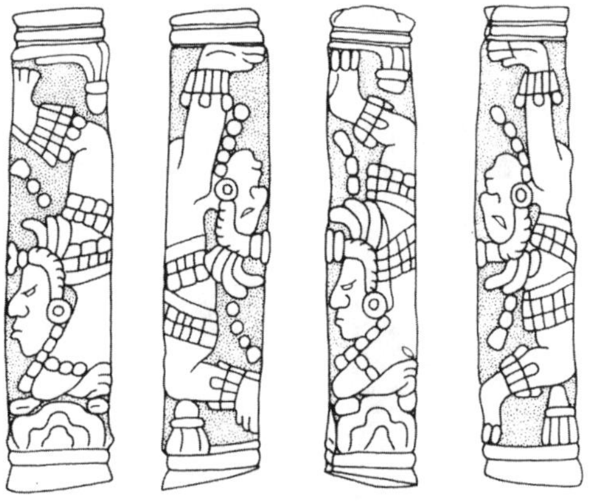

FIGURE 3.11. Four sides of a jade bead showing the Contortionist–Diving Maize God that was placed below a *Spondylus* shell at the head of Burial 196 (originally published in Moholy-Nagy, *The Artifacts of Tikal*, fig. 118a, image courtesy of the University of Pennsylvania Museum of Archaeology and Anthropology).

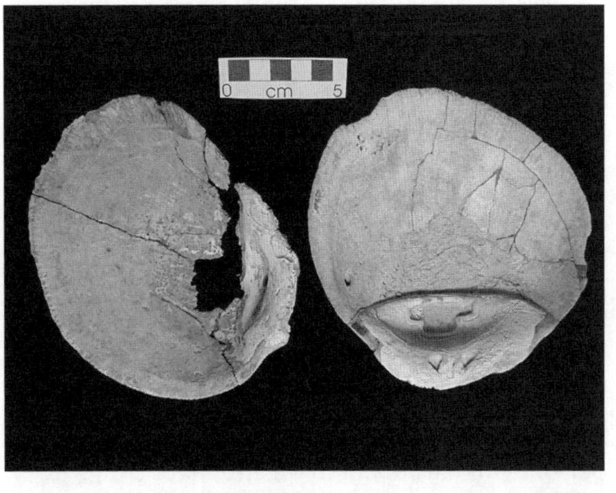

FIGURE 3.12. *Spondylus* shell with perforation and *ik'* sign from Piedras Negras Burial 13 (photograph courtesy of Stephen Houston).

head, a second lay at the feet, and three were laid over the body, to situate the decedent within an assemblage of shells (see fig. 2.27).[54] In other examples the bodies of the dead were enveloped by shells, with bivalves placed around and over the body, as in Tikal Burials 23, 24, 116, and 196 (see figs. 2.17 and 2.50). As with the figurines and jade beads that were placed within the shell, these arrangements seem to imply that the dead, as seeds of life, will be reborn from within the husk of the *Spondylus*. A *Spondylus* shell from Burial 13 at Piedras Negras was incised with the sign for *ik'* (wind or breath), underscoring its animate value (fig. 3.12).

The vast majority of *Spondylus* shells that were placed above the head of the deceased had large, central perforations (figs. 3.12, 3.13). These openings in the *Spondylus* are distinct from the typical small, paired holes drilled into the base of valves to facilitate suspension on clothing. In some instances these perforations are square, in others they are irregular and jagged, and in a few cases they are perfectly circular. The possible significance of these large, central perforations can be gleaned from a few examples that were further embellished. Two *Spondylus* shells from Burial 23 at Tikal had large perforations in their centers that were surrounded by four circular incisions, forming the glyph for *bih* (road or path) (fig. 3.13b).[55] Recall that one of the most common euphemisms for death in Classic Maya texts was *och bih* (to enter the road), suggesting the postmortem

FIGURE 3.13. Perforated *Spondylus* shells: (a) large perforation from shell at head of Piedras Negras Burial 5 (photograph by author); (b) *bih* sign from Tikal Burial 23 (originally published in Moholy-Nagy, *The Artifacts of Tikal*, fig. 165d, image courtesy of the University of Pennsylvania Museum of Archaeology and Anthropology); (c) possible *uh* sign from Tikal Burial 24 (originally published in Moholy-Nagy, *The Artifacts of Tikal*, fig. 165e, image courtesy of the University of Pennsylvania Museum of Archaeology and Anthropology). Note that the external spines on Fig. 3.13(a) were scraped off in antiquity.

FIGURE 3.14. Incised alabaster vessel from Bonampak showing the emergence of a youth from a shell in a watery scene (drawing by Stephen Houston).

journey of the spirit. As another variant, the perforated shell placed over the face of the woman from Yaxchilan Tomb 2 in Structure 23 (possibly Lady K'abal Xook) had two small circles drilled on either side of a larger circular perforation.[56] A shell with a similar pattern of perforations was found in Tikal Burial 24 (fig. 3.13c). Iconographically, this distinctive sign appears within crescents as the signs for moon (*uh*) and an untranslated glyph that semantically relates to "hole."[57] The meaning implied here is not fully clear, but the signs may mark the body as underground in anticipation of its emergence, much as the sign is used as a prefix affixed to the glyph for *wayeb*, the five days that are "in the hole" at the end of the solar year. Alternatively, the *uh* sign might imply lunar ascent of souls, as, for example, on the Berlin Vase (as noted in chapter 2). Other evidence suggests that the drilling of shell was closely related to emergence: an evocative scene from an alabaster vessel from Bonampak shows a young woman pulling a youth from a conch shell while another youth drills the back side of the shell (fig. 3.14).[58] In sum, the Maya practice of placing *Spondylus* shells over the heads and bodies of the dead parallels the placement of jade bead-seeds and effigy figures within the shells and alludes to the potential for rebirth and resurrection of the soul.

The Maya widely understood burial, rebirth, and resurrection as occurring within the heart of turtle-earth.[59] According to Houston, a passage on a looted panel from the region of Cancuen describes the arrival of a ruler at a place that is described as "within/inside the pool, the three turtle island, within/inside the heart of the turtle."[60] The meaning of the passage on the Cancuen area monument is vague, and Houston interprets it as a reference to a journey to the sea. A similar passage appears on an altar from El Perú–Waka', however, where a ruler "finishes his fifty-two years *tu-yohl, ahk* (within the heart of the turtle)," likely a reference to the amount of time that had passed since the burial of the king depicted on the monument.[61] Early representations of the heart of turtle-earth (or crocodile-earth) are evident on Izapa Stela 8 and Takalik Abaj Altar 48, where enthroned lords emerge from quatrefoil openings that are surrounded by reptilian limbs (figs. 3.15a and b).[62] Thus "three turtle island" may be a reference to the earth itself. What might be a "three turtle island" is shown in the Madrid Codex, where three hearthstones mark the center of the back of a turtle as an axis mundi (fig. 3.16).[63] In a similar fashion the Maize God is shown encased within a turtle's shell on the San Bartolo murals (see fig. 1.21b). This concept relates directly to Maize

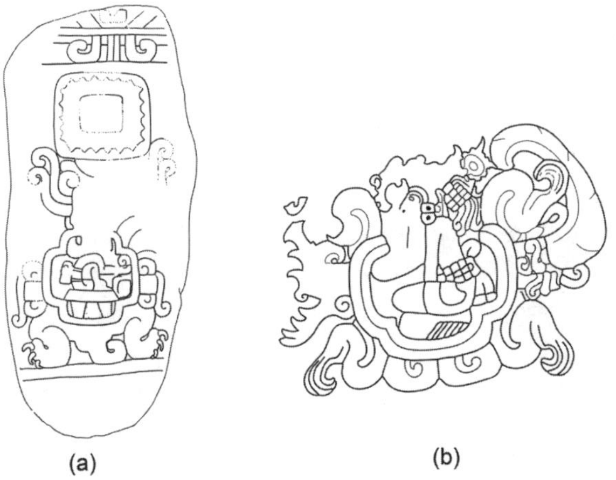

FIGURE 3.15. Beings within turtle/crocodile earth: (a) Izapa Stela 8 (drawing by author after Kaplan, "El trono incienso y otros tronos de Kaminaljuyu, Guatemala," fig. 12); and (b) Takalik Abaj Altar 48 (drawing by author after Schieber and Orrego, "El descubrimiento del Altar 48 de Tak'alik' Ab'aj," fig. 1).

God imagery, where the youthful supernatural is shown reborn from a water lily skull that resides in the heart of the turtle (see fig. 1.12). Houston, Stuart, and Taube also call attention to a series of Classic period monuments that show dead kings seated either within turtles or on the backs of crocodiles.[64]

Despite the imagery that shows the Maize God and Maya lords within or emerging from turtles and turtle shells, I am not aware of any situation in which the dead were placed beneath a turtle shell. Nevertheless, other mortuary practices are suggestive of mythic emergence from the earth. One such practice is the placement of the dead beneath perforated ceramic dishes, as evident in ancient graves throughout Mesoamerica and the American Southwest (fig. 3.17; for these and similar dishes in situ, see figs. 2.45 and 2.46). It may be that the practice of placing perforating *Spondylus* shells near the head was related to this.

In some cases the perforation within these ceramic dishes was carefully drilled (fig. 3.17c), while in others it was punched out (fig. 3.17a, b, and d). Popular belief holds that perforated ceramics were "ritually killed" in order to release the spirit of the vessel.[65] This assumption, however, does not explain why inverted vessels, usually found over the face, are invariably perforated, whereas other grave ceramics are rarely treated in such a manner.

FIGURE 3.16. Turtle with three stones on its back from page 71 of the Madrid Codex (drawing by author).

Nor does it account for the absence of perforated dishes in nonburial ritual contexts, such as cache deposits. Resting over skyward-facing skulls, these ceramic dishes more likely represent the surface of the earth, drilled to establish the axis mundi within the burial space and to recall the split turtle carapace from which the Maize God emerges during his resurrection.

Early examples of the use of mortuary dishes to symbolize ritual centering, quadripartite divisions, and spiritual travel include a series of seven Late Preclassic ceramic bowls with distinctive cross markings that have been recovered at K'axob, Belize.[66] Rather than being perforated, the interiors of the K'axob dishes were painted with crosses, denoting the quadripartite division of space (fig. 3.18). On six of these vessels the cross was created by two sweeping, perpendicular brushstrokes

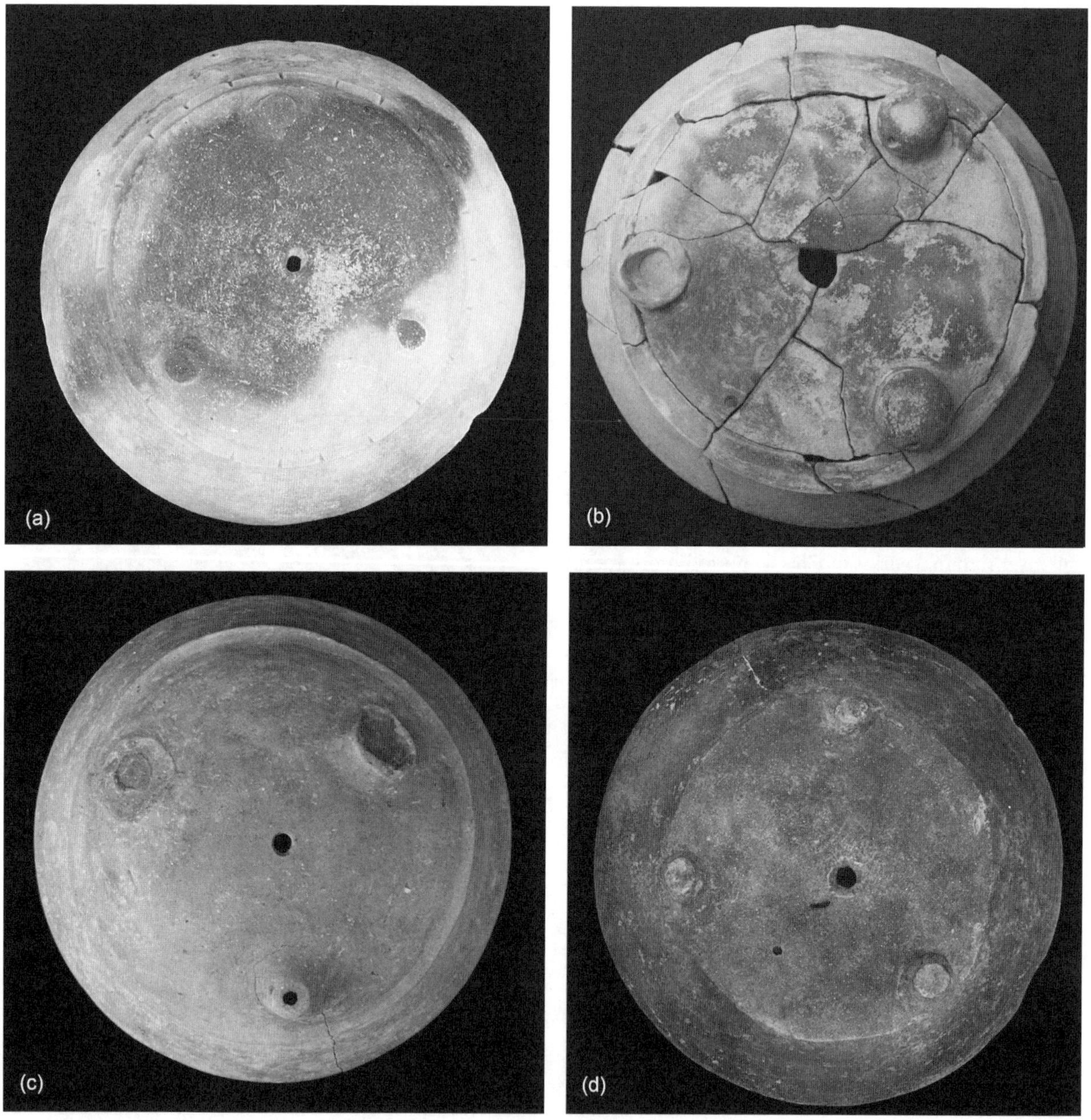

FIGURE 3.17. Perforated dishes from burials within the kingdom of Yaxchilan: (a) El Kinel Burial 11; (b) El Kinel Burial 12; (c) El Kinel Burial 10; (d) Tecolote Burial 6 (photographs by Charles Golden). Note the postperforation burning in (a) and (b).

FIGURE 3.18. Sierra Red dish with cross motif found over the head of the deceased in Late Preclassic K'axob Burial 1-25 (photograph courtesy Patricia McAnany).

across the exterior base of the vessel. On the seventh vessel the cross was painted on the interior surface of the vessel and a large dot was painted at its center. All of these vessels were found within burials, and four of them were placed over the heads of the deceased.[67] Annabeth Headrick suggests "that the people of K'axob not only saw the painted crosses on their vessels as representative of the four cardinal and horizontal directions but that they also thought of these crosses as emerging from these bowls to symbolize the world tree and its three cosmological levels."[68]

Michael Coe proposed some of these concepts years ago when he suggested that "the clay bowl represented the underside of the surface of the earth . . . the bowl was pierced to release the soul of the defunct."[69] As Mary Miller cogently points out, the polychrome perforated dishes that have been recovered from elite mortuary contexts often depict the actions of the Maize God—either his resurrection or his dancing—establishing a material connection between the perforated mortuary dishes and the myths of this supernatural being that embodies the concepts of fertility, vitality, and ascension from the underworld.[70] Thus the perforated vessels—so many of which have been looted and now populate the art museums of the world—are best understood as quintessential symbols of rebirth and resurrection of the Classic Maya dead.

Nearly all the Classic Maya perforated mortuary dishes are tripod vessels (see fig. 3.17). The three vessel legs may symbolize the three hearthstones of creation, underscoring the perforation in the vessel as an axis mundi.[71] Recall the three stones that mark the center of the back of the cosmic turtle from the Madrid Codex (see fig. 3.16). The placement of three objects to replicate the hearth and thus properly center place and space is pervasive in Maya worldview. For example, one of William Hanks's informants indicated that in laying marker stones around the milpas of Oxkutzcab, Yucatan, a person ideally "grabs three stones, like the kitchen fire."[72] An unprovenanced vessel with tripod legs, presumably

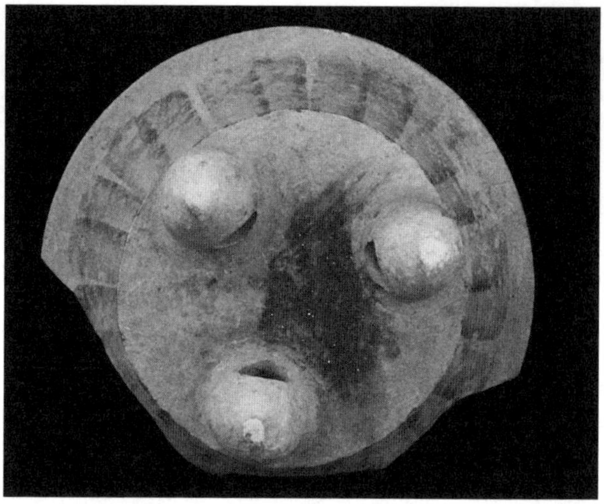

FIGURE 3.19. (*above*) Unprovenanced vessel recovered near Piedras Negras with its center painted with red hematite (photograph by author).

FIGURE 3.20. (*left*) A woman's head placed within three stones, a possible symbolic hearth, in Piedras Negras Burial 81 (drawing by Zachary Hruby). Also note the metate fragment placed at the foot of the burial.

from a looted burial, was shown to me in a modern community near Piedras Negras. Relevant to this discussion, the center of the dish's underside was painted with red hematite in a variation of the center perforation (fig. 3.19). In this context the red may symbolize the fires of creation at the center of the sacred hearth or perhaps solar rebirth, akin to the sun in the east.

Two perforated vessels from the Yaxchilan kingdom show burning that occurred after the dishes were placed over the heads of the deceased, evidenced by deep blackening on the underside of the vessel (see fig. 3.17a and b). In the case of the vessel shown in figure 3.17a, the smudging extends into the hole, evidence that the burning occurred *after* the perforation was drilled. The smudging also extends into the broken tripod supports, indicating that this burning was not conducted when the dishes were still functional but only after they had been modified for placement in the burials. Presumably copal or some other substance was burned over the face of the dead as part of the funerary rite, perhaps to facilitate the departure of the soul, to "feed" the deceased, or to replicate the sacred three stone hearth. In a similar fashion, the so-called head-cists of the Classic Maya may have been intended to situate the head within a three-stone hearth (fig. 3.20).[73]

The placement of perforated vessels over the head of the deceased, as opposed to other parts of the body, suggests a number of interpretive possibilities, none of which are mutually exclusive. Recall that the ancient and modern Maya understand the head as the physical manifestation of self and identity.[74] The head was the seat of the senses and the locus for vitality.[75] The spirits of the dead depart from the head as a final flowery breath.[76] Further communication with the dead is also focused on the head, as implied in the fiery breath imagery that travels upward from the nose of the interred ancestor on Stela 40 at Piedras Negras (see fig. 2.39). The emphasis on the head also recalls the pervasive Maya metaphor of skulls as seeds and heads as fruit. Many scenes of the Maize God's rebirth show him rising from a skull. Centering the head is evident in a diverse range of ancient and modern Maya practices. For example, when a contemporary Tz'utujil woman puts on her huipil she is situating herself in the center of the world, with her head emerging from a break in the earth's surface, much like the Tzotzil women noted in chapter 2.[77]

Curiously, the use of perforated dishes and shells within burials seems restricted to adults. For example, at El Kinel, Guatemala, my team and I excavated four burials with perforated dishes, all graves of adults.[78] We also excavated five burials of children, none of which contained such dishes. The significance of this distinction is not entirely clear but may relate to the undeveloped nature of the children's souls or differences in the postmortem afterlives of children relative to those of adults.

POLES AND NAVELS

Today ritual practice throughout Mesoamerica reflects the importance of sacred trees as world centers. This is especially evident in a broad group of acts that are typically referred to as "pole rituals" and involve climbing, descending, and performing atop what were historically wooden poles (metal poles are used in some contexts today). The best-known of these is the *danza de los voladores*, a performance that involves the planting of a pole from which dancers descend on spinning ropes. Nowadays the *voladores* perform widely across Mexico at festivals and for tourists, with much of the underlying symbolism forgotten.

In the modern K'iche' community of Momostenango preparations for the feast of Santiago begin with the erection of a large pole in the central plaza of the town, anchoring it as the center of the world (fig. 3.21).[79] The pole is stabilized by a guy-rope that extends from the top of the pole to the plaza floor. In the Momostecan dance performers dressed as monkeys, jaguars, and pumas (*leones*) first execute a series of acrobatic maneuvers atop the pole and then descend the guy-rope to the plaza floor below, performing additional acrobatics along the away. In Momostenango the top of the pole is reached by going up through the church. The ascent is not part of the spectacle. Instead the performance begins at the summit of the pole and continues during the performers' descent. As long as the dancers have properly performed *costumbre* (proper religious observances) to summon their *nawal* (animal companion spirits), they will be protected and will not fall to the plaza floor below. As Garrett Cook and Tom Offit have documented, the base of the hole into which the Momostenango pole is set is divided into a quadripartite space using copal and candles, ritually centering the space.[80] When Cook probed his informants about some of the symbolism of the dance, however, he found that much of the original significance was no longer understood by the performers. He suggests that in this matter Momostecan ritualists are "explicit practitioners but vague theologians," having long since lost an understanding of the original meaning behind some of their rituals.[81] At least in practice, the Momestecan pole ritual echoes important aspects of the *danza de los voladores* in the erection of a pole as ritual center and descent from that pole by rope.

Christopher Beekman has compiled information on pole rituals from across Mesoamerica and notes that

FIGURE 3.21. The Monkey Dance performed at Momostenango, Guatemala, in July 2008 (photographs by author). At lower left is a puma dancer balancing atop the pole. At lower right is a monkey dancer suspended over the crowd.

FIGURE 3.22. Ceramic diorama of a Nayarit pole ritual showing a central figure balancing atop a pole erected in the center of a village (image © Yale University Art Gallery 1959.55.18).

many of these rituals are linked to agricultural fertility.[82] Ceramic tableaus from the Nayarit (western Mexico) show that such rituals have deep antiquity, going at least as far back as the Late Preclassic period (fig. 3.22). Some depictions of the Nayarit poles show their tops divided into quadripartite spaces, similar to the setting of the pole in the K'iche' performance, establishing these early poles as world centers. Moreover, Beekman demonstrates that these pole rituals were linked to authority and rulership, echoing the earlier discussion of political and ritual officials as likened to trees and sacred centers in Mesoamerica. Indeed the Classic period Maize God of the Maya not only is an embodiment of the axis mundi but is also often depicted as a contortionist and acrobat who dives to the earth from the celestial realm above (see fig. 3.11). A variant of this theme in Postclassic art is the Diving Maize God, who appears descending from a celestial realm, foreshadowing the descent of the modern pole rituals noted above.[83]

Aztec pole-raising ceremonies emphasize the importance of rope, not only in raising and stabilizing the tree but also in removing and transporting it from its original location.[84] As noted earlier in this chapter, the path between the earth and the celestial realm was seen by the conquest-era Yucatec as a ladder made of vines in some conceptualizations. Arthur Miller suggests that the Maize God descends with a rope, specifically a celestial umbilical cord that he uses to bridge the earth to the heavens, leading some scholars to speculate that many of the twisted chords in Maya iconography may be celestial umbilical cords.[85] Yet, as Taube has shown, some of these cords may otherwise have been birth ropes, twisted fabric that is still used in Maya houses, suspended from the ceiling and used to support women as they give birth. The woman who "births" the youth from the shell shown in figure 3.14 has a rope running around her arms, perhaps an allusion to the birth ropes. As Taube points out, birth ropes were conceived as similar to umbilical cords, both essential emblems of birthing in the Maya worldview.[86]

Complicating matters, other twisted cords appear in Precolumbian iconography, flowing from the abdomens of Maize Gods and skeletal beings. Some scholars have confused these objects with umbilical cords, though they are more likely intestines.[87] Yet it is entirely possible that the Maya saw parallel symbolism in all of these objects. As umbilical cords twisting from the navel may be understood as the rope associated with the vital start to life, intestines distending from the abdomen may have symbolized life's end. Whatever the case, ropes and chords, as with crosses and poles, may have

FIGURE 3.23. Unprovenanced vessel from the Museum of Fine Arts, Boston, showing an enthroned lord, a sacrifice, and a dancer grasping the rope or tether of an entombed lord (K1377 © Justin Kerr).

been important symbols in Classic Maya conceptions of world centering and the passage between earthly and otherworldly realms.

The Maya have a deep tradition of likening the world center to a navel. The concept is anchored in their understandings of body and space. The contemporary Maya attribute the body's center to the navel. For the Tzeltal, this center is embodied by a small organ, referred to as *sme' winik* (mother of man), that they believe is located beneath the navel and is conceived as the center of the body.[88] The Yucatec Maya recognize a similar organ and refer to it as *tipte'*.[89] As the corporal center this organ is conceived as the point of origin of the body's blood vessels and is closely linked to the heart via the circulatory system.

The contemporary Maya map this navel-as-center metaphor onto their conceptions of the landscape. Thus center places, particularly the center of their natal communities, are likened to earthly umbilici. As Gossen observes, "most basic to Chamula [a Tzotzil community] spatial orientation is the belief that they live at the center of the earth, *smisik banamil*, or 'the navel of the earth.'"[90] The Tz'utujil of Santiago Atitlán have a nearly identical notion of the world and conceive of their community as the central "umbilicus of the world."[91] The Atitecos understand the navel of the earth's surface to be located in the floor of the town's church, where it is conceived as a portal to the underworld.[92] The colonial period Yucatec Maya "believed that a sacred umbilical cord, through which nourishment flowed in both directions, linked heaven and earth."[93] The protagonist of the *Rabinal Achi* describes the Rabinal fortress as the "navel of the sky, navel of the earth."[94] In a number of the west Mexican pole ritual tableaus, the performer is shown balancing atop the pole on his abdomen (see fig. 3.22). Such a display was obviously meant as a show of bravado. Nevertheless, it is worth considering whether these acts were also meant to suggest a tether running between the navel of the performer and the earth. K'iche' performers strike the same pose in their acrobatics atop the pole (see fig. 3.21).

Linguistically, there is a curious linkage between interment and the navel. In colonial Ch'olti' and modern Ch'orti', the term *muc* or *muhk* is used for things buried (including corpses) and for the umbilicus.[95] *Muhk* is used in the Classic period inscriptions as the term for burial or tomb.[96] As Raphael Girard observes for the Cho'rti': "Like the pit of the temple and of the milpa, the grave is a receptacle of sacred remains and is located in the Center of the World. The symbolic equivalence is expressed linguistically by a common word: *muhk* or

FIGURE 3.24. A being emerges from the underworld on La Venta Altar 4, assisted by another being grasping a rope (photograph by author).

muh, that designates the grave, the navel or pit of offerings, which also have a mythical common denominator: *Pucbal chaj*, the grave of Ahpú."[97]

Kerry Hull has suggested that one of the means by which Classic period souls journeyed from the tomb was along ropes that connected the earth to the celestial realm.[98] The concept is evident in the fiery rope-breath shown emerging from the tomb on Piedras Negras Stela 40 (see fig. 2.39). A Late Classic period vase (K1377) in the Museum of Fine Arts (MFA), Boston, shows a dancing figure holding a chord that links him to an enthroned and entombed lord, situated within the mouth of the earth crocodile that doubles as his sepulchral throne (fig. 3.23).[99] Notably, the MFA vessel is effectively a more elaborate depiction of the Middle Preclassic Altar 4 from La Venta, complete with the lord situated within the mouth of a reptilian creature (fig. 3.24). In the earlier twentieth century, in the Yucatec village of Chan Kom, a knotted cord tied to the belt of the deceased was understood to represent the rosary.[100] The practice, however, may have earlier roots related to the notion of ropes and cords as channels for the movement of the soul.

Within the polity of Yaxchilan the Maya drew specific attention to the deceased's navel as the center and focus within the mortuary space. As a variant to the positioning of inverted perforated dishes over the faces of the deceased, a number of burials excavated at the Yaxchilan-affiliated sites of Tecolote and El Kinel revealed dishes positioned in the middle of the burial crypts (fig. 3.25). Although it may seem that the practice was intended simply to place the dish-as-center in the actual center of the burial chamber, evidence from recently excavated royal burials at El Perú-Waka' suggests otherwise. El Perú-Waka' Burials 39 and 61 each contained royal personages with perforated dishes located over their abdomens.[101] Other vessels within the

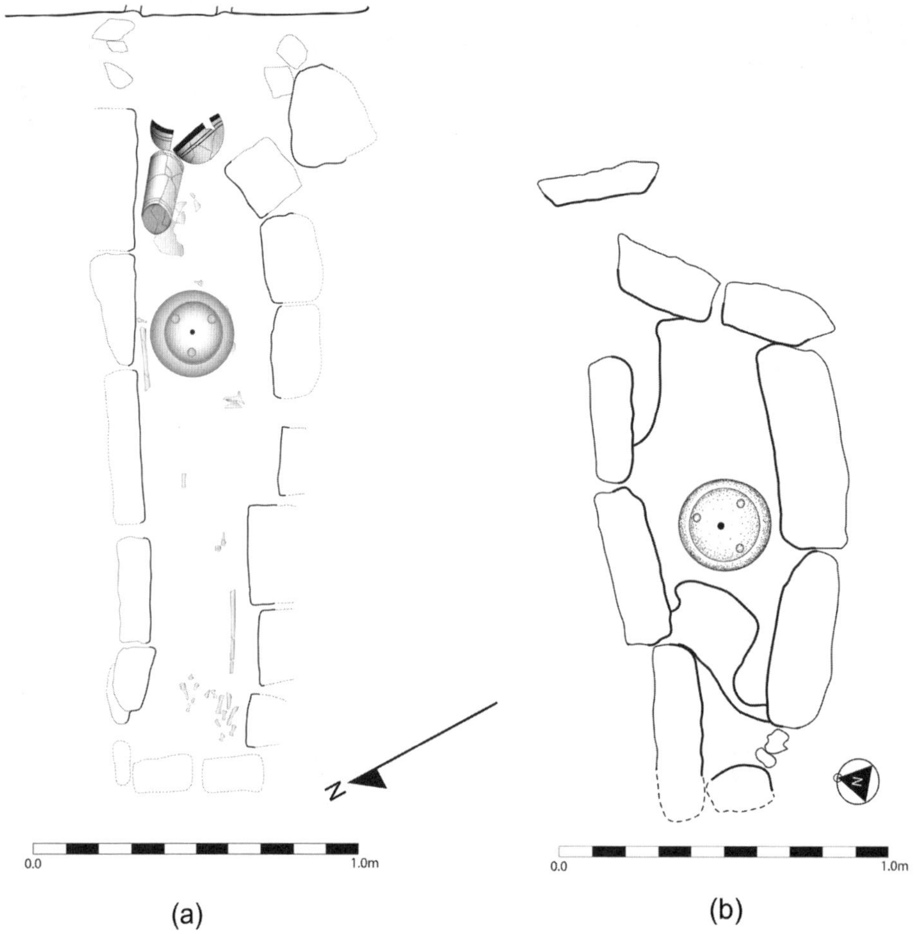

FIGURE 3.25. Perforated dishes at the center of the burial chamber: (a) El Kinel Burial 11 (drawing by Charles Golden); and (b) Tecolote Burial 6 (drawing by author).

tombs were unperforated. At least in the case of Burial 39, the body lay on a bench and was not centered within the tomb, so that the dish was centered on the body, not the tomb. Perhaps in these instances the departure of the soul was envisioned as through the umbilicus, not through the head. Nevertheless, some mortuary contexts hint at the centering of the mortuary space itself. For example, Tikal Burial 23 had a red dot, 15 cm in diameter, painted on the middle slab of the burial chamber that may have represented either the center of the burial or a portal through which souls departed, reminiscent of the openings made in the roofs of the houses of the Tzotzil.[102]

FIRE AND THE JAGUAR SUN

Performers descending on ropes and Maize Gods diving from the celestial realms are both reflections of a broader mythic concern with ascension and descension of celestial bodies, supernatural beings, and ancestors. In Classic period thought the jaguar was the embodiment of the setting and nighttime sun, depicted as a poorly understood being known as the Jaguar God of the Underworld (fig. 3.26). He is recognized as a jaguar supernatural (for example, he has a projecting central tooth instead of the normal canines of "natural" jaguars), he is often marked with *k'in* signs, and his symbolism overlaps with that of the Sun God. Unlike the Sun God, however, the eyes of the Jaguar God of the Underworld have spiral, underworld pupils. He is apparently emblematic of the sun's journey through the underworld and

FIGURE 3.26. Jaguar God of the Underworld from the west stairway of the East Court of Copan (photograph by author). Note his "cruller," projecting center tooth, spiral eyes, and the star (*ek'*) signs that flank him.

is associated with ferocity and war. GI of the Palenque triad, although a distinct being, seems to embody the transition of the Jaguar God of the Underworld into the Sun God during solar ascent in the east. The Baby Jaguar appears to be his newborn manifestation. The Water Lily Jaguar seems to be an utterly distinct entity; he has "normal" jaguar canines and appears on a number of vases with the Baby Jaguar.

One of the telltale attributes of the Jaguar God of the Underworld is the so-called cruller, a cord that is strung under his orbits and twisted between his eyes. Though likely not an umbilical cord, the object nevertheless may relate to the concept of movement between the earth and otherworldly places. The ends of the cruller worn by what is either the Jaguar God of the Underworld or GI on the recently uncovered stucco façade at El Zotz terminate within his earspools, which are marked with *bih'* (road) (fig. 3.27).[103] In light of the Jaguar God of the Underworld's association with fire, the rope may be the twisted cord used for fire drills, as suggested by Taube.[104] In Mesoamerican ritual practice, new fire ceremonies, involving the act of fire drilling, were linked to symbolic dawning.[105] Maya lords and priests are shown dressed in the guise of the Jaguar God of the Underworld to conduct fire drilling rites (as on Tikal Altar 5; see fig. 2.58), leading Stuart to propose that the Jaguar God of the Underworld "was the supernatural patron of fire and fire making."[106] As Taube observes, it is particularly difficult to disentangle umbilical cord, birth rope, and fire drill symbolism in Mesoamerican art, perhaps because they are not necessarily mutually

FIGURE 3.27. Jaguar God of the Underworld as a stucco mask on the El Diablo Pyramid, El Zotz (image courtesy of Stephen Houston, El Zotz Archaeological Project).

exclusive concepts. Aztec thought equated fire drilling with both copulation and conception, connecting human creation with fire starting.[107]

Functionally and symbolically, fire is deeply important to the contemporary Maya. At the start of the agricultural season the Maya burn the overgrowth in their milpas in preparation for planting. Cutting and burning is a necessity for Maya farmers. Nonproductive crops are cut to minimize competition with the crops. That cut vegetation is burned in order to introduce nutrients into the soil. In light of its importance to the agricultural process, clearing the milpa also has important spiritual significance. Vogt reports that for the Tzotzil maize plants, like humans, have souls. If the milpa is not kept clear of competing vegetation and "weeds are allowed to grow, 'souls' of the maize plants will move to a 'clean' milpa to grow."[108] Stuart demonstrates that the Classic Maya used fire ritually to prepare and regenerate built space, including domestic residences and funerary contexts.[109] Classic period texts describe the rite as *och k'ahk'* (fire enters). The ideological importance of fire relates to the association of heat with life and especially power, as derived from the sun. As discussed in chapter 1, warmth is associated with vitality, the heart, and blood, whereas cold is linked to illness and death.[110] Guiteras-Holmes describes the view of the Tzeltal: "as the *wahyal* joins man to nature to make possible his life on the surface of the earth, 'heat' relates him to the gods or powerful beings that preserve and destroy life. Through 'heat' man becomes godlike."[111] Fire not only purifies space but vivifies it, as burned places are imbued with animate spirit.

Epigraphic evidence for mortuary fire-entering rituals and a variant practice, *el-naah* (house censing), is evident at Seibal, Piedras Negras, and Tonina.[112] Piedras Negras Panel 3 indicates that Ruler 7 oversaw an *el naah umukil* (house burning at the burial) of Ruler 4, twenty-five years after the king was buried and one year after Ruler 7 came to the throne.[113] In 1997 Héctor Escobedo and Tomás Barrientos excavated Burial 13, the probable tomb of Ruler 4.[114] Inside the chamber they found the remains of an adult male accompanied by two youths. Many of the remains are browned and blackened, demonstrating exposure to heat or flame (fig. 3.28). The pattern of burning, however, indicates that the bone was burned after the body had already skeletonized.[115] For example, the parietal bones of one of the youths is blackened while adjacent areas of the frontal bone show little to no thermal alteration, indicating that the cranium was disarticulated prior to thermal exposure. Similarly, the roots of one of the youth's maxillary teeth are blackened though the enamel is relatively unaffected, which suggests that the teeth had already fallen out of the cranium prior to heat exposure. Some long bone fragments show similar blackening on both the endosteal and periosteal surfaces, indicating that the bone had already fragmented prior to heat exposure. Other objects in the tomb, such as a bone bloodletter handle, are also blackened. Nevertheless, the focus of

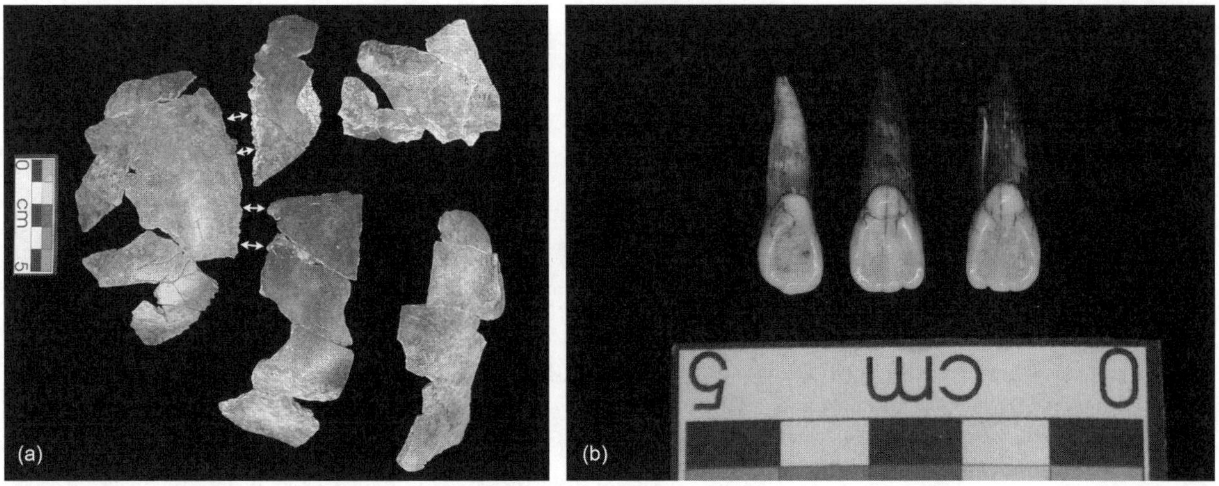

FIGURE 3.28. Burned subadult human remains from Piedras Negras Burial 13: (a) cranial vault showing differential burning to the frontal (*left*) and parietal bones which articulate along the coronal suture (*arrows*); and (b) anterior maxillary teeth showing burning of roots (photographs by author).

the flame seems to have been the human remains. The floor of the burial consisted of a very thick, dense, gray-blue clayey sediment.[116] It is likely that this matrix was a mix of scorched floor and burned organic material, the contents of the tomb.

Though we do not know the nature of their relationship, Ruler 7 made little mention of his predecessors, Rulers 5 and 6; nor do we really know the fate of these intervening kings. By reentering and censing Ruler 4's tomb (along with a variety of other memorial acts), Ruler 7 sought to legitimize his reign through the legacy of Ruler 4 even though he was not the lord who oversaw his original interment. In this sense the burning and censing of Burial 13 revivified the king's remains. Much as Maya farmers order the landscapes of their milpas through fire, Ruler 7's *el-naah* cleansed Ruler 4's burial chamber of the legacy of Ruler 5. The censing of the tomb at Piedras Negras can be understood as not only a mortuary rite but as a political act significant enough to be memorialized on a stone monument (see chapter 4).

The *el-naah* rites are just one example of a rich and varied tradition of fire and censing associated with Maya funerary rites. Recall the apparent burning of copal or other substances on the ceramic vessels placed over the face of the dead in the kingdom of Yaxchilan. Today burned offerings serve as the food for the deceased and supernatural beings. Censers and censer stands are shown widely in Maya imagery and are frequently found in the archaeological record (fig. 3.29). Many have spikes that replicate the youthful trunk of the *ceiba* tree, which is among the largest tree species in the Maya area and often conceptualized as a sacred world tree. On El Cayo Altar 4 a lord scatters something in front of an *incensario* stuffed with faggots and a ball (or balls) of copal (or a human heart) (fig. 3.29b). A bundle of feathers (or feathered objects) is inserted into the copal/heart. Recalling the discussion of trees and crosses in chapter 2, the censer-as-tree facilitated the communication and exchange of the offering with otherworldly places. The censer depicted on the El Cayo altar is marked by the face of a supernatural being. The jaguar ear would be consistent with the Patron of Pax (the embodiment of trees), but the visage and especially the murky underworld eyes suggest some other being. Altars, like the one shown on the El Cayo monument, are invariably three-legged. With

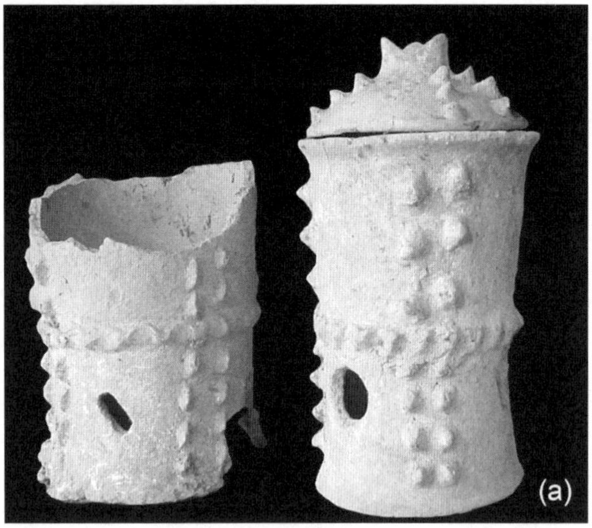

FIGURE 3.29. Classic period *ceiba* tree-style *incensarios*: (a) unprovenanced from the western margins of the Piedras Negras kingdom (photograph by Charles Golden); and (b) atop El Cayo Altar 4 (drawing by Peter Mathews).

incense burning atop them, they may have been likened to sacred hearths, also symbolic of the axis mundi.

Arguably the most impressive assemblage of censing devices is the roughly 100 ceramic *incensarios* (actually censer stands) recovered from the Palenque Cross Group. The Cross Group was effectively a monumental three-stone hearth that was activated when such censers held burning offerings (see chapter 4). Most of the censer stands at Palenque were fashioned into the likeness of GI (the ascending sun) and the Jaguar God of the Underworld (the setting sun). Other censer stands depict human beings, presumably Palenque's elites that were buried in crypts and tombs throughout the Cross Group.[117] As Taube suggests, the stucco supernatural beings that adorn Maya pyramids were effectively monumental, immovable censer stands, architectural elements that supported burning braziers in fire ceremonies enacted at the *wahyib*, the sleeping places of gods and ancestors.[118]

SOLAR PORTALS

Although their related symbolism is less obvious, iron ore mirrors (pyrite, hematite, and so forth) were also used in rituals of fire throughout Mesoamerica and are frequently found in elite mortuary contexts. Such objects were usually crafted as a mosaic of pyrite pieces, though occasionally monolithic pyrite mirrors and smaller discs are also found (fig. 3.30). They are depicted in Mesoamerican iconography as objects hanging over the lower back of warriors, as ornaments worn over the chest and abdomen, and as devices that the nobility gazed into. Although archaeological examples of iron ore mirrors are relatively rare in the tombs of the Maya lowlands, they are more common in the highlands, perhaps due to the proximity to iron ore sources or a greater degree of material and cultural exchange with central Mexico.[119] Excavations by the Carnegie Institute at Kaminaljuyu produced thirty-five mirrors, distributed across eleven Early Classic tombs.[120] Only one tomb excavated by the Carnegie team at Kaminaljuyu did not produce a pyrite object. Pyrite mirrors have been recovered in

FIGURE 3.30. (*opposite top*) Pyrite objects: (a) mosaic mirror from Kixpek, Guatemala (photograph courtesy of the University of Pennsylvania Museum of Archaeology and Anthropology, Object Number NA 11613); and (b) small disc (diameter about 4 cm) showing the severed head of a lord from Hix Witz found in Piedras Negras Burial 13 (photograph courtesy of Stephen Houston).

FIGURE 3.31. (*opposite bottom*) Piedras Negras Burial 5. Note the pyrite mirror below the feet of the primary occupant at number 17 and the pyrite mosaic bands in Niche 1 (originally published as Coe, *Piedras Negras Archaeology*, fig. 64, image courtesy of the University of Pennsylvania Museum of Archaeology and Anthropology; note that Coe had the north arrow incorrect in his illustration, which I have corrected here).

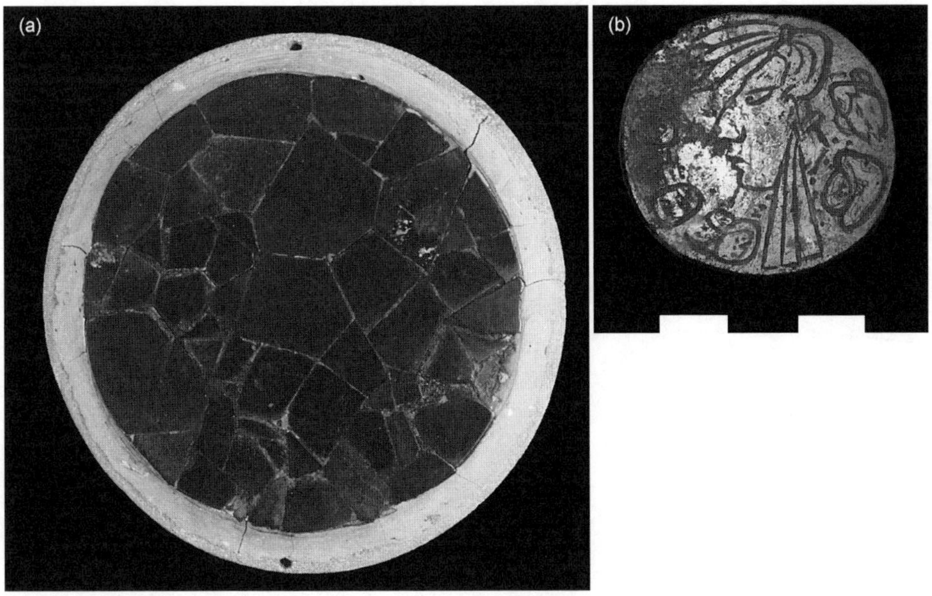
(a)
(b)

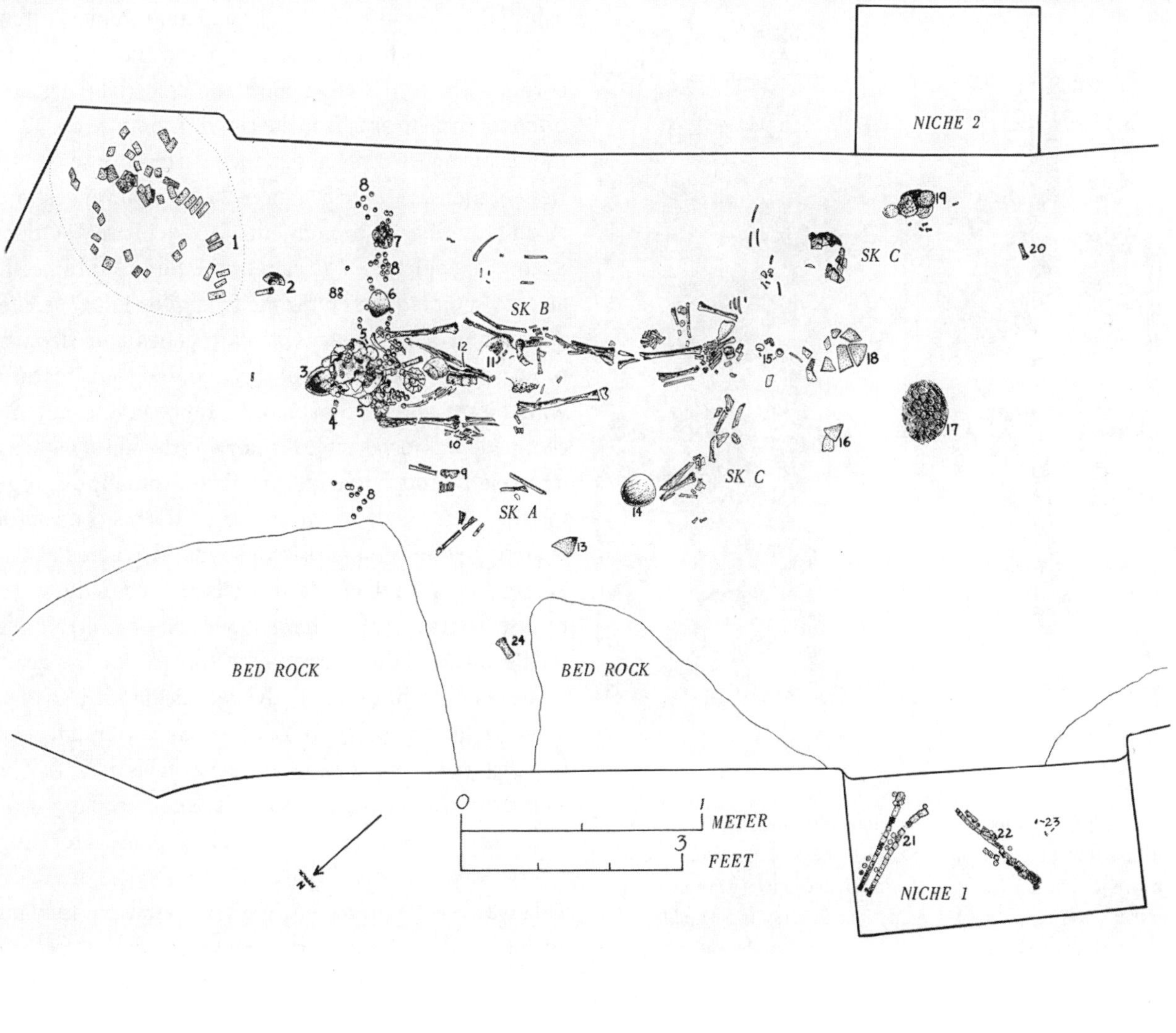
NICHE 2
SK C
SK B
SK A
SK C
BED ROCK
BED ROCK
METER
FEET
NICHE 1

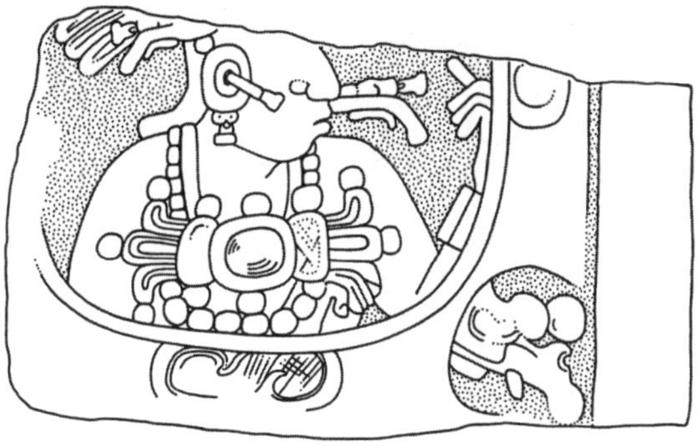

FIGURE 3.32. Ancestor within solar cartouche on Yaxchilan Stela 30 (drawing by Peter Mathews).

(a) (b)

FIGURE 3.33. Mirrors as an element of Mesoamerican backracks: (a) Yaxchilan Stela 11 (note the eroded ancestor cartouche above) (drawing by Linda Schele, © David Schele, courtesy Foundation for the Advancement of Mesoamerican Studies, Inc.); and (b) warrior from the Pyramid B at Tula (photograph by author).

tombs in the lowlands at Piedras Negras (Burial 5), Tikal (Burial 116 and 196) (fig. 3.31; see also figs. 2.17 and 2.50), and Bonampak (Building 4, Burial 1).[121]

Taube has shown that mirrors had a diverse range of symbolic associations at Teotihuacan, including eyes, faces, flowers, fires, water, webs, shields, the sun, and caves.[122] At least some of these associations can also be demonstrated for the Classic Maya. The link between mirrors and the sun and fire is especially obvious: highly polished pyrite mirrors capture the heat and glare of the sun and other sources of light and reflect them back at the viewer. The Maya often depicted the sun and other celestial bodies as hard, bright punctate frames, often with centipede heads emerging from the disc's four corners (fig. 3.32). These solar and celestial discs are often referred to as cartouches in the literature because they show figures reflected in their surfaces or seated in front of them. In some instances they contain ancestors, as on a number of monuments at Yaxchilan.[123] Other examples depict the *k'in* sign or the Sun God himself, as on a sculpted bench throne from Dos Pilas.[124] The association of centipedes with cartouches stems from a more general connection of this many-legged creature with the sun and the Sun God. As Taube points out, the centipede was responsible for carrying the sun at night as it traveled through the underworld, eventually disgorging the solar disc at dawn in the east. This concept is evident on the Palenque sarcophagus lid, where Pakal's emergence from the underworld is likened to the sun's release from the jaws of the centipede (see fig. 2.5).[125] On the Berlin Vase a solar cartouche floats above the dead body, likening the ascent of the deceased's spirit to that of the sun (see fig. 2.15). On Yaxchilan Stela 11 Bird Jaguar IV wears a solar or ancestor cartouche on his back, visible as a mirror marked by centipedes at its corners (fig. 3.33a). Back mirrors were a common costume element for warriors in Mesoamerica. The Atlantean warriors of Toltec Tula wear such objects, each mirror with what may be an ancestral head marking its center (3.33b).

A remarkable vessel from the American Museum of Natural History (AMNH) shows the celestial ascent of a solar disc (fig. 3.34), set within the mouth of a square-nosed fire serpent, a being emblematic of solar vitality. The square-nosed serpent emerges from the mouth of a second serpent whose body winds across the vessel's surface. The second serpent emerges from the underworld, rendered as the mouth of an earth crocodile, similar to the one shown on the MFA vase noted earlier (see fig. 3.23). In addition to the solar cartouche that is being disgorged, the Moon Goddess is seated in a lunar cartouche on the serpent's back. Overall the scene seems to depict the ascent of celestial bodies.

In Maya iconography crocodilian beings represent not only the earth and entrances to the underworld but also the night sky of the underworld. David Stuart identifies the underworld sky crocodile as the "Starry Deer Crocodile." Inscriptions at Palenque describe his decapitation, which is followed by a passage that seems to name him as a "fire-drill entity."[126] In central Mexico serpents were understood to emerge directly from mirrors as a result of fire drilling. Taube has shown that the Aztec fire serpent, Xiuhcoatl, both emerges from mirrors and wears them on his body. A number of Postclassic and colonial depictions of the Aztec fire serpent show beings drilling fire on mirrors that are attached to Xiuhcoatl's body. The Aztec fire serpent seems to have been derived from the Classic period war serpent of central Mexico. This creature has an upturned snout similar to the upturned or square nose of the Maya fire serpents.[127] Both the central Mexican war serpent and the feathered serpent are shown emerging from a mirror on the Temple of the Feathered Serpent at Teotihuacan.

FIGURE 3.34. The emergence of a mirror-cartouche from the mouth of the square-nosed serpent on a cylinder vase from the American Museum of Natural History (drawing by Linda Schele, © David Schele, courtesy Foundation for the Advancement of Mesoamerican Studies, Inc.).

Similarly, square-nosed serpents emerge from mirrors on the tips of the cross on the Palenque sarcophagus lid (see fig. 2.5). A number of mirrors have been recovered at Chichen Itza that show the square-nosed serpent around their rims. The recovery of burned mirrors at Chichen Itza led Clemency Coggins to suggest that the Maya, like the Aztec, used these objects in fire drilling.[128] It may also be that ancient mirrors, by reflecting the hot glare of the sun, were used to start fires. Robert F. Heizer and Jonas E. Gullberg report that Gordon Ekholm of the AMNH was able to start a fire with a Mesoamerican mirror in the museum's collection.[129]

Aside from their association with fire drilling and fire serpent conjuring, mirrors also appear in courtly scenes on Maya polychrome vases, where lords gaze intently into their surfaces. Though perhaps these are simply scenes of royal vanity, it is more likely that the lords are searching the face of the mirror for something more significant. Considering that mirrors are used as conduits to otherworldly places, it is probable that Maya lords employed the devices for scrying: by gazing into their shiny surfaces, lords may have been able to gain insight by seeing into other realms of existence.[130] As Taube notes, Mesoamerican mirrors have eyelike qualities: the dark mirrors are similar to the pupil of a human eye.[131] In this sense mirrors were both lenses and actual portals into other places, particularly solar or celestial realms.

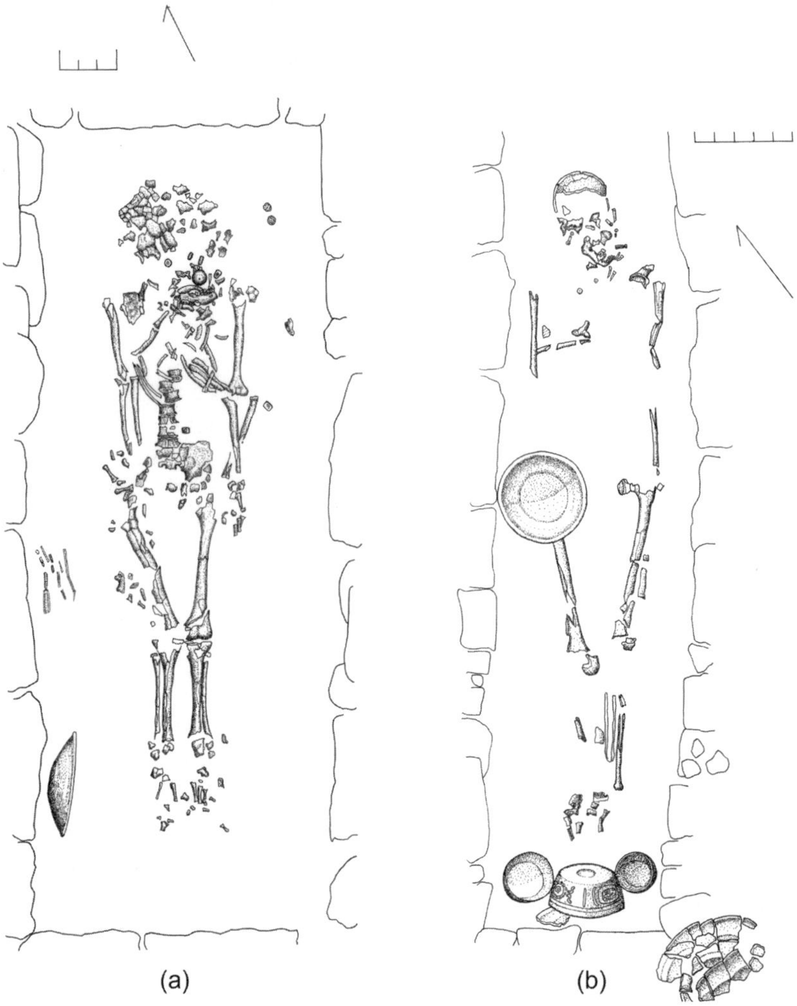

FIGURE 3.35. (*above*) Piedras Negras burials with tilted vessels: (a) Burial 82; and (b) Burial 77 (drawing by Zachary Hruby).

FIGURE 3.36. (*right*) Field photograph of tilted bowl with smoking/breathing mirror (*black arrow*) and "*ajaw*-face" (*white arrow*) from Piedras Negras Burial 77 (photograph courtesy of Stephen Houston).

The mirrors at Piedras Negras, at Bonampak, and in Tikal Burial 116 were located below and to the right of the deceased, near the foot-end of the burial chamber. Mary Miller and Simon Martin suggest that the Bonampak mosaic mirror may have been meant to suggest the earth turtle, circular in shape and composed of composite polygons that resemble the carapace of the turtle's shell and suggest the Maize God's emergence from turtle-earth.[132] Yet there is little association between mirrors and turtles in Maya imagery. Many of the tombs at Kaminaljuyu that contained mosaic mirrors also contained turtle carapaces; such redundancy suggests that this ideological link may be incorrect. Four mirrors were placed to the right (west) of the head and legs of the individual in Tikal Burial 196 (see fig. 2.50). The westward position of the mirrors may have been meant to suggest the setting sun, recalling the association of mirrors, portals, and solar travel.[133] Mirrors may have been placed in Maya graves to suggest or even facilitate the ascent or descent of the souls of the dead, in accord with the ancestor solar cartouche imagery. Burial 5 at Piedras Negras contained a mirror at the foot and slightly to the right (west) of the primary occupant (see fig. 3.31). In other elite mortuary contexts ceramic vessels were substituted for mirrors. A vessel inscribed with a square-nosed serpent was located near the feet and to the right (westward) of the noble youth interred in Burial 82 (fig. 3.35a; see also fig. 2.35). I have suggested elsewhere that this dish was used in bloodletting rituals.[134] In this sense dishes (containing burning, blood-soaked paper) and mirrors may have both been used to facilitate the movement of supernatural essences. In a similar fashion a noble lord (a probable *sajal*) was interred in Piedras Negras Burial 77 with a vessel at his feet with both a smoking/breathing mirror and an "*ajaw*-face" sign decorating its sides (figs. 3.35b, 3.36). Coe notes that the mirror in Burial 5 was "set on angle." Similarly, the vessels in both Burials 77 and 82 were set on their sides.[135]

In addition to mosaic mirrors, smaller pyrite discs were recovered from Piedras Negras Burials 5 and 13 (see fig. 3.30b).[136] A pair of pyrite mosaic bands made of tesserae cut into circles and rectangles and possibly worn as a headband, part of a headdress, or some other object of clothing was also found in Burial 5 (see fig. 3.31). Worn by Maya nobility, such objects would have reflected light and hint at the solar nature of the Piedras Negras kings. This may also explain the pyrite dental inlays used by the Maya. For example, Ruler 3 of Piedras Negras had a pyrite inlay set into at least one of his lower incisors (see fig. 1.17). Such dental decorations allude to the solar vitality of Maya nobility, whose breath could be rendered as the square-nosed fire serpents, the very beings shown to emerge from pyrite mirrors. In this sense the Piedras Negras kings fully embodied their royal title: *k'inich ajaw* ("radiant lord" or "lord of the sun").

FIERY CATALYSTS

The Precolumbian and early colonial Maya pierced their tongue, cheeks, and lower lips in order to let blood ritually. Men also pierced their genitals, specifically the foreskin, at the time of the conquest, according to Landa's *Relación*.[137] The *Relación* also mentions that string or rope was pulled through the incisions, an act that is also shown on Yaxchilan Lintels 17 and 24, the Bonampak murals, and polychrome pottery, including a vase on display at Dumbarton Oaks (fig. 3.37).[138] In rites of bloodletting, blood was dripped onto paper that was then burned. The resulting conflagration brought forth fiery snakes and other serpentine forms that served as conduits from otherworldly places (fig. 3.38).

The piercing of the body and the splattering of blood on paper must have been a dramatic spectacle. Yet monuments that depict bloodletting, such as the panels from the Cross Group at Palenque or the lintels of Yaxchilan, were housed in the interior of buildings, suggesting that the ritual was an intimate affair involving only members of the royal family, their ancestors, and patron spirits. The ritual itself likely took place in closed contexts, perhaps inside temple chambers or sweatbaths. It is likely that Maya bloodletting, much like the Sun Dance

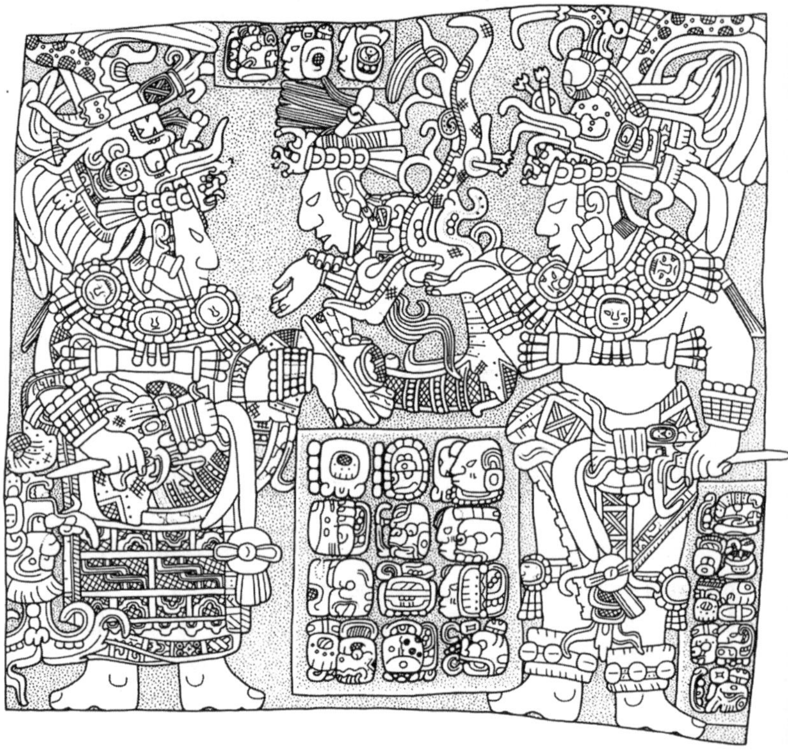

FIGURE 3.37. (*above*) Bloodletting by pulling a rope through the tongue depicted on an unprovenanced vessel at the Dumbarton Oaks Research Library and Collection (drawing by author after K2783).

FIGURE 3.38. (*right*) Lady Great Skull and a *sajal* conjure a vision serpent through bloodletting on Yaxchilan Lintel 14 (drawing of Lintel 14, by Ian Graham, from Graham and von Euw, *Corpus of Maya Hieroglyphic Inscriptions*, vol. 3, part 1, *Yaxchilan*, 3:37 © 1977 by the President and Fellows of Harvard College, Peabody Museum of Archaeology and Ethnology, Harvard University, PM# 2004.15.15.1.144 [digital file# 125510030]). Note the blood-soaked paper in the bowl and the paper wrapped around the bloodletters.

and other autosacrifices practiced by the Plains tribes of North America, was preceded by sweats and fasting.[139] Coupled with the consumption of intoxicants, Maya bloodletting surely induced hallucination, adding an intense psychological edge to the ritual invocation.

The importance of autosacrifice lay in the value of blood as a catalyst for triggering communication with or appeasement of supernatural beings. The hot, solar nature of Maya nobility was embodied in their blood, which was understood to be *k'uhul*, a sacred or spiritually charged essence. A bloodletter shown on the Temple of the Cross panel gushes a liquid marked as *k'an* (precious), presumably a graphic representation of the *k'uh* that infused royal bodies (fig. 3.39a). The fiery quality of royal blood is suggested by the symbolic associations with the square-nosed serpent, as suggested, for example, by the incised vessel from Piedras Negras Burial 82, which was likely used by the prince in his own bloodletting rites (see fig. 2.35).

Spines from various stingray species, members of the cartilaginous marine fish family Dasyatidae, were the preferred tool for bloodletting (fig. 3.40). The serrated barbs range in length from 5 to 35 cm and in living stingrays are coated in venom that is not preserved after the spines are removed from the fish.[140] As described in Landa's *Relación*:

> There is another fish on this coast, which they call *ba*—broad and round and good to eat, but very dangerous to kill and to meet, since it also does not know

> how to go into deep water, and likes to go into the mud where the Indians kill it with bow and arrows. And if they are careless in going near it or treading on it in the water, it at once has recourse to its tail, which is long and thin and stabs with a saw, which it has, so seriously that it cannot be taken out from where he puts it without making the wound large, since its teeth are backwards. . . . The Indians used these little saws in cutting their flesh in the sacrifices of the devil and it was the duty of the priest to keep them, and so they had many of them. They are very nice, for they are of very white bone and curiously formed in the shape of a saw so sharp and fine that it cuts like a knife.[141]

Elizabeth Benson points out that sharks' teeth, also found in some Maya burials and caches, resemble stingray spines in basic form.[142] The edges of the teeth from certain shark species, such as the tiger shark (*Galeocerdo cuvieri*) and the great blue shark (*Prionace glauca*), are serrated, much like the edges of stingray spines. In light of the physical similarities, Benson suggests a functional and metaphorical overlap between sharks' teeth and stingray barbs: both were presumably used for bloodletting.

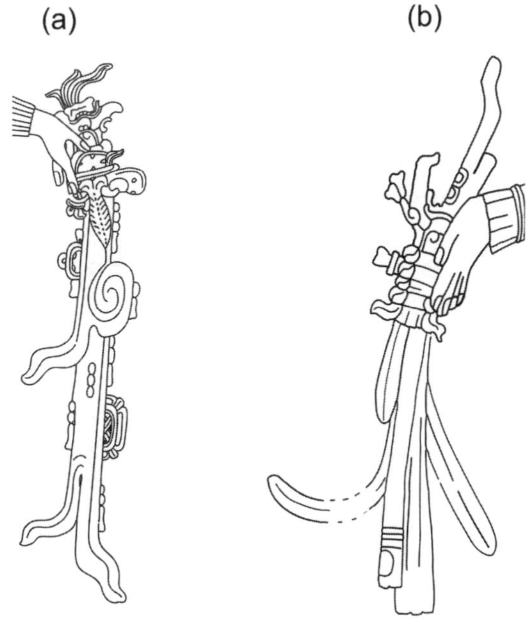

FIGURE 3.39. (*above*) Bloodletters in Palenque iconography: (a) Quadripartite Badge on the west door jamb of the Temple of the Cross; and (b) from the Tablet of the Foliated Cross (drawings by author). The bloodletter in 3.39a is streaming blood marked by the *k'an* sign and floral crossbones.

FIGURE 3.40. (*below*) Incised stingray spines from the Piedras Negras royal burials: (a) select incised spines from Burial 5 (originally published as Coe, *Piedras Negras Archaeology*, fig. 56, image courtesy of the University of Pennsylvania Museum of Archaeology and Anthropology); (b) in situ spines from Burial 13 (drawing by Stephen Houston); and (c) in situ spines from Burial 82 (drawing by Zachary Hruby).

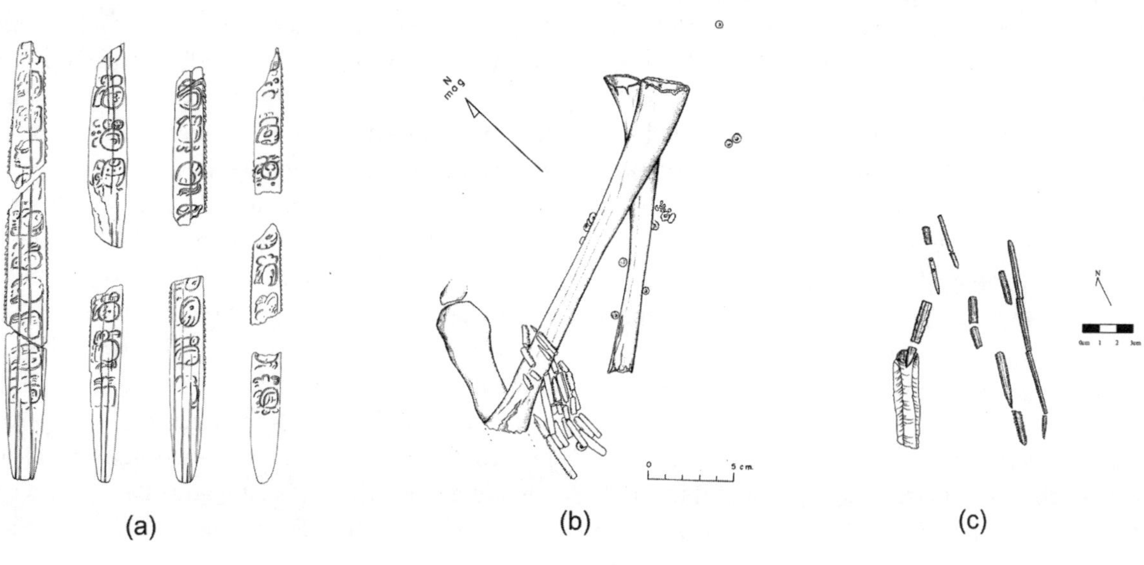

Stingray spines have been recovered in royal tombs across the Maya area, including Piedras Negras, Yaxchilan, Altun Ha, Kaminaljuyu, Uaxactun, and Tikal.[143] Despite the diversity of Maya elite mortuary ritual, the ubiquity of bloodletters in royal graves suggests that they were requisite for royal interment. At Tikal the tradition was established at least in the Late Preclassic period with Burial 85 and continued for centuries, including the eighth-century tombs of Burials 116 and 196. At Piedras Negras the royal tombs attributed to Ruler 3 (Burial 5), Ruler 4 (Burial 13), and a *ch'ok ajaw* (unripe lord) all contained what can be described as bloodletter bundles (fig. 3.40). Near the right arm of Ruler 3 lay four inscribed stingray spines, eleven other stingray spine fragments, three fragments of animal bone, and a carved jaguar ulna (see figs. 3.31, 3.40a).[144] Intermingled with the bones of the right arm of Ruler 4 were six stingray spines, two jade imitation stingray spines, two bone needles, and a carved jaguar ulna (fig. 3.40b).[145] Near the right arm of the *ch'ok ajaw* in Burial 82 were six stingray spines, a stylized jade stingray spine, a bone needle, an obsidian blade, and, near the right femur, a carved jaguar ulna (see figs. 3.35a, 3.40c). In each case the items were originally bundled or packaged in a perishable container. One possibility is that they were kept in a cloth bundle and that the bone needles were used to pin the bundle shut. Alternatively, the bloodletting kits may have been kept in wooden boxes similar to ones shown on Yaxchilan Lintel 43 and found in a cave to the north of Piedras Negras at Alvaro Obregón, Tabasco.[146] In either case the containment of these items corresponds to broader Maya concerns with the wrapping or bundling of spiritually potent objects.

The prevalence of royal bloodletters in mortuary contexts suggests that these objects were closely tied to the royal body. Like headdresses, they may have been inalienable objects that were removed from use following the death of their owner. One possibility is that interment with these objects points to continued ritual obligations for Maya lords and ladies even in death. Yet we must also consider the intimacy of these objects, both in terms of the anatomy penetrated and their infusion by royal blood. Classic Maya spines frequently were inscribed with the names of their owners, as was the case with the Piedras Negras spines. Yet it seems unlikely that the same spines were used throughout a person's life; they were probably retired periodically, as evidenced by the shifting titles on the objects. For example, the spines from Burial 82 name the youth as a *ch'ok ajaw*. Had he completed his first *k'atun* he would presumably have acquired new spines that no longer named him as a youth. Similarly, caches with stingray spines are likely the remains of spines and other ritual accouterments that were retired after their use in a particular ritual event. As potent objects, bloodletters and other cached items had to be carefully discarded and sealed in ceramic or stone, much as the royal body is carefully contained within a tomb.

The handles of bloodletters were shown to be crafted in the likeness of supernatural beings, though the identities of these entities are poorly understood (figs. 3.38–3.41). Following work by David Joralemon, the beings that embody bloodletters are typically glossed as "perforator gods" in the Mayanist literature.[147] In particular this name was applied to bloodletter handles depicted with a hook-nose (or beak) on monuments at Palenque and on Yaxchilan Lintel 14 and to similar bloodletters that appear on a vase from Huehuetenango (K1362). The identity of this being remains enigmatic. In some sense the nose resembles the beak of birds shown on Palenque monuments, atop the cross-trees (see figs. 2.5, 2.6). Yet the nose/beak of the "perforator gods" also resembles the elongated snouts of some serpents or crocodiles in Maya imagery, particularly the water (*witz'*) serpent, which is shown to gush blood from its mouth.[148] The aquatic symbolism of this creature would certainly accord with the use of stingray spines and sharks' teeth as bloodletters, each perhaps representing the fang of this serpent being. Other bloodletters are shown as the "Quadripartite Badge," especially at Palenque (see fig. 3.39a). The Quadripartite Badge

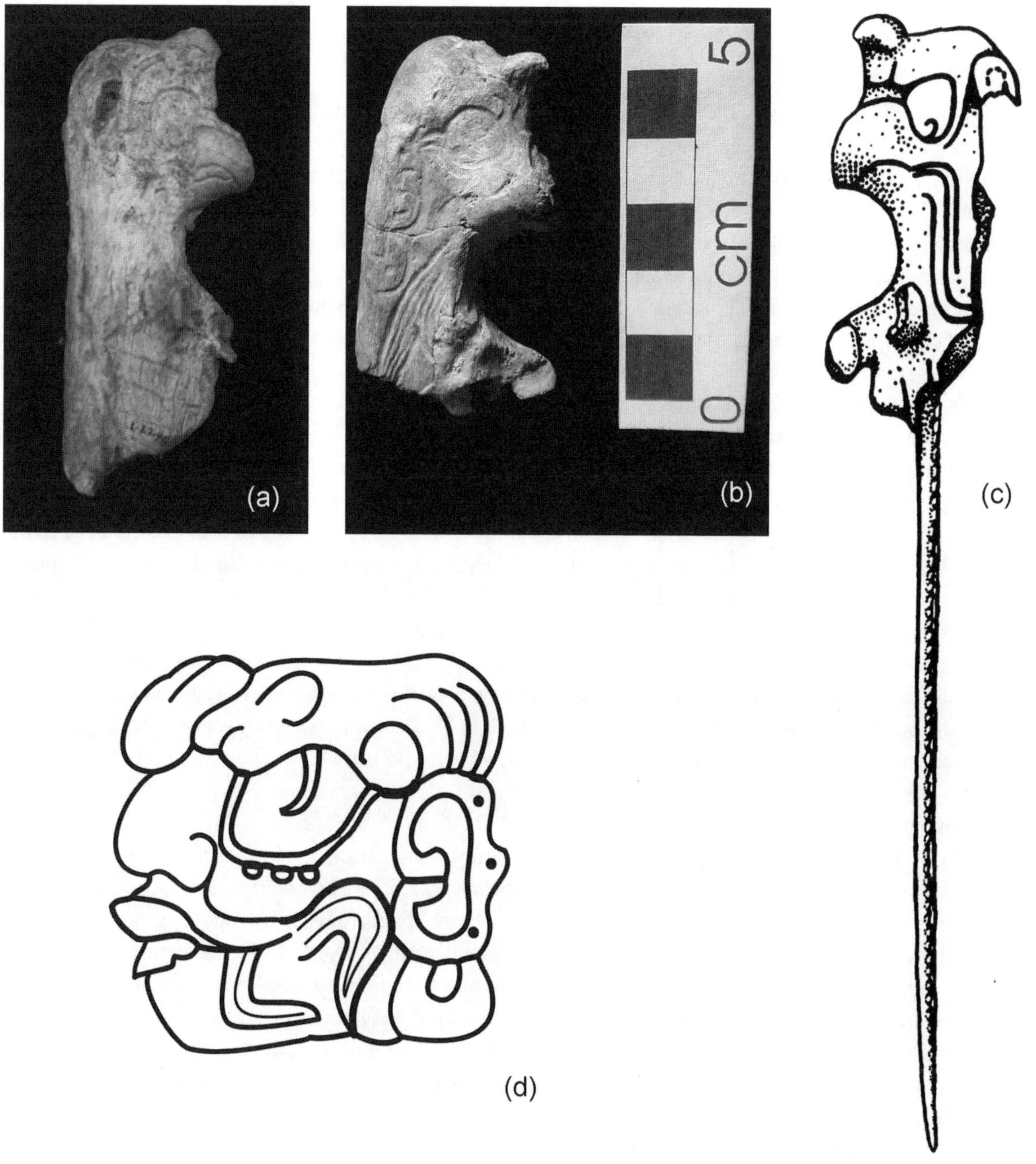

FIGURE 3.41. Piedras Negras bloodletter handles: (a) Burial 5 (photograph by author); (b) Burial 13 (photograph courtesy of Stephen Houston); (c) Burial 82 (drawing by James Fitzsimmons); (d) compared to GI glyph from the Temple of the Foliated Cross (drawing by author).

depicts a supernatural head (often with a skeletal jaw) whose forehead is a *k'in*-marked receptacle that holds a stingray spine, a *Spondylus* shell, and a crossed-band floral element.[149]

Unfortunately, the archaeologically recovered bloodletter handles from Piedras Negras do little to clarify the identity and nature of the beings that embodied bloodletters. Each of the royal bloodletting kits from Piedras Negras Burials 5, 13, and 82 included a carved jaguar ulna that served as the handle for the stingray spines (fig. 3.41). Each stingray spine was subtly modified to fit its associated handle. These handles generally have been assumed to depict Chahk because of the *Spondylus* ear ornament that is faintly visible on the handle from Burial 13 (a more visible *la* sign hangs below).[150] Unfortunately, the images from Piedras Negras contain no depictions of bloodletters; nor are any bloodletters depicted at other sites that show Chahk.[151] Nevertheless, Chahk is an important supernatural patron at Piedras Negras. Some of the attributes of the Burial 13 handle do not match depictions of Chahk, however, particularly the finlike barbel incised around the trochlear notch.[152] This distinctive element is more characteristic of GI (3.41d; see also fig. 1.20b). The Burial 5 handle lacks the barbel and the Burial 82 handle has a long barbel of three incised lines. Burial 5 and 82 ulnae also have incisions that modify the olecranon process to suggest either a beak or long snout, perhaps in parallel to depictions of the "perforator god" noted above. There certainly is good reason to suspect that GI is depicted on the bloodletter handles at Piedras Negras, at least on the ulna from Burial 13. GI is repeatedly shown in Classic period imagery to have a central fang that is either a stingray spine or a serrated shark's tooth, which suggests a link to bloodletting. Moreover, he frequently wears the Quadripartite Badge on his head, the quintessential symbol of bloodletting (see fig. 1.20b).[153] GI is not depicted at Piedras Negras, however, which complicates this interpretation. Whichever supernatural is represented, it is clear that we are to understand the bloodletters as the very embodiment of a supernatural blood-drinker whose tooth was a stingray spine inserted into the anthropomorphized handles.

Although bloodletting was a royal obligation, the inclusion of bloodletters in the graves of nonroyals suggests that other members of society were also obligated to make blood offerings. At Piedras Negras 10 out of 122 (8.2 percent) of the burials contained stingray spines, including 7 from nonroyal interments. In some cases those burials were clearly elite, such as the remains of a possible *sajal* (Burial 77; shown in fig. 3.35b, though his bloodletter is not in the drawing). In other instances the burials were very modest and the stringray spine was the only item found within the grave, including the burials of at least two women (Burials 28 and 66). In Burial 28 the stringray spine was found near the woman's pelvis. In two cases (Burial 63 and 91) only the tip of the stingray spine was present. Such an object may not have been used for bloodletting but was included in the grave because of its symbolic value. Alternatively, it is possible that the tip had broken off inside of the person during an act of bloodletting shortly before her death.

MORTUARY SACRIFICE

For the Classic Maya, ritual violence was directed against not only a person's own body but also the bodies of others in a range of acts that can be roughly grouped into two types of sacrifice. The first was the torture and eventual slaying of enemies of war.[154] The second is less well understood but seems to have involved ritual killing, particularly of children, during episodes of transition and rupture, including royal accession and especially the death of an important royal person.[155]

Burial 9 at El Zotz, an Early Classic period tomb, contained the remains of an adult individual accompanied by six children (fig. 3.42).[156] The children were placed in a series of lidded cache vessels along the central axis of the tomb. Skeletal fragments of the adult lay scattered atop these vessels and on the surrounding floor. Though they are poorly preserved, the distribution of the adult remains indicates that the body had been interred in

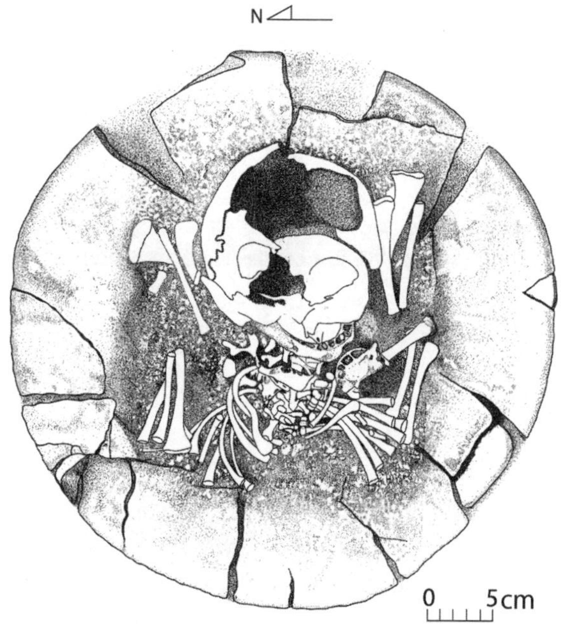

FIGURE 3.42. (*left*) El Zotz Burial 9 (photograph courtesy of Stephen Houston, El Zotz Archaeological Project). The child sacrifices were located within the lidded cache vessels that run through the center of the image.

FIGURE 3.43. (*above*) A child (Burial 15) sacrificed outside of the El Zotz tomb (Burial 9) (drawing by Yeny Gutiérrez).

a primary supine position atop a perishable funerary bier, still detectable by fragments of stucco that once covered this wooden structure. When the bier collapsed, the skeletal remains cascaded to the floor. My analysis of the skeleton suggests a probable male, at least forty years old at the time of death, presumably the founder or one of the earliest kings of the El Zotz dynasty.[157] The vessels containing the children were located directly underneath the bier. The youngest child was only a few months old, the oldest no more than five. Outside the tomb archaeologists found two other caches containing the remains of children (fig. 3.43). The bone in all of these deposits demonstrates evidence of thermal exposure. Three questions come to mind. Why was the taking of human life and the offering of body parts necessary for the interment of this king? Why were these corporal elements exposed to flame? How do these practices relate to Classic period conceptions of the soul?

FIGURE 3.44. Sacrifice of the Baby Jaguar on an unprovenanced vase (K0521 © Justin Kerr). The depiction of Chahk is unusual in that he is given a human face.

In Classic Maya worldview humans were quite literally the food upon which some supernaturals feasted.[158] This belief is related in part to the understanding that humans were crafted from maize. Imagery suggests that vile underworld beings seemed to prefer limbs and other random body parts. The Sun God, however, dined on human hearts, likened to tamales, one of the favorite dishes in Classic period cuisine. Human sacrifice and bloodletting ended long ago in the Maya area, replaced by chickens, tortillas, tamales, maize foods, copal, and other offerings.[159]

Copal is a tree resin that has long been the favored incense among Mesoamerican peoples. The hard resin nodules are placed on charcoal or a similar slow-burning device, producing a tremendous amount of acrid smoke as the resin is slowly consumed. Among the contemporary K'iche', blood and tree sap are known by the same word, *kik'*. The link between tree resin and offerings may explain why *incensarios* are fashioned to resemble the young *ceiba*, as noted earlier. In the K'iche' *Popol Vuh* a ball of copal is offered in place of an underworld maiden's heart in order to trick the lords of the underworld.[160] As this section shows, copal was also burned in antiquity, including during ritual human sacrifice. Today copal continues to be understood as food for ancestors and supernaturals.

Sacrificial offering is referred to today as *k'ex* (exchange or substitution) in most Maya languages. Bruce Love observed that Yucatec ritual specialists offer tamales as *k'ex* to placate dangerous winds that cause illness.[161] Ruth Bunzel noted the sacrifice of a chicken as a substitution among the contemporary K'iche', and Vogt made similar observations for the Tzotzil.[162] As Taube shows, infant sacrifice was a particularly powerful form of *k'ex* in the Classic period: children offered to trick or placate supernatural beings who may otherwise have hungered for the souls of more powerful humans.[163]

Although we may find the notion of child sacrifice disturbing, the killing of children in the interest of adults is widespread in human history. For example, Kathy Stuart has documented 116 cases of child homicide in northern Europe during the eighteenth century in what she calls "suicide by proxy." Catholic and Protestant men and women, wishing to commit suicide but fearing the damnation of their souls, instead killed innocent children and immediately turned themselves into law enforcement, forcing their own executions. By confessing their sins, these murderers saved their souls from hell. Children were selected as their victims because they were understood to be without the taint of sin. As Stuart explains, "it was murder as a religious offering . . . child-victims of suicide by proxy were constructed as martyrs," and all souls involved in the transaction went to heaven.[164]

For the Classic period Maya, the killing of children seems to have been given mythic charter in the sacrifice of the Baby Jaguar (*unen bahlam*).[165] Although the

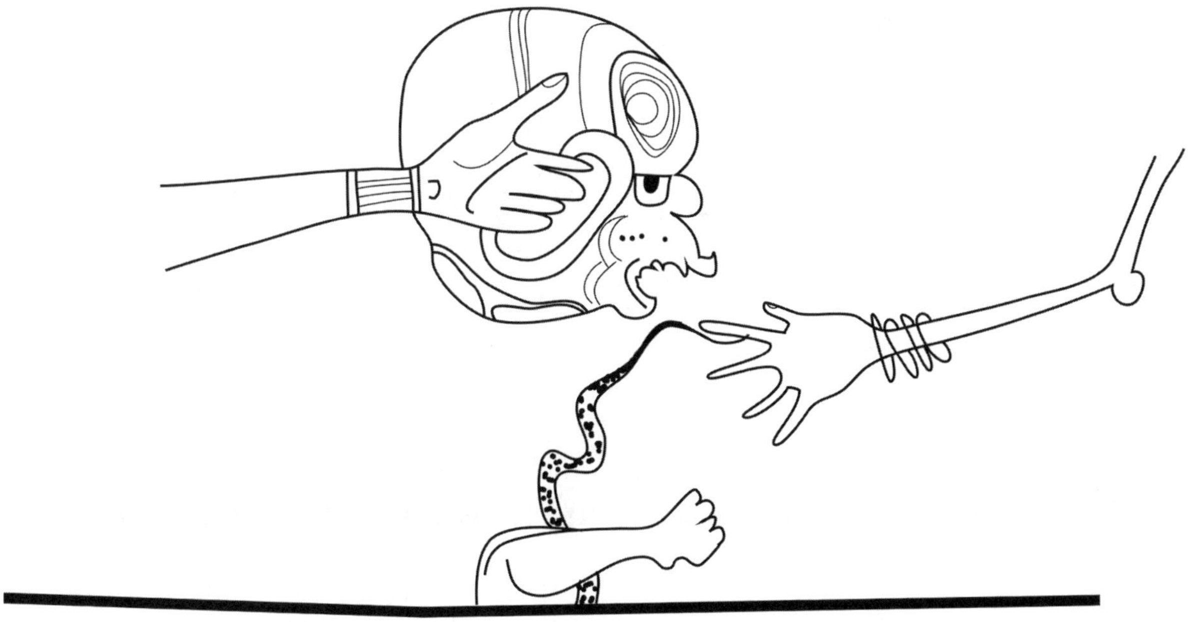

narrative has no equivalent in contact period documents, at least sixteen Late Classic period vessels have been documented that depict the sacrifice of the Baby Jaguar, suggesting the myth's importance and offering some clues as to its meaning (fig. 3.44).[166] The absence of the myth by the time of the Spanish conquest suggests that it may have been intimately linked to the institution of divine kingship of the Classic period.[167] Although the majority of the Baby Jaguar vessels are without provenance, most were produced in the same workshop or closely related workshops in the Central Petén. The Baby Jaguar also appears in inscriptions and is especially prominent in Early Classic period texts at Tikal.[168]

The three primary actors in this drama of sacrifice are the Death God (God A), Chahk, and the Baby Jaguar.[169] The Baby Jaguar has a central fang and is often shown wearing a cruller, two attributes that link him to the Jaguar God of the Underworld and distinguish him from other jaguars in Classic period art. Although Classic period imagery contains many versions of the Death God, the one that appears in the Baby Jaguar myth is shown with a bound topknot and usually with a backrack. Depending on the particular pot, the backrack may be a large bone or a bundled "bone throne" (which may be understood as an underworld throne) from which hang pieces of cloth or paper. Most versions of the myth are depicted as taking place at an animate *witz* (hill or

FIGURE 3.45. The Baby Jaguar plunges into the underworld, dropped by the Death God (*right*) as Chahk (*left*) strikes with his "knuckleduster" (drawing by author after K1199).

mountain), usually shown horizontal (looking up at the sky). Swirling water is visible among the protagonists' feet. The combination of *witz* and turbulent water led Simon Martin to suggest that the sacrifice of the Baby Jaguar took place at a "rocky island in a primordial sea."[170] The Death God tosses the Baby Jaguar at the *witz* as Chahk dances or strikes with either his axe or "knuckleduster," depending on the particular vase. In some depictions the infant slides below the scene as he falls into the ground or water, descending into the underworld (fig. 3.45). The horizontal orientation of the *witz* may suggest that it represents an opening into a world below, perhaps a cave.

As the infant form of the Jaguar God of the Underworld, the Baby Jaguar's descent into the jaws of the underworld may be a metaphor for the setting sun. The baby's reclining position on most of the vases (for example, see fig. 3.44) is characteristic of birth in Maya imagery. Yet impending death is also suggested on a number of vessels: the Baby Jaguar makes the gesture of woe, holding its hand to its forehead, as figures typically do in scenes of death and mourning.[171] More overtly, death is implied by the presence of the Death God and

FIGURE 3.46. Presentation of an infant for sacrifice on an unprovenanced vase (K4384 © Justin Kerr). The vase has some overpainting.

the string of eyeballs that Chahk wears on most vases of the sacrifice drama. In sum, the scene seems to represent both the birth and sacrifice of the infant form of the Jaguar God of the Underworld as he descends into the underworld. During the rainy season in the Maya lowlands, tempestuous storms usually begin in the late afternoon, hours before sunset, which may explain Chahk's presence in the scene. As Martin notes, the particular version of Chahk that appears in these scenes is named "First Rain Chahk," suggesting that he is the onset of the rainy season.[172] He may be there as the sacrificer or to crack open the earth to allow the Baby Jaguar to enter. The *k'ex* sacrifice of the Baby Jaguar might suggest a mythic duality: the western descent of the sun at dusk is balanced against solar rebirth in the east.

At least four vessels (K1200, K4384, K5855, and K8655) have been documented that show a man offering an infant to a seated lord in a scene that vaguely recalls the mythic sacrifice of the Baby Jaguar (figs. 3.46 and 3.47).[173] The style of painting indicates that most of these vases were produced in the same workshops that generated the Baby Jaguar vessels. On at least two known vases the infant is garbed as the Baby Jaguar (see fig. 3.47). Imminent sacrifice is strongly implied on one vessel by an attendant holding a long obsidian knife (see fig. 3.46). On some of these vessels the sacrificial victim is presented on a leaf bed, recalling the leaf cloaks worn by participants in bloodletting, as on the Palenque Temple XXI platform and Dos Pilas Panel 19 or the leaf bed that contains the severed head on the Bonampak murals and the heads surrounded by leaves on the Tonina stucco frieze. Tamales and other sacred breads are prepared in leaf wraps. The leaf beds may have framed the corporal offerings—someone engaged in bloodletting, a ritually killed infant, or a severed head—as things of sustenance for otherworldly beings. The lord who presents the infant for sacrifice in both vessel scenes wears a distinctive costume that partly draws on imagery of the Death God and is generally worn by officiates in sacrifice and other mortuary rites, as seen, for example, on Tikal Altar 5 and Piedras Negras Stela 5 (compare with figs. 2.58 and 3.44).[174] The hair of these ritual specialists is worn in a topknot, pulled through a conical sheath or hat of cloth or paper. The bound hair replicates the topknot and backrack of the Death God, and the flaring brim is perhaps reminiscent of the flaring cloth or paper that hangs from the Death God's backrack (compare figs. 3.44, 3.46, and 3.47). As in many depictions of the Death God, black paint covers the eyes of the impersonator in figure 3.47.

On what grounds can the infant sacrifices from the El Zotz tomb be linked to the scenes of mythic (and possibly real) infant sacrifice depicted on the polychrome vessels? One important clue is the placement of the child sacrifices within ceramic dishes. Although the dishes are not

FIGURE 3.47. Presentation of an infant dressed as the Baby Jaguar (K8655 © Justin Kerr).

present in the mythic scenes, a polychrome vase depicts a Baby Jaguar (or an impersonator) in a dish and offers a tentative link between the myth and the actual practice of child sacrifice (fig. 3.48). Similar scenes of actual infant sacrifice are shown on other ceramic vessels, monuments at Piedras Negras, and Yaxha Stela 13 (figs. 3.49, 3.50; see also fig. 2.14). The infant carved on Piedras Negras Stela 11 lies on a pelt, presumably that of a jaguar. As Taube shows, a careful examination of Teobert Maler's original photograph of Piedras Negras Stela 14 shows that the infant's left arm is clad in a jaguar skin shirt, suggesting that at least in this scene the child may have been dressed as the Baby Jaguar.[175] Although these connections are admittedly tenuous, the repeated infant-jaguar imagery suggests a loose thematic link between these scenes of child sacrifice and the mortuary archaeological evidence recovered at El Zotz and elsewhere.

If I am correct in characterizing these child sacrifices as *k'ex*, what then was actually exchanged for the death of the infant? Unfortunately, the texts on the ceramic vessels are of little help. One possibility is that the mortuary infant *k'ex* offerings were made as exchanges to placate the dead. Recall that the contemporary Maya are concerned that the souls of the recently deceased must be encouraged to depart, lest they linger and attempt to bring the souls of their loved ones with them (see chapter 1). As Guiteras-Holmes explains of the recently dead among the Tzotzil: "The *ch'ulel* of a dead person is dangerous to the living because it generally does not wish to leave the earth unaccompanied. . . . It is not good to seek relations with the dead for fear that they might want to take the living with them."[176] Could the mortuary sacrifices be made in exchange for the souls of the royal family and courtiers, as a means to assuage the recently dead and encourage their peaceful departure?

The vases that show infant sacrifice may suggest that a dead lord was the recipient of these precious offerings. On one vase the receiving lord is shown with a cloth strand breath ornament, a motif that is usually characteristic of the breath of dead beings (see fig. 3.46). His regalia lie before him on the ground, much as they would be situated on the floor of a tomb. On another vase the lord sits on a throne within the open maw of an earth crocodile, mirroring the entombed lord visible on the MFA vase (compare fig. 3.47 with fig. 3.23). Yet this interpretation does little to explain the possible connection to the myth of the Baby Jaguar.

FIGURE 3.48. (*top*) Unprovenanced vase from the Museum of Fine Arts, Boston, showing a Baby Jaguar (or its impersonator) within a bowl (K1184 © Justin Kerr).

FIGURE 3.49. (*bottom*) An infant burns before an enthroned bundle on an unprovenanced vase (drawing by Linda Schele after K3844, © David Schele, courtesy Foundation for the Advancement of Mesoamerican Studies, Inc.).

(a) (b)

FIGURE 3.50. Child sacrifice at Piedras Negras: (a) Stela 11 (drawing by Linda Schele, © David Schele, courtesy Foundation for the Advancement of Mesoamerican Studies, Inc.; note that Schele's drawing excluded the pelt, which I have added based on Teobert Maler's original photograph of the monument); and (b) Stela 14 (drawing by John Montgomery, © Foundation for the Advancement of Mesoamerican Studies, Inc.).

Alternatively, the *k'ex* sacrifice of infants may have been conducted to placate greedy underworld beings. By offering the infant, the deceased lord or lady's soul was allowed to ascend to celestial realms or at least not be captured, contained, or otherwise harmed by these denizens. We know from imagery on pottery that underworld beings were voracious consumers of human body parts. Among the modern Maya no sacrificial practices accompany death, to my knowledge. But the Maya do have a very real concern with being trapped or enslaved by capricious underworld spirits. For example, Pedro Pitarch explains that the *ch'ulel* of the Tzeltal of Cancuc are in danger of being kidnapped by mean-spirited *lab* and also the *yajwal witz* (Mountain Lords), spiritual owners of the significant mountains that surround Cancuc. These spirits kidnap souls for a variety of reasons, including the "hope to obtain something in exchange for its liberation (basically incense, liquor, tobacco, and flattering words of recognition that the shaman serves up through the cross on the household altar); and at other times—these are the worst and most difficult cases to resolve—they aim to ensure the body's death and obtain the deceased's Bird of the Heart in advance." The *lab* may simply imprison the *ch'ulel*—for example, stashing it in the earth, inside the hearth, or in a rock at a cave entrance. The *yajwal witz* may, however, enslave the *ch'ulel*, which causes the body to die.[177] Guiteras-Holmes's informant similarly explains the dangers of the Tzotzil souls, which are targeted by Pukuh, an evil spirit or demon:

> The Pukuh carries the soul away in the same manner that one carries away a hen, a little animal; that is *ch'ulelal*, and [in order to cure *ch'ulelal*] it must let it go. It sold it; it extracted it and sold it like a hen, in the world, among other people. That is why a hen is killed. The *ch'ulel* of the hen goes to the place where it [a human's soul] has been sold, so that man's *ch'ulel* return to his body; the hen is its [the soul's] *kexol*, given in exchange for man's life.[178]

Considering the ubiquity of *k'ex* sacrifice as offerings to underworld denizens among the modern Maya, the practice almost certainly has some historical depth. The question is how such *k'ex* offerings might relate to the solar descent suggested by the Baby Jaguar imagery and to mortuary contexts more generally.

A possible clue is the Palenque sarcophagus lid. Pakal lies atop a sacrificial receptacle (shown as the Quadripartite Badge) in a pose that is well recognized as a metaphor for birth but is also a position that mirrors the depictions of infant sacrifice on ceramic vessels discussed previously (see fig. 2.5). The general consensus among scholars is that the sarcophagus scene shows *ascent* from the underworld, not *descent*, as is clearly implied on the Baby Jaguar vases. Nevertheless, Pakal and the Baby Jaguars share the same pose and both lie in sacrificial receptacles. Pakal's solar ascent is shown

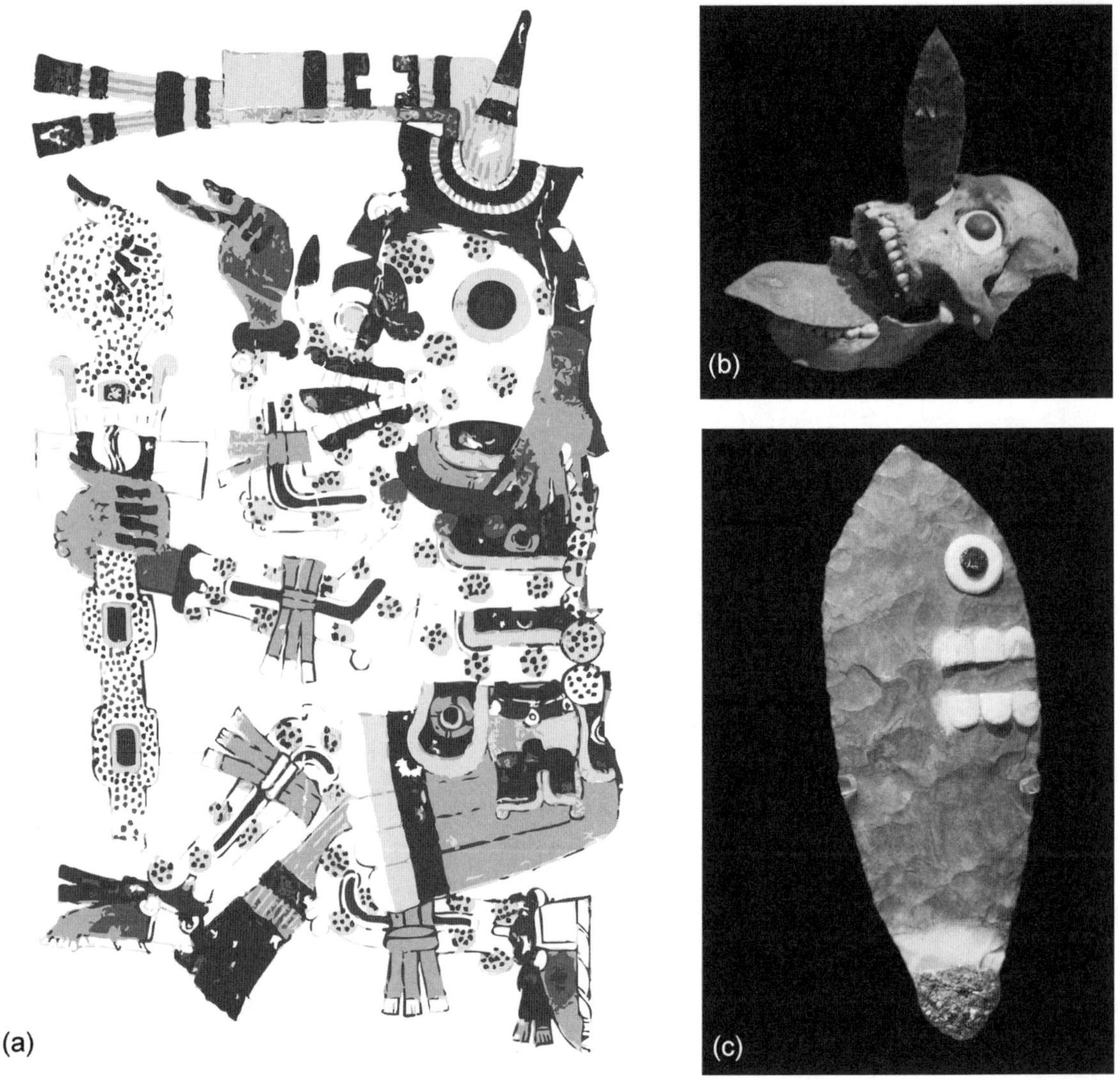

FIGURE 3.51. Aztec Death God (Mictlantecuhtli): (a) from the Codex Borgia; (b) as a skull mask from the Templo Mayor; (c) as a sacrificial knife (*tecpatl*) from the Templo Mayor (photographs and drawing by the author). Note the traces of copal on the base of the blade in 3.51c.

in a fashion similar to the solar descent implied by the Baby Jaguar sacrifice (compare figs. 3.44, 3.46–3.47). Consider especially the scene of the Baby Jaguar (or his impersonator) that seems to take place in the underworld (see fig. 3.48). The Baby Jaguar lies in the birthing/sacrifice position inside of a dish supported by serpents that emerge from a head with a skeletal jaw. A serpent rises from the infant's naval, disgorging a solar cartouche (compare with fig. 3.34). The implication may be that the death of the infant facilitates the emergence of the solar disc, presumably emblematic of a rising soul.[179]

In addition to the mythical scenes of infant sacrifice, actual depictions are also known. Most relevant for understanding the El Zotz tomb is the imagery on an unprovenanced vessel that shows an infant burning atop an *incensario* in front of an enthroned bundle (fig. 3.49). The bundle was likely meant to suggest the wrapped body of a dead lord (compare with fig. 3.23). Note that amidst the flame in figure 3.49 are the signs *yax* (blue-green, unripe, new) and *k'an* (yellow, ripe, and precious). The scene vaguely recalls the burning of the infant sacrifice on Yaxha Stela 13 by a ruler dressed as the Jaguar God of the Underworld (see fig. 2.14).[180] Child sacrifices are also shown on Piedras Negras Stelae 11 and 14, where sacrificed children lie in bowls and are about to be burned (see fig. 3.50). Stela 11 shows an infant, with its abdomen cut open and a bundle stuffed inside, lying on kindling under which a pelt is draped. The sacrificed infant is barely visible on the right of the scene in Stela 14, obscured by erosion. The child is lying on its back, with head hanging down, limp, gazing out at the viewer. As on Stela 11 a bundle is inserted into the victim's abdomen. Here the lady assumes the role of the sacrificer. She wears the Death God in her headdress and holds what may be a sheathed sacrificial knife. Her huipil is adorned with the same smoking flower motif that marks the cloak of the ritual specialist that presents the infant for sacrifice in figure 3.47. As Taube suggests, the bloody steps that ascend the scaffolds on Stela 11 and Stela 14 underscore these sacrifices as essential for royal accession at Piedras Negras. At least one such infant sacrifice has been discovered at Piedras Negras. The skeleton of a child was found in a lidded cache vessel associated with the R-3 pyramid.[181]

A wrapped feathered item is inserted into the bulbous object that fills the dead children's abdomens on the Piedras Negras stelae and is also visible on the El Cayo altar (see fig. 3.29b). These bulbous objects are balls of copal, a human heart, or perhaps a combination of both, to be burned as an offering. As Taube suggests, the wrapped, feathered object is likely the hilt of the sacrificial knife.[182] The wrapped handle of this object superficially resembles the paper wrapping that surrounds the topknot of the Death God (and his impersonators) and some representations of his backracks in scenes of infant/Baby Jaguar sacrifice (see fig. 3.44).[183] In this sense the Death God may be the embodiment of the sacrificial knife (and vice versa). Postclassic Mesoamerican Death Gods are shown to have knives extruded from their noses. The Aztec likened their sacrificial knives to supernaturals, especially their version of the Death God, Mictlantecuhtli (fig. 3.51). Quite a few of their sacrificial blades were decorated with his face, and many such blades were found embedded in balls of copal (fig. 3.51c).[184] Much as some Maya bloodletters represented G1 (or Chahk or some other supernatural being), some of their sacrificial blades seem to have been the Death God.

With this symbolism explored, we can return to the remains of the eight infants and children from the El Zotz tomb (six inside and at least two outside the tomb) that were sacrificed upon the death of the king entombed within Burial 9 (fig. 3.52). The bodies of the youngest two children (aged six months to a year and a half old and one to two years old) were nearly complete and were laid across their vessels (one lidded, the other not) in a fashion similar to that of the children shown on the Piedras Negras stelae (see fig. 3.50). The next set of children (aged one and a half to two and a half and two to four years old) were only partially complete, represented by fragments of their skulls, thorax, and legs. The absence of certain bones may simply be a matter of poor preservation or may indicate that their bodies were dismembered to accommodate the limited space within the vessels. The final pair of children (aged four to five years old) were represented only by rows of articulated teeth, all that remained of what once were severed heads. Excavations outside of the tomb to date have revealed two other sets of lidded cache vessels, one containing the complete skeleton of a child two to four years old and the other the skeleton of a child three to twelve months old (see fig. 3.43). Additional *k'ex* sacrifices are likely still buried in the area around the El Diablo pyramid.

Each of the children's skeletons exhibits thermal alteration. The vessel interiors show blackening, indicating that the children were within the vessels when they were exposed to heat or flame. None of the skeletal remains are calcined, as is expected in cremations where bodies are exposed to high heat for a prolonged period. The bones exhibit only slight discoloration and some surficial cracking, consistent with bodies that were briefly exposed to flame or a smoldering material that consumed the flesh in isolated locations. Thermal alteration was most significant for the dorsal aspect of

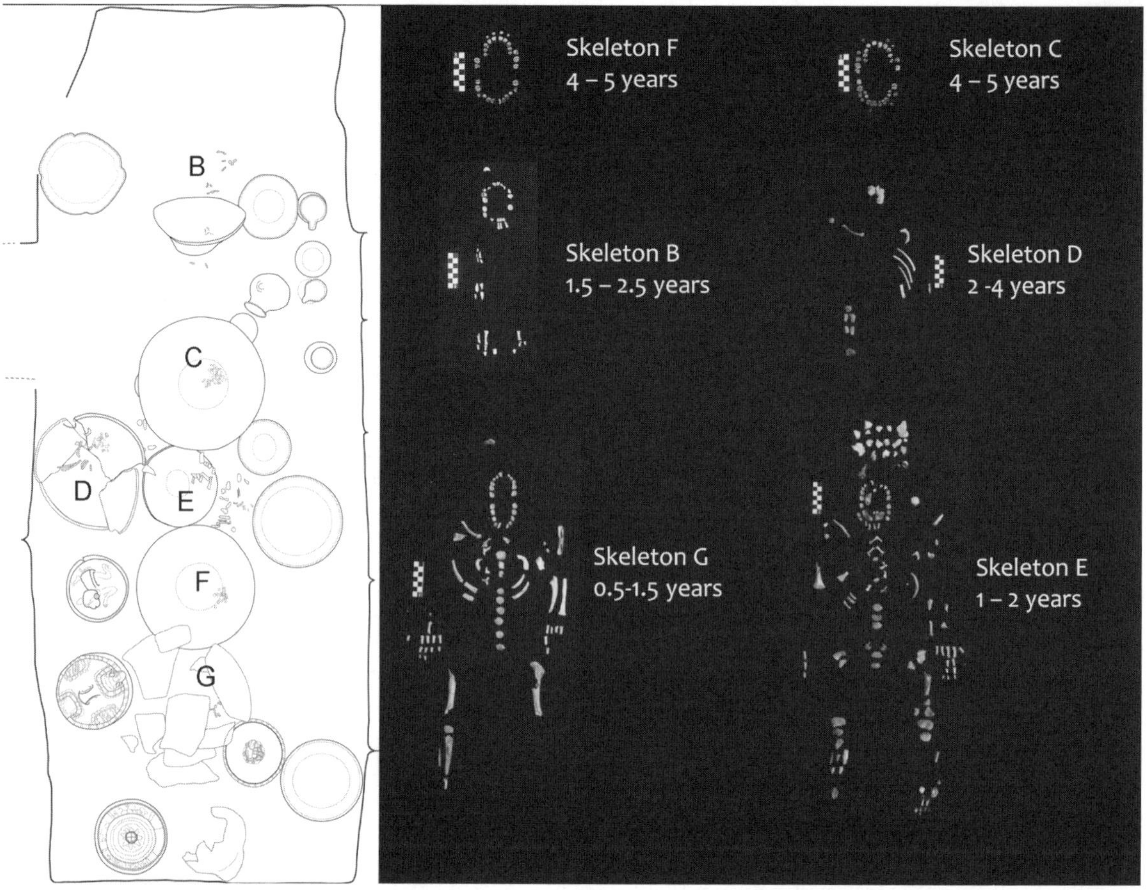

FIGURE 3.52. Remains of the infants and children found within El Zotz Burial 9 (photographs by author, drawing by Stephen Houston). Skeleton designations correspond to field numbering, with their locations shown on the plan. Skeleton A, the adult, is not shown.

the body as well as the thorax. It is likely that some sort of kindling was placed underneath the children, similar to what is shown on Piedras Negras Stela 11 (see fig. 3.50a). Discoloration of all of the vertebral bodies was evident in the cases where the thoracic and lumbar vertebrae were present. Notably, the body of a lumbar vertebra of Skeleton G was more significantly burned than the posterior aspect. In other words, the source of greatest heat was *inside* the abdomen, consistent with the burning of copal or another substance within the thoracic cavity. As suggested by the parallel imagery discussed above, the burning and censing was apparently necessary to release the essence of the child as an offering for supernatural beings.

A blackened substance adhering to bones of both Skeleton E and Skeleton G may be carbonized copal or some other resinous material that was burned within the vessels. Two of the child's skeletons exhibit heat exposure that targeted the face and lower body. In regard to the burning of the face, these children may have been wearing masks, perhaps a composite of wood and other combustible materials (compare with fig. 3.48, a possible masked Baby Jaguar impersonator). That would explain why this part of the skeleton demonstrates greater thermal alteration. The costuming of sacrificial victims has precedent in Mesoamerica, as is well attested among the Aztec.[185] The bone of a felid claw was found among the scattered pelvic remains of the primary occupant in the El Zotz tomb, probably either the remains of a pelt that he wore or perhaps part of the infant's costume. In sum, the sacrificial *k'ex* offerings at El Zotz closely mirror depictions of actual burning of infants shown in Classic period imagery and may be further understood through the myth of the Baby Jaguar.

PRECIOUSNESS OF YOUTH

Considering the association of child sacrifice with accession and death, the Classic Maya must have understood such offerings as necessary during turbulent times. Presumably such deaths placated temperamental supernatural beings during such periods of crisis and transition. The value of these offerings was in the preciousness of youth. The age of the sacrificial victims was a primary consideration in most mortuary sacrifices enacted by the Maya. In a sample of twenty-two elite tombs with mortuary sacrifices from Kaminaljuyu, Tikal, Piedras Negras, and Palenque, three of the victims were infants (zero to two years), seven were children (two to ten years), thirty-four were adolescents (ten to eighteen years), and ten were adults. Children and adolescents were generally preferred over adults, with adolescents selected more often than infants and children. Notably, of those twenty-two tombs, only El Zotz Burial 9 contained more than a single infant (six children all under the age of five), underscoring the singularity of that deposit.

Only a few cases of traumatic injury have been documented on the remains of sacrificial victims from mortuary contexts. In part this likely relates to poor preservation in the Maya area. Andrea Cucina and Vera Tiesler have demonstrated trauma on the remains of victims sacrificed at Palenque and Calakmul, some of which relates to heart extraction, consistent with the imagery of *k'ex* sacrifices discussed above.[186] Otherwise, the absence of trauma may reflect the manner by which most victims were killed. In the ancient Andes, where skeletal preservation is generally much better, sacrificial remains in mortuary contexts rarely demonstrate evidence for traumatic violence. Rather, the recovery of ligatures around the necks of individuals killed in high-status Moche burials at Pachacamac and El Brujo indicates that sacrifice was accomplished by strangulation.[187] Notably, the Inca act of *capacocha* (child sacrifice), like Classic Maya *k'ex* sacrifice, seems to have been enacted at a range of liminal times, including an emperor's accession or death and the birth of a royal son.[188] *Capacocha* involved methods that would not mar the child's body, including strangulation and drugging, and the children were carefully deposited on mountaintops or buried within the earth.[189] The emphasis in many of these Andean sacrifices was on perfection and beauty, almost certainly a concern for the Maya in the selection of youth for mortuary sacrifice.

The practice of youth sacrifice in the Central Petén largely occurred during the Late Preclassic and Early Classic periods. An Early Classic looted tomb, Burial 1 at El Zotz, contained fragments of both an adult individual and a child aged five to nine years. A series of Late Preclassic and Early Classic tombs at Tikal produced infant and child sacrifices, including Burials 10, 48, 160, 162, and 167.[190] Burial 10, the probable tomb of Yax Nuun Ayiin I, contained the remains of at least nine human sacrifices, eight of which ranged in age from five to six years old to adolescence (ca. fifteen years old). The ninth seems to have been an adult.[191] At Tikal only the Late Preclassic Burial 167 and Early Classic period Burial 162 contained infants, both apparently placed within bowls, similar to the El Zotz tomb discussed in the preceding section. In Tikal Burial 167 a cross was etched on the vessel near the infant's head, perhaps to facilitate the exchange of the sacrificed child's soul (fig. 3.53). The child skeleton from Burial 160 (paired with an adolescent) is reported to demonstrate thermal exposure similar to the El Zotz tomb.[192]

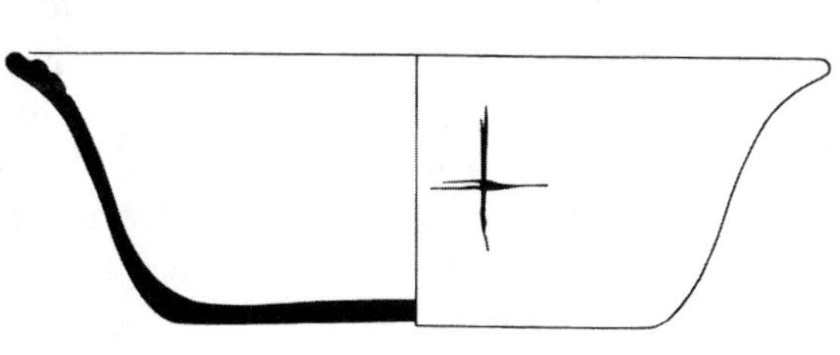

FIGURE 3.53. Late Preclassic bowl from Tikal Burial 167 with a cross incised at the location of the child's head (originally published as Culbert, *The Ceramics of Tikal*, fig. 13A, image courtesy of the University of Pennsylvania Museum of Archaeology and Anthropology).

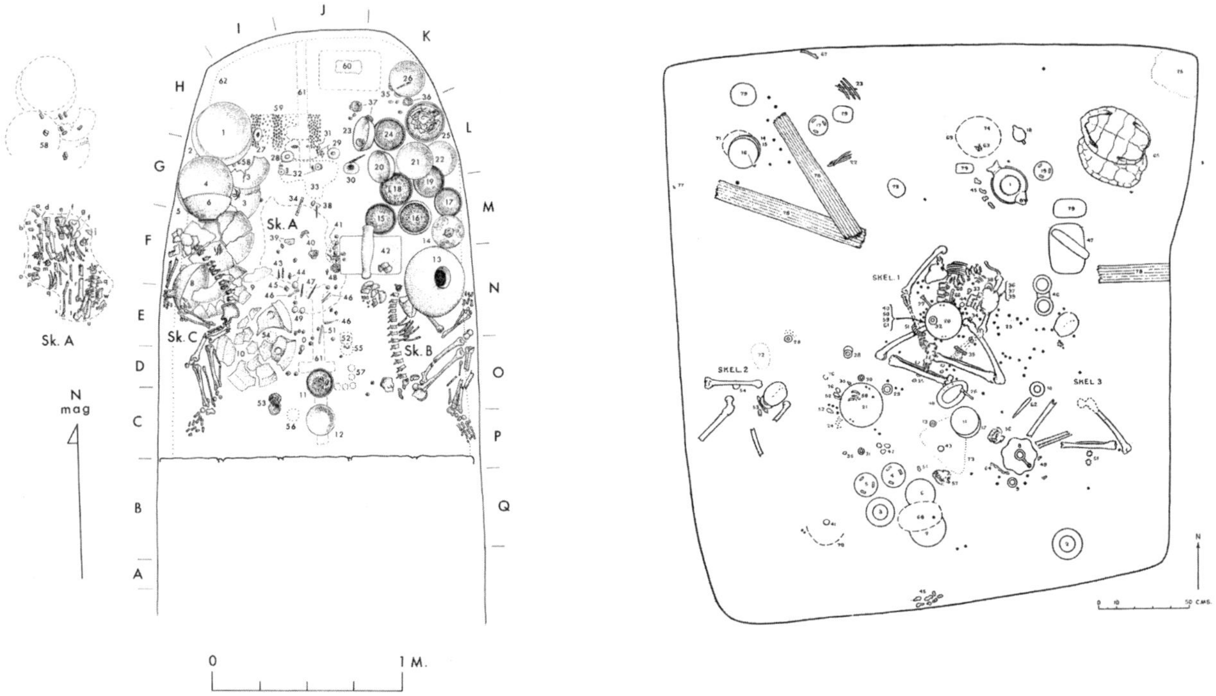

(a)

(b)

FIGURE 3.54. The sacrifice of youth pairs: (a) Tikal Burial 48 (originally published as Coe, *Excavations in the Great Plaza, North Terrace, and North Acropolis of Tikal*, fig. 174, image courtesy of the University of Pennsylvania Museum of Archaeology and Anthropology); and (b) Kaminaljuyu Tomb A-IV (originally published as Kidder, Jennings, and Shook, *Excavations at Kaminaljuyu, Guatemala*, fig. 26).

Tikal Burial 48, identified as the tomb of Sihyaj Chan K'awiil II and dated to AD 457, presents a distinctive arrangement of sacrificed youths (fig. 3.54a). The primary occupant lay bundled in the center of the tomb, with two adolescents lying on either side of the king in semiflexed positions.[193] Tomb A-IV of Kaminaljuyu, roughly contemporaneous if not a few decades later than Tikal Burial 48, contained a primary occupant thirty-five to forty-nine years of age, accompanied by two adolescents, nine to fifteen years old and ten to thirteen years old (fig. 3.54b). Similar pairings of youth are evident in the tombs of Piedras Negras Rulers 3 and 4. The youths in Burial 5 (Ruler 3) were aged between six and ten at the time of their deaths, and those from Burial 13 (Ruler 4) were both aged between nine and fifteen at the time of their deaths. Strontium isotopic testing of Ruler 4 and one of the sacrificed youths from his tomb indicates that both were from the region around Piedras Negras. Although no evidence for trauma was found on the remains, the consistency of their ages suggests that they were selected based on their status as adolescents—an age category otherwise rare in the Piedras Negras mortuary series.

The pairing of adolescents in the tombs at Piedras Negras, Tikal, and Kaminaljuyu not only accords with this general theme of youthful sacrifice but, more speculatively, may relate to the trials of the Hero Twins, Juun Ajaw and Yax Baluun, of Classic period mythology. These beings appear widely in Classic period iconography, and one of their most important roles was attending the Maize God during his resurrection. On the dish from the Museum of Fine Arts, Boston, Juun Ajaw reaches out to assist the Maize God as Yax Baluun pours water on the crevice from which he sprouts (see fig. 1.12). In a curious parallel to the MFA dish, one of the adolescent skeletons from Burial 48 was flexed around a large water vessel (see fig. 3.54a). Although the *Popol Vuh* is not a perfect window into earlier mythology, the K'iche' Hero Twins help us understand the role of the Classic period Hero Twins. In the K'iche' myth the twins descend into the underworld and through a series of tests and trials and

FIGURE 3.55. Caches of adult body parts at El Diablo, El Zotz: (a) bones of a finger in Cache 8; and (b) mandibular incisor in Cache 7 (photographs courtesy of Stephen Houston, El Zotz Archaeological Project).

their own sacrificial deaths are able to defeat the lords of the underworld and ensure the resurrection of their father, Hun Hunapu. The myth is effectively an elaborate *k'ex* sacrifice, which allows all of the protagonists to escape the underworld. Unfortunately, sex cannot be determined for the skeletons of these adolescent pairs from Tikal and Kaminaljuyu, which would help validate or reject this hypothesis.

Notably, child and infant sacrifice declines in the Central Petén at the start of the Late Classic period and may reflect a shift in local ideology and belief. Following more than a century of royal mortuary rites involving the sacrifice of youths at Tikal, around the end of the sixth century Animal Skull (Burial 195) was entombed alone. Although we cannot be certain that youths were not still sacrificed upon the death of Tikal kings, no later royal tombs contained their remains. The apparent decline in youth sacrifices also corresponds to a decline in textual references to the Baby Jaguar at Tikal.[194] Burial 39 at El Perú-Waka' contained the remains of a child approximately seven years old, placed atop a series of ceramic dishes, with the head oriented to the south, in opposition to the orientation of the primary occupant.[195] Dating to the first half of the seventh century, the El Perú-Waka' tomb is one of the latest examples of child mortuary sacrifice in the central Petén. Nevertheless, the practice continued in the west, at Palenque and Piedras Negras, and elsewhere in the Maya area throughout the Classic period.

MUTILATION AND GRIEF

In addition to the remains of infant and child sacrifices, lidded cache vessels have been found with a diverse range of offerings, including fragments of adult bodies, especially manual phalanges (finger bones) and anterior teeth. Such deposits are restricted to sites in western Belize and the central and eastern Petén, pointing to a regionally circumscribed ritual tradition.[196] At El Zotz eleven lip-to-lip caches containing phalanges and mandibular incisors were found outside of Burial 9 (fig. 3.55). At Tikal fingers and teeth were discovered in extremely complex cache assemblages in association with tombs and stelae and contained bloodletting accouterments, shell, lithic eccentrics, and fingers and teeth.[197] These Tikal deposits were found throughout the royal precinct, suggesting that these body parts not only functioned as offerings to supernaturals but were essential to the ritual processes that activated the sacred center of the site.

Diane Chase and Arlen Chase note that deposits of phalanges and teeth span the Late Preclassic through Late Classic periods at Caracol and were placed primarily in lightly fired bowls, apparently crafted for the express purpose of caching human remains.[198] They also

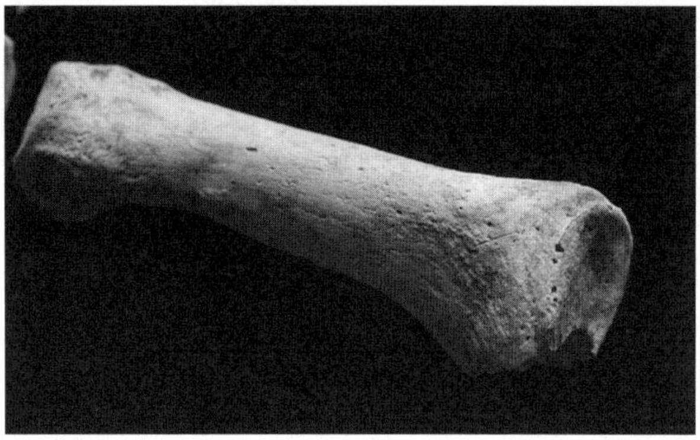

FIGURE 3.56. Cut mark visible on the dorsal-proximal aspect of a proximal manual phalanx from Cache 5 of El Diablo, El Zotz (photograph by author).

note that by the Late Classic period bowls with finger bones and teeth were located in both elite and nonelite contexts at Caracol and were associated primarily with eastern structures that house human interments.[199] David Cheetham reports a deposit of 200 small bowls at Cahal Pech that contained 225 proximal, medial, and distal manual phalanges associated with an entombed stela. At the base of the stela 36 permanent mandibular incisors were deposited.[200]

Cut marks are rarely reported on these remains. In the El Zotz assemblage, three of the nineteen phalanges bear an obvious cut (fig. 3.56). The general absence of visible trauma has led some scholars to suggest that the phalanges and teeth must have been collected from burials and other deposits of already skeletonized remains.[201] Yet removal of the fingers was likely not done by cutting but instead by chopping with an axe, large knife, or some other heavy stone implement. The removal of fingers by machete in a modern forensic case documented by John Verano resulted in angular fragmentation of the proximal and intermediate phalanges, similar to patterns of fragmentation noted on three of the El Zotz phalanges.[202] Furthermore, these deposits consistently contain only anterior teeth (never posterior) and almost always manual (as opposed to pedal) phalanges.[203] This suggests either that the Maya had superior anatomical knowledge and specifically selected these elements from within the grave or, much more likely, that these were corporal elements extracted from living people. Of the teeth, mandibular incisors are among the easiest to avulse in living people and the least likely to impair masticatory function if missing. Moreover, fingers may be severed without seriously maiming a person, especially if some fingers are left intact. The severance of fingers presumably makes for a better ritual spectacle than the dismemberment of toes, which explains their preponderance in the caches. Cache 14D at Tikal contained thirty-nine manual phalanges, of which all but two were distal. A preference for fingertips (not the bases of the finger) further demonstrates that these were body parts cut from the living.[204]

The removal of fingers and the extraction of teeth are reflective of a more general Classic Maya preoccupation with pain and bleeding through the faces and hands.[205] Such acts produced pain, blood loss, delirium, and, most importantly, spectacle, without actually taking the life of the victims. Yet it is not necessary to presume that the corporal offerings of teeth and fingers were extracted from unwilling ritual participants. Among tribes of the North American Great Plains, bloodletting, flesh-cutting, and even finger sacrifice were all noted by European and Euro-American explorers.[206] James Beckwourth, a nineteenth-century mountaineer, observed of the Crow: "Their mourning consists in cutting and hacking themselves on every part of the body, and keeping up a dismal moaning or howling for hours together. Many cut off their fingers in order to mourn through life, or, at least to wear the semblance of mourning; hence the reason of so many Western Indians having lost one or more of their fingers, and of the scars which disfigure their bodies."[207] William MacLeod notes that finger sacrifices and other flesh offerings were sometimes made as a ritual obligation in mortuary settings, while in other cases they were a form of reciprocity between the living and the dead. For example, among the Crow, if the living were indebted to the dead, an offering was needed. For the Dakota, a friend who offered flesh at the death of a friend "will be met in the other-world by the person whose death he so remembered and this person will give him presents." MacLeod also notes that fingers and other flesh offerings were often made directly to the

FIGURE 3.57. (*above*) A water lily jaguar serves a cache vessel with a human arm and eyeball to a Death God (K1380 © Justin Kerr).

FIGURE 3.58. (*right*) Detail of figure missing the distal end of the second digit on his left hand from the stone panel of the Palenque Temple XIX pier (photograph by author).

sun among some of the Plains tribes, which, in parallel to Classic Maya belief, was perceived as either an eater or receiver of human flesh.[208]

The finger and tooth caches at El Zotz demonstrate evidence for heat exposure. It may be that these fragments of the human body were placed atop or inserted within balls of copal or some other substance (note the organic remains that surround the tooth in fig. 3.55). In this fashion these corporal offerings were sustenance for the Sun God (recall his taste for human hearts) and other supernatural beings as part of Classic period mortuary rites. Numerous polychrome vessels depict scenes of flared cache vessels filled with human body parts presented to supernaturals (fig. 3.57). Recalling the connection between seeds and bones, a possible contemporary parallel is the frequent inclusion of squash seed paste as layers within the sacred maize breads that are fired in the *pib* and offered as *k'ex* in contemporary Yucatec ritual.[209]

To my knowledge Maya iconography contains no realistic depictions of auto-dismemberment. Certainly other acts of self-inflicted pain, such as bloodletting, are shown. Yet the absence of auto-dismemberment imagery may suggest that precious royal bodies, the

(a) (b) (c)

FIGURE 3.59. Akan and related beings on Classic period vases: (a) decapitating himself; (b) smoking a cigar, vomiting, and holding an enema; and (c) releasing a jar of bees (drawings by the author after K1230, K0927, and K2284). Note the darkness sign on the chest of (b) and on the jar of (c).

focus of Classic period Maya art, were exempt from the practice. But there are a few examples of noble lords missing body parts. A kneeling figure carved on one of the piers of Palenque Temple XIX is missing the ends of the fingers on his right hand (fig. 3.58). Another figure (perhaps the same person) is shown on the Temple XIX platform, missing the ends of digits two through five of his right hand.

Quite a few mythic scenes of auto-dismemberment, however, show the supernatural being Akan engaged in self-decapitation (fig. 3.59a). As with other Maya supernaturals, Akan imagery is diverse, suggesting a series of closely related beings with overlapping attributes (all referred to here as Akan for simplicity) (see fig. 3.59). Among Akan's most prominent characteristics are an **AK'AB** darkness sign and the "percentage" markings on his skin.[210] Depending on the particular depiction, he may share a variety of underworld attributes with the Death God, such as black paint around his eyes, knotted cloth breath ornaments, a fleshless jaw, an eyeball necklace, crossbones, headdresses that consist of bones or centipedes, and a hair topknot wrapped in paper.

Akan is frequently shown intoxicated, smoking, vomiting, and holding an alcoholic enema. His name means "wasp" in many Maya languages, though the flying insects he is often associated with are more likely the bees needed to make fermented honey (*balche*), the favored alcoholic beverage of the Maya. Akan, or a being closely related to him, is sometimes named *mok chih*, (pulque sickness).[211] He is quite literally the embodiment of drunkenness. In contemporary Maya culture alcohol is not something drunk casually with an evening meal. When alcohol is consumed, it is in large quantities with the goal of utter inebriation, understood to be a state in which the co-essences of the heart are allowed to come to the fore.[212] Classic period imagery shows drunken individuals who have lost control of their physical bodies, achieved both by imbibing alcoholic beverages and by using alcoholic enemas.[213] The linkage between alcohol and the release of the spirit made it an essential tool in Maya ritual contexts. Akan is also a regular smoker of tobacco, another ritual intoxicant.

The self-decapitation of Akan may be metaphoric, suggesting the splitting headache that follows an evening of drunken revelry. Yet considering the ritual significance of alcohol for the Maya, the meaning is likely more complex. Akan is not a threatening denizen of the underworld but instead appears in underworld scenes in contexts of intoxication and self-mutilation. He seems to be the embodiment of the mind-altering substances and practices that were used to access the otherworldly places. In some depictions it is clear that the "percentage" markings are idealized lacerations and punctures in the skin (see the right arm in fig. 3.59b). In one scene a bloated figure holds a knife in one hand and a three-pronged eccentric (an idealized jaguar paw) in the other (fig. 3.60). He gestures at a woman in the guise of Akan who bears "percentage" markings corresponding to trauma to the skin.[214] Rather than simply being a curious drunkard who decapitates himself, Akan is perhaps better understood as the embodiment of ritual transformation through intoxication and mutilation.

FIGURE 3.61. The Maize God in a watery place attended by the Hero Twins, two maidens, and two Akan impersonators wearing masks and with skin marred by the "percentage" markings (K6979 © Justin Kerr).

On one Classic Maya vase two female Akan impersonators accompany the Hero Twins and two maidens as they prepare the Maize God for his resurrection in a watery place (fig. 3.61).[215] This vase, along with other depictions of the Maize God cycle, highlights the importance of women in the myth and perhaps also actual royal mortuary rites (see figs. 2.4, 2.15, 2.16, and also the collapsed woman in front of the sacrificial victim in fig. 3.23). As has been well recognized, women dressed the Maize God in what may be a metaphor for the preparation of the corpse. They were also depicted in expressions of extreme grief, however, an emotion that is otherwise rarely shown in Maya art. The vessel in figure 3.61 captures both aspects of this dynamic. A maiden and an Akan impersonator are on either side of the Maize God. Behind the first impersonator is a maiden holding the Maize God's regalia. Behind the Hero Twins is another impersonator, wrenching her hair in an expression of grief. This combination of violent grief and self-mutilation attributed to Akan and his impersonators recalls the accounts of mortuary rites among North American Plains tribes during the nineteenth century as noted above and allows us to make better sense of the finger and tooth caches found throughout the Petén and Belize.

FIGURE 3.60. Female Akan impersonator with centipede headdress and "percentage" markings presumably made by the stone claw held by the bloated figure to her right (K2286 © Justin Kerr).

GROUND AND BAKED

Throughout the Maya area manos and metates were placed within burial chambers. Nearly every one of the tombs excavated by the Carnegie Institution at Kaminaljuyu contained a mano and metate pair (see fig. 3.54b).[216] A mano and metate were also placed in Tikal Burial 48, the tomb with the pair of sacrificed youths noted above (see fig. 3.54a). Metates were also used in the construction of burial chamber walls and lids. Along the Usumacinta River metate fragments were placed at the foot end of the graves of women in Piedras Negras Burial 81 and El Kinel Burial 12 (see fig. 3.20). El Kinel Burial 3 was the grave of a six- to twelve-month-old infant, capped by a metate fragment (fig. 3.62). In El Kinel Burial 4 a probable male was buried with a mano placed over his abdomen (see fig. 2.49a), and in El Kinel Burial 9 an eight- to sixteen-month-old child was buried with a small mano resting near the left os coxa. The inclusion of grinding implements with infants and children indicates that these were not objects that the decedents

FIGURE 3.62. Metate as lid on El Kinel Burial 3 (photograph by the author).

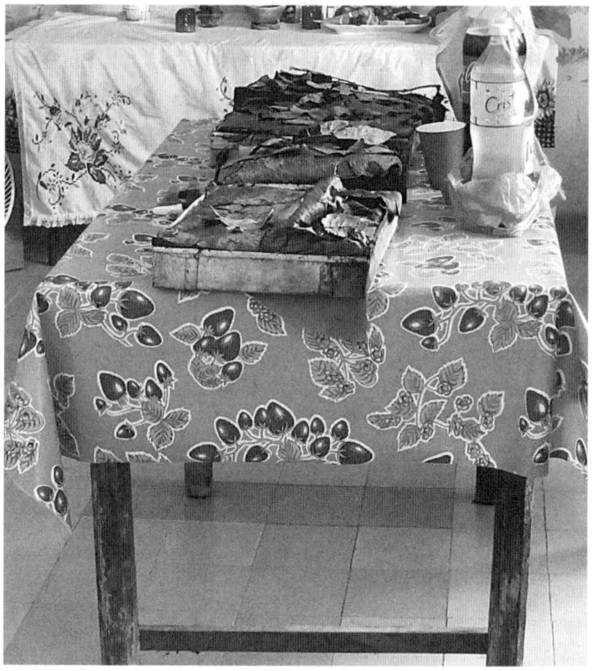

FIGURE 3.63. Maize breads cooked in the *pib* and served for Día de los Muertos in Campeche, 2012 (photograph by the author).

had used in life. Rather, the objects likely had symbolic value relating to the pan-Maya belief that humans are made from maize and that the human life cycle mirrors the cycle of the plant.[217] The association of grinding stone implements with the dead finds resonance in a passage from the K'iche' *Popol Vuh* where the lords of the underworld plot the death of the Hero Twins: "It is good that they should die. And it would be good if their bones were ground upon the face of stone like finely ground maize flour."[218] There may be a connection between the sacrificed youths from Tikal Burial 48 and the mano and metate in the tomb, considering that this particular pairing (twin sacrifice with a mano and metate) is otherwise absent from the site. In similar respects a sculpture on display at the Baluarte de San Miguel, Campeche, shows a sacrificial victim splayed out on a metate.[219]

When the lords of the underworld finally settle on a strategy to kill the Hero Twins in the *Popol Vuh*, they lure them "into the pit oven" where both of them die, their bones ground and strewn into the river.[220] As noted, some contemporary Yucatec Maya communities bury their dead for only a temporary period, a process that is likened to cooking in the *pib*. After exhumation the bones of the dead are placed in mausoleums where they can be visited by the living, especially on Día de los Muertos. Throughout the Yucatán Día de los Muertos is celebrated with a feast whose central food is a maize bread cooked in the *pib*, similar to a large tamale and stuffed with meat, squash seeds, and various other foods (fig. 3.63; see also fig. 2.61). In the early twentieth century Redfield and Villa Rojas noted that "seven kinds of sacred breadstuffs are made in the earth-oven," served as offerings "suitable for the gods."[221] Recall that such sacred breads also serve as *k'ex* offerings in Yucatec Maya ritual. In this sense the placement of wrapped bodies into the earth may have paralleled the placement of wrapped foods in the *pib*.

Epigraphically, the sweatbath of the Classic period was identified as a *pib naah* (oven house).[222] Throughout the indigenous Americas, sweatbaths are essential for both physical and spiritual health. Among the Maya, sweatbaths underscore the curative power of heat, returning balance to the ill whose body has become too cold. Sweatbaths are and were especially associated with childbirth in Maya communities, and the afterbirth is

FIGURE 3.64. Bobwhite (*Colinus virginianus*) (drawing by the author).

FIGURE 3.65. The remains of two bobwhites inside a dish from El Zotz Burial 9 (photograph courtesy of Stephen Houston, El Zotz Archaeological Project).

often interred under their floors. Thus there is a duality of sorts between sweatbaths, as ovens for the living, and burials and tombs as ovens for the dead. Recall that mortuary spaces are also likened to navels. Conceptually, the Maya have viewed both sweatbaths and burials as liminal spaces, entrances to the underworld and other supernatural places from whence the living enter the world and the dead depart. The parallels are intriguing, though epigraphic or iconographic evidence is needed to demonstrate that the Maya conceived burials as like-in-kind to the *pib*.

BIRDS, RODENTS, AND TURTLES

Mukbil pollo (literally "chicken that is buried") is among the most common maize bread dishes served at contemporary Día de los Muertos celebrations in the Yucatán. The name reflects the way the meal is prepared, in the *pib* oven. Chicken and other fowl are often served during celebrations related to the dead in the Maya region, either stuffed inside the ground maize loaf or as an accompaniment to such sacred breads.[223] Among the Tzotzil, chickens are sacrificed as *k'ex* to supernaturals that molest the souls of the living. Vogt suggests that chickens and similar domesticated animals are selected because they are food items but also because they mediate between nature and the wild.[224] Recall that the Tzeltal believe that one of the souls of the heart is a bird, described as a grackle, pigeon, or, most commonly, rooster or hen.[225] Various supernaturals in the Tzeltal worldview seek to eat this heart-bird, thus causing death. Pitarch suggests that in particular the hen is linked to the heart soul, because it was a delicacy favored by Spanish priests and was thus especially coveted by the malicious *palé* (priest *lab*). Pitarch describes the Bird of the Heart as "jumpy and skittish . . . when confronted with danger or bodily exertion, becomes agitated and flutters," likening the bird's behavior to the palpitations of the heart. Yet domesticated chickens are not particularly temperamental, certainly not in the manner that Pitarch describes for the Bird of the Heart, and of course they did not exist in Precolumbian Mesoamerica. Such flighty behavior is more characteristic of wild game birds indigenous to the region, particularly one that was very important to the Precolumbian Maya: the bobwhite (New World quail: *Colinus* sp.).[226]

Bobwhites, as members of the order Galliformes, are distant cousins of chickens and roosters (fig. 3.64). Bobwhites and other unidentified birds have been widely recovered in tombs and caches throughout the Maya area. El Zotz Burial 9 and El Perú Burial 24 each contained a dish holding two bobwhites (fig. 3.65). A pair of birds was placed in Cache J-29–21 at Piedras Negras, and bird bones were recovered in other cache contexts at the site as well as in the tomb of Ruler 3 (Burial 5).[227] Skeletons of birds were found in Early Classic period royal tombs at Tikal, including Burials 10, 160, and 48 (which contained the remains of at least five bobwhites as well as a turkey).[228] In Tomb 1 from Structure 3 of Group B at Palenque a lidded vase contained the remains of a bird along with a polished piece of meteorite.[229] It seems

FIGURE 3.66. Pair of birds on an incised vessel from Tikal Burial 48 (originally published in Culbert, *The Ceramics of Tikal*, fig. 31a, image courtesy of the University of Pennsylvania Museum of Archaeology and Anthropology).

highly unlikely that the bird bones found in Maya tombs simply represent a feast for the dead, as the remains of similarly preferred animals, such as deer, peccary, or even turkey, are rarely found in mortuary contexts.

Rather, the sacrificed bobwhites in Maya tombs may be evidence of the Precolumbian origins of certain bird species as appropriate sacrifice, appropriate emissaries between the earth and supernatural realms, and perhaps even a belief in birds as like-in-kind to aspects of the human soul. Gifted with the power of flight, birds had the ability to travel to places and see things that no human, confined to the earth, would be capable of matching. Such supernatural avatars include God D's raptor (often glossed as the Principal Bird Deity) and the owls of God L, a pan-Amerindian messenger of death and ill fortune.[230] Taube calls attention to the frequent portrayal of precious birds associated with the celestial paradise of "Flower Mountain," beings that are likely emblematic of human souls.[231] Bird imagery permeates the Early Classic period assemblage of mortuary ceramics at Tikal and other Maya sites. For example, an incised vessel from Burial 48 shows a pair of birds framing a central cartouche, with speech scrolls suggesting their calls (fig. 3.66).[232]

Vogt notes that for the *k'ex* conducted during Tzotzil healing ceremonies chickens are generally sacrificed specifically in pairs, one for the "ancestral gods" and the second for reasons not entirely clear but somehow operating as a "means of establishing communication between men and gods." Vogt also notes that pairs of chickens are sacrificed in house ceremonies: one is consumed by the ritual officiate and the other by the gods, thus establishing communication between the two.[233] It may be that the pairs of birds in Classic period tombs were used to establish links between the dead and other supernatural beings or places, perhaps to help guide the deceased on their way. Among the Tzotzil, a rooster's head is placed within the grave in order to help the deceased navigate the correct paths of the afterlife.[234]

Much as jaguars embodied the sun in its underworld guise and birds represent aspects of the human soul, a host of animals of the Maya world have been ascribed important spiritual roles and attributes. Today dogs are understood to guide Maya souls to the underworld. Although dogs in Mexico and Guatemala are treated with the sort of indifference that would trouble a North American or European pet owner, they are nonetheless valued as companions for travel, as they are useful for spotting snakes and other potential hazards. A collection of vases seems to display the journey of a lord in a palanquin in the company of a dog, which some have interpreted as a trip to his grave. On one vessel a procession of musicians and an incense bearer precede a black-faced lord with a smoking celt or torch in his forehead, perhaps indicating that he is indeed dead (fig. 3.67).[235] He is followed by another courtier carrying the lord's jaguar cushion, which could be understood to seat him in the world of either the living or the dead (compare with the cushion and bundle shown on fig. 3.23). Whatever the case, the remains of canids are for the most part absent in Maya mortuary contexts. Exceptions include the Early Classic

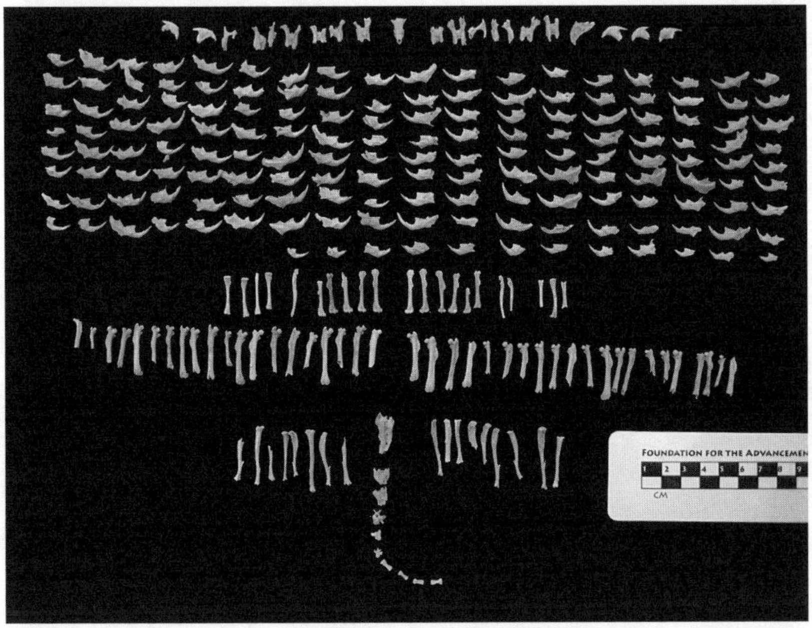

FIGURE 3.67. (*above*) Lord carried in a palanquin and accompanied by a dog on an unprovenanced vase (K5534 © Justin Kerr).

FIGURE 3.68. (*left*) The remains of 81 Toltec cotton rats from El Kinel Burial 11 (photograph by author).

tombs at Kaminaljuyu, a number of which contained the remains of sacrificed dogs.[236] Placed at the edges of the tombs, these companions may have served to guide the souls of the dead.

Even more common than dogs, rats and other rodents appear regularly in Classic period imagery, particularly as denizens of the underworld. On the MFA vessel rats are shown scurrying underneath the throne of a dead king, contrasted to the dwarves that serve in the living court (see fig. 3.23). Similarly, an underworld rodent is shown on the fifth terrace stucco mural from Tonina, scurrying behind a version of the Death God, carrying a bound vessel or ball.[237] Their underworld association is unsurprising; many rodents of the Maya world, such as the *tepescuintli* (paca), are nocturnal. Other rodents are burrowing animals that live in the earth, such as the pocket gopher, or even make their home within burial chambers. At the western Preclassic site of Rancho Búfalo, I excavated a burial (Burial 1) that was incorporated into part of the burrow of a pocket gopher. The animal had tunneled throughout the burial, disturbing bones and artifacts. Although I never saw the creature, it returned at night after I had left the partially excavated burial and even absconded with some of the bones and potsherds from the grave. At the site of El Kinel Charles Golden and I found the remains of 81 Toltec cotton rats (*Sigmodon toltecus*) within Burial 11, the remains of creatures that had converted the crypt

FIGURE 3.69. Tikal Burial 10 (originally published as Coe, *Excavations in the Great Plaza, North Terrace, and North Acropolis of Tikal*, fig. 160, image courtesy of the University of Pennsylvania Museum of Archaeology and Anthropology).

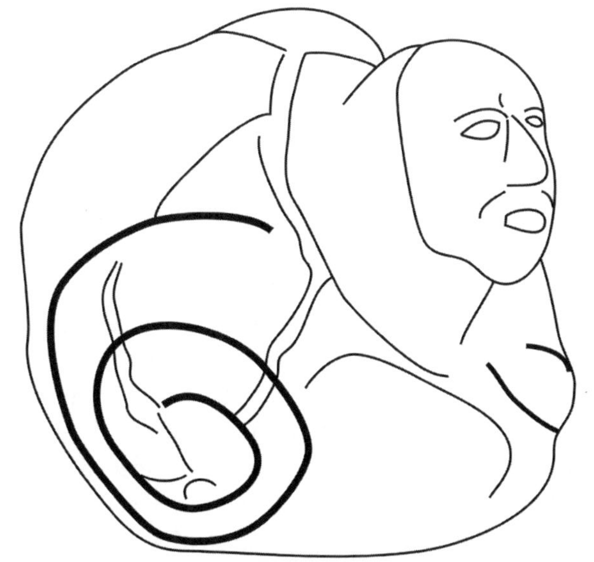

FIGURE 3.70. Sketch drawing of alabaster vessel from El Perú–Waka' Burial 61 (drawing by author based on a photograph by the El Perú–Waka' Regional Archaeological Project).

FIGURE 3.71. Spirals incised on the underside of the lid from Piedras Negras Burial 103 (photograph courtesy of Stephen Houston).

into their home (fig. 3.68).[238] We frequently find rats' nests within crypts. Sarah Newman even excavated a burial at El Zotz that contained a nest full of living baby rats. Crypts provide a convenient subterranean shelter for rats. Unlike those of humans, rodent incisors grow perpetually and must be worn down by habitual gnawing; human bone, potsherds, and other artifacts provide convenient objects to gnaw, much to the chagrin of the archaeologist. And, of course, there is the unpleasant possibility that hungry rats were also attracted to both the offerings of food and the decaying corpse itself in the weeks immediately after interment.[239]

As emblematic of the earth from which the Maize God is resurrected, turtles also appear on occasion within Maya mortuary contexts. Prior to his resurrection, the Maize God of the San Bartolo murals dances within turtle-earth and beats a turtle carapace (see fig. 1.21b). As Taube suggests, the turtle drum is associated with rainmaking, symbolizing either the rumbling thunder or rain pelting the earth, and is an appropriate symbol for resurrection in Maya mythology.[240] Turtle shells are most common in the tombs of Kaminaljuyu, where four burial chambers excavated by the Carnegie project produced turtle shells.[241] The roughly contemporaneous Burial 10 at Tikal contained five complete turtle shells, one of the many parallels between that burial chamber and those at Kaminaljuyu (fig. 3.69; compare the location of the turtle carapaces with fig. 3.54b).[242] In Classic period iconography the cranial and caudal ends of turtle shells are shown to be points of emergence for K'awiil, God N, and other supernaturals. These beings also emerge from openings in marine gastropod shells (see fig. 3.14). A recently recovered alabaster jar from El Perú–Waka' shows a human figure emerging from a gastropod shell, represented as a stylized spiral (fig. 3.70).[243]

Spirals are also carved on the rocks near Piedras Negras, perhaps suggesting the shells or, more likely, the whirlpools in the violent current of the Usumacinta River. Similarly, a spiral was carved on the underside of one of the stone slabs that covered Piedras Negras Burial 103, likely meant to suggest emergence from a watery, underworld place (fig. 3.71). Recall also that watery and underworld beings have swirls for pupils. In general, many of the spiral motifs in Classic Maya art are likely meant to suggest watery places. In the Early Classic period Tomb 1 at Río Azul, the mortuary space was rendered as an aquatic place by a swirling water band painted on the walls of the burial chamber.

VASES AND OTHER VESSELS

Thus far I have suggested that the animal remains and symbolism associated with Maya mortuary contexts were for the most part not related to food but had other significances connected to Maya views of life, death, and the supernatural. Nevertheless, we must at least consider that some foodstuffs were placed within Maya mortuary spaces, especially in view of the large number of ceramic vessels in Maya graves. Ethnography indicates that the number of vessels, the objects they contain, and their placement within the graves are largely contingent on the particular traditions of each Maya community. At Zinacantán Vogt noted that villagers placed bowls and gourds within the graves in order to provide food and drink for the souls of the dead.[244]

Most mortuary ceramic vessels seem to have been used prior to their placement within funerary contexts. Houston suggests that many of the very finest Maya funerary vases—the codex-style cylinder vessels of the Central Petén—were crafted as instructional devices, given to male youths before their passage into adulthood.[245] The pictorial subject of such vessels is primarily supernatural. Many show scenes of the underworld and the *wahy* that dwell there, as is evident in many of the illustrations in this book. Moreover, the iconography and imagery of certain vessels may have been relevant to the mortuary context, as was the case for the mirror motif noted on the vessel from Piedras Negras Burial 77 (see figs. 3.35b, 3.36).

It is perhaps notable that in the western region of the Maya lowlands, where codex-style vase painting was rarely if ever executed, there is also a notable reduction of the quantity of ceramic vessels in mortuary contexts.[246] Of 121 burials at Piedras Negras, only 19 contained ceramic vessels (15.7 percent). Even the royal tombs at Piedras Negras generally contained only a single vessel, likely linked to bloodletting and conjuring, as noted above. The most elaborate tombs at Palenque, those of Pakal and the Red Queen, contained only four and three vessels, respectively. In contrast, the Early Classic Burial 10 of Tikal contained thirty vessels, and the Late Classic Tikal Burial 116 held nineteen (see figs. 2.17 and 3.69). Ceramic vessels generally are more common than not in the burials of Tikal. This is not to say that Piedras Negras was impoverished in ceramics; rather, they chose not to place these objects within the burial chambers. Thus it is unlikely that the "number of vessels included as offerings proves to be a significant definer of the wealth of a burial," as suggested by Estella Krejci and Pat Culbert.[247] More likely, the number and arrangement of ceramic vessels relate to the execution of certain rituals or the enactment of particular symbolism within the mortuary space.

The material from which vessels were constructed may also have been relevant in some instances. Although most mortuary vessels were ceramic, archaeologists have also detected wooden bowls and gourds (used to contain water and other beverages).[248] Alabaster vessels have been found in Tikal Burials 48 and 160 and in a number of tombs in the Yaxchilan kingdom, including Tomb 2 of Structure 23 and the recently discovered tomb from Structure 1 at Bonampak.[249] In Structure 1 the head was removed and replaced by the alabaster vessel. Judging by the frequency with which alabaster was used to depict shell motifs, it may have been viewed as a like-in-kind material, perhaps also seen as a material that facilitated movement between earthly and otherworldly places. Alternatively, the gleaming whiteness of the material may have suggested bone: hence the replacement of the cranium with an alabaster vessel at Bonampak.

The inclusion of rare and unique items such as alabaster vessels or headdresses (as discussed in chapter 2) also hints at issues of inalienability in Maya mortuary practice. Certain objects were linked to the persona of the deceased and could not be inherited by their descendants.[250] Indeed ethnohistoric and ethnographic reports indicate that the dead were buried with their personal possessions as a means of removing them from circulation. Vogt reports that a wide variety of objects may be placed in and around the coffin at Zinacantán, many of which are personal objects "said to possess the soul of their owner."[251] Some of these goods are intentionally damaged so that they will not transform into an object that might harm the deceased or the living. Similarly, Bartolomé de Las Casas reports for the graves of the Pokom: "With the men laborers they placed their gear and implements for farming, and with the hunters, for hunting their bows and arrows, and so with the rest. With the women were the stones for grinding maize, pots for cooking and preparing food, the vessels for drinking, plates, and bowls. The lord placed in the tomb, his face toward the south (held by them as more fortunate than the north) they put around him all the dead servants."[252]

The passage from Las Casas reminds us that sometimes the objects within Maya graves may simply be the possessions of the dead.

THRONES, BENCHES, AND BIERS

The vast quantities of ceramic vessels transformed the tombs of the kings and queens of the Central Petén into subterranean versions of the courtly places that they enjoyed in life. Other accouterments of royal life were included to equip their tombs as regal chambers, including one of the most important symbols of authority: thrones. Houston has identified the Classic period word for this object as *teem*, a term that is evident in more recent Maya languages as "seat."[253] Royal *teem* are simply elaborate versions of benches found in structures from all social contexts, much as European thrones are more elegant versions of the chairs of the common people. One of the common Classic period inscriptions for accession is "he sits in kingship."[254]

In describing the interment of a Pokom lord at the time of the conquest, Las Casas describes the "seating" of a royal corpse within its mortuary space:

> All the things that everyone had brought were put together and they placed them over those he already had, and thus, well dressed, with many folded mantles and decorated with his jewels, they placed him in a box made of stones or wood, with a cover, in which he was carried seated cross-legged. This box they placed in the tomb, which was a grave or hole made in the top of a mountain or on the mountain heights. Then, placing the lord within, they killed the male and female slaves that they had to go to serve.[255]

Similarly, the Kaqchikel lords at Iximche were seated within their tombs, and the bodies of the dead in many of the tombs of Early Classic period Kaminaljuyu were seated (see fig. 3.54b).[256]

Yet the nobility of the Classic period lowlands were not seated in their tombs. Rather, as shown on the Berlin Vase, royal bodies were wrapped and laid horizontally on thrones that doubled as funerary biers (see fig. 2.15). The supine positioning of the body in the Maya lowlands likely relates to more general Maya concepts of horizontality and soul travel discussed in chapter 2. Although iconographic depictions of the interior of royal tombs are rare, those that exist invariably show the body bundled atop a mortuary *teem* (see figs. 2.39, 3.23, 3.49).

Wooden biers have been found across the Maya area, including El Zotz Burial 9; Tikal Burials 10, 48, 24, 23, 160; and Calakmul Structure III, Tomb 1.[257] Most of these tombs date to the Early Classic period. The bier in the El Zotz tomb was identifiable by the thin veneer of green painted stucco that covered it. The wooden bier in the Tikal tomb had "pink-red round stuccoed supports."[258] In El Zotz Burial 9 and Tikal Burial 10 the primary individual was placed on the bier and the sacrificial victims were placed underneath (see figs. 3.52 and 3.69).[259] When the thrones in El Zotz Burial 9 and Tikal Burial 10 rotted out, the skeleton and costume of the primary occupant collapsed onto the ceramic vessels and tomb floor below. This positioning of the king over the sacrificial victims replicated the spatial arrangement of the living court. The king always sat above his subjects, as is evident in so much Classic period imagery. Moreover, these tombs replicated the splendid display

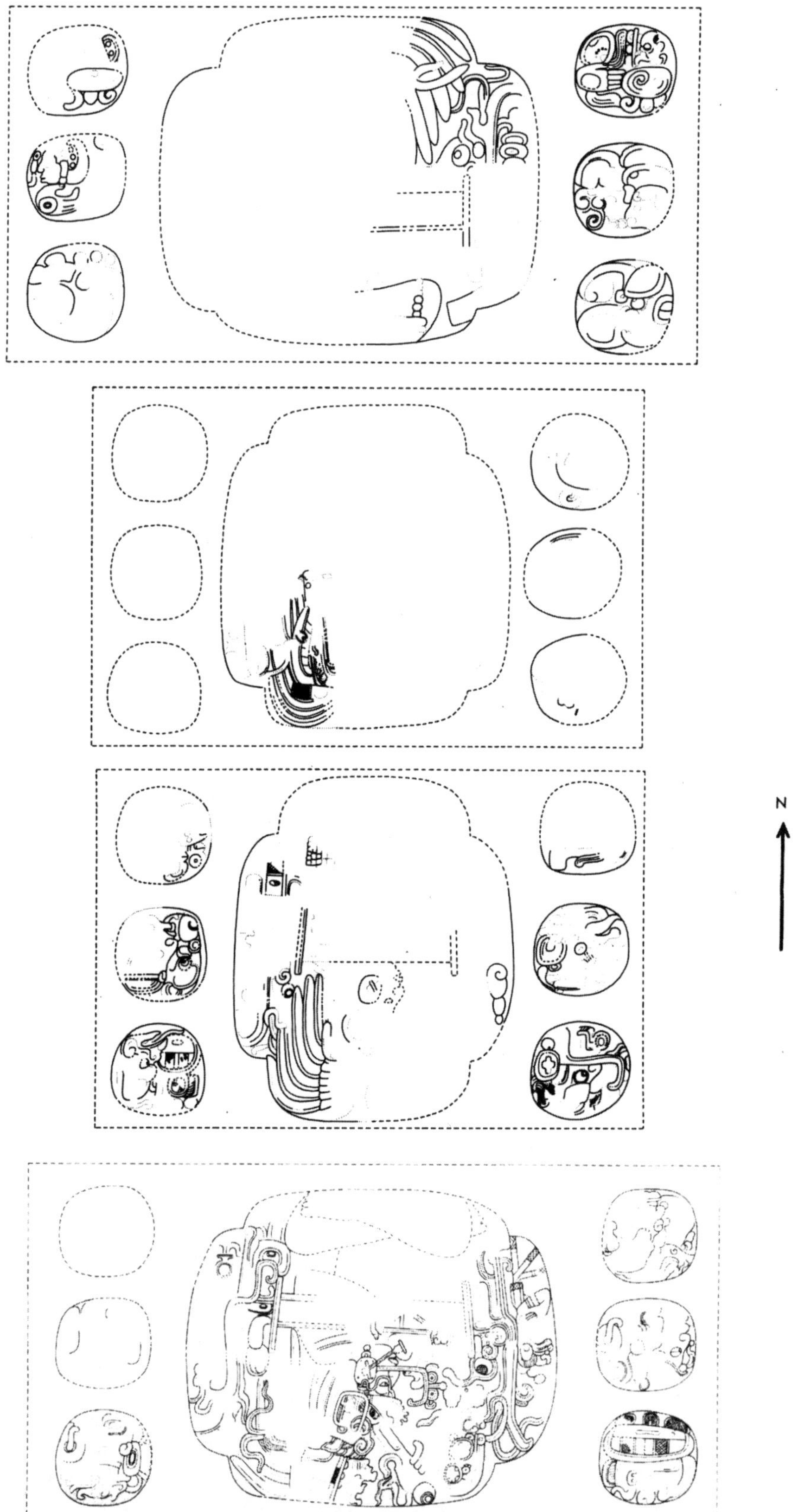
N

of wealth expected in Maya throne room scenes. It may be that the Maya king in his regal mortuary chamber was understood to receive his final tribute and preside over one last feast. Yet the elevation of the corpse may also have had specific implications in the mortuary space. The cache vessels were likely placed in the El Zotz tomb while copal still smoldered within, covering the king's corpse in a thick layer of smoke and carrying his soul to otherworldly places.

A distinctive funerary bier was discovered in Burial 195 of Tikal, identified only because of the rapid sedimentation of the tomb and careful excavation by the Penn team (fig. 3.72; see also fig. 2.27).[260] The primary individual, presumably the king Animal Skull, was laid out on a series of four boards, with the two largest at the head and foot ends, the smaller boards in the middle. The boards were painted green, like the bench from the El Zotz tomb. These green *teem* signified verdant fertility and centered the lords as the axis mundi. Though the precise construction of the Tikal bier is unclear, the boards were likely part of a larger bench, perhaps placed on a supporting scaffold that disintegrated. The boards were carved with a series of figures set in quatrefoils, presumably depictions of the deceased or his ancestors seated in otherworldly places. The top of the head in the northernmost cartouche pointed north, while the heads of the remaining figures pointed south. The significance of this arrangement is not clear. Also included in the tomb was a small, stuccoed wooden bench, painted green with red details, similar to the other benches noted here. This smaller bench may have been a seat used by the Tikal lord in life, a portable throne to journey with him as he moved throughout the community and beyond.

FIGURE 3.72. *(opposite)* The remains of a funerary bier from Tikal Burial 195 (originally published as Moholy-Nagy, *The Artifacts of Tikal*, fig. 229, image courtesy of the University of Pennsylvania Museum of Archaeology and Anthropology; the boards are reoriented here to reflect their original position within the tomb).

FIGURE 3.73. A lord on a portable litter from an unprovenanced vase (drawing by author after K5456).

Functionally, Classic period biers overlap with a range of similar devices from colonial and contemporary times. Yucatec Maya from Chan Kom are reported to have placed bodies on tables during the 24-hour period of visitation prior to inhumation.[261] Girard notes that for the Cho'rti' "the rough stretcher of wood or reeds on which the dead rests is a replica of the altar: both symbolize the cosmic plane in the center of which is placed the body or the image."[262] In the Classic period leaving a corpse unwrapped and unburied would have been risky business, as putrefaction is accelerated in such a hot and humid environment. Corpses with complex funerary wraps would have required some sort of device to carry the dead to the grave. The litter of a living king shown on a polychrome vessel is probably a reasonable approximation for the sort of device used to transport corpses in the Classic period (fig. 3.73). The transport of royal corpses was not merely a functional event, however. As Barbara Arroyo and Lucia Henderson have recently suggested, it is quite likely that Maya lords and ladies were deposited in their tombs only after a lengthy ceremonial procession, facilitated by their funerary biers.

As James Fitzsimmons points out, the *och bih* (road entering) death event may occur as much as a week after the actual death. Although it is generally assumed that this expression is a euphemism for the postmortem journey of the soul, it may also have specific connotations regarding the actual procession that brought the corpse to its final resting place.[263]

Wooden biers are more common in Early Classic period funerary chambers than in the Late Classic period. Instead the Maya fashioned mortuary benches from masonry and bedrock at that time, especially in the Central Petén. In Tikal Burial 116 the floor of the tomb chamber was excavated into bedrock as two levels, an upper primary floor and a sunken alley to the west, 40–50 cm below the level of the body (see fig. 2.17). The king was arrayed on a mat that lay on the plastered upper level that may have functioned as a bench or throne.[264] The alley floor was filled with ceramic vessels, pyrite mirrors, and other offerings. In this regard the burial chamber was organized like Early Classic tombs where kings were supported on perishable *teem*, with their wealth spread below. Tikal Burial 196 and El Perú-Waka' Burial 39 were similarly constructed, with the dead nobility laid on masonry mortuary thrones.[265]

At other sites nobility were interred on stone slabs that were functionally equivalent to thrones and benches. The founder of the dynasty at Copan, K'inich Yax Kuk Mo', was placed on a massive stone slab, presumably understood to be his eternal throne.[266] A tomb within the Temple of the Skull at Palenque contained the body of a noble on a stone throne consisting of three large slabs of limestone.[267] The practice was echoed in elite residential groups at Palenque, where individuals were entombed atop benches that were constructed of large limestone slabs set atop stone blocks. For example, a chamber located below the sanctuary of Structure 3 of Group B, north of the site core, contained the remains of two females, each lying on a bench within the same sealed chamber.[268] Frans Blom and Oliver La Farge report human remains atop a low stone platform in Group E of Palenque.[269] They also note that a tomb opened at the site of Comalcalco to the west contained four low pillars, presumably supports for a wooden platform upon which the body lay.[270]

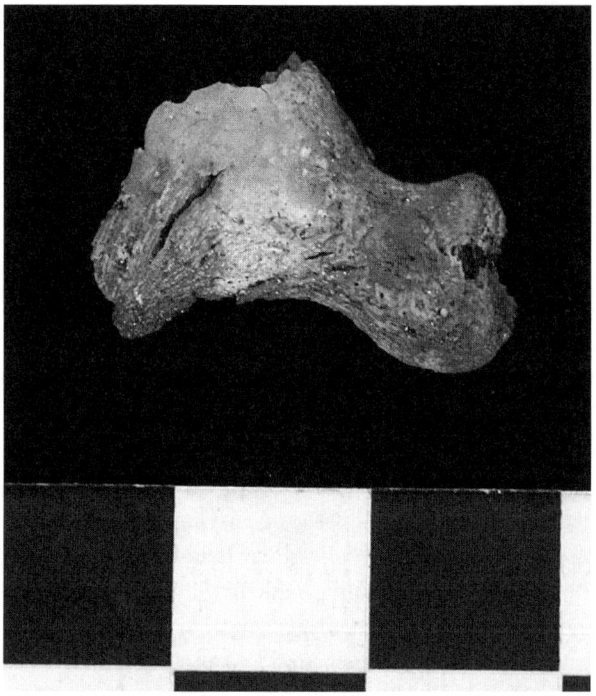

FIGURE 3.74. Felid phalanx from El Zotz Burial 9, likely part of a pelt or cape (photograph by author).

Whether set atop benches or on the floor of the tomb, royal bodies were laid out on a variety of mats, including jaguar and other felid pelts, as evidenced by the frequent recovery of the phalanges of cats' paws within mortuary spaces (fig. 3.74). For example, the skeleton from Tikal Burial 196 was surrounded by seventy-two felid phalanges located in sixteen different locations around the body, presumably the claws of at least four different jaguars (see fig. 2.50).[271] Similarly, eight sets of felid phalanges were found in Tomb 4 of Structure II of Calakmul, indicating at least two jaguar pelts.[272] Burial 39 at El Perú-Waka' contained six distal phalanges from the front paws of a jaguar, corresponding to a cape worn by the deceased or a pelt upon which the dead lay. Roberto García Moll reports that Tomb 3 of Structure 23 of Yaxchilan contained twenty-seven jaguar phalanges clustered in six groups, and nineteen jaguar claws in three groups were recovered in neighboring Tomb 2.[273] Jaguar phalanges were also recovered in an Early Classic tomb, Burial 110 from Piedras Negras. The single claw recovered from the royal tomb at El Zotz may have been part of a pelt associated with the sacrificed infants or perhaps a cape worn by the king (see fig. 3.74). Considering that Maya thrones are generally shown to be covered by jaguar skins, it is unlikely that such pelts had any meaning particular to the mortuary context beyond the animal's more general symbolic association with power and the exclusivity of such objects.

POLYSEMY IN MAYA MORTUARY SPACE

This chapter only scratches the surface of the complex symbolism and ritual activity evident in Classic Maya mortuary spaces. Burials were conceptualized as entrances to the underworld, points of access to celestial realms, courtly chambers, and perhaps even ovens. Other meanings have only vaguely been addressed (some of which I reserve for chapter 4). Scholars have long noted that, as entrances to the underworld, Maya graves were likened to caves and watery places. Since the Preclassic period such passages were shown as quatrefoils that contained figures such as revered ancestors and the Maize God (see fig. 3.72).[274] The prevalence of shell and marine objects within Maya mortuary contexts indicates that they may also have been understood as watery places or, as suggested by Fitzsimmons, watery surfaces, an appropriate parallel to the rebirth of the sun in the eastern sea.[275]

In this regard Maya mortuary spaces reflect broader aspects of Maya cosmology. If the earth is a turtle or crocodile floating on a great cosmic sea, it is only logical that the underworld is a watery place. It is also a place of darkness where the sun descends at the end of each day, dipping into the western sea only to rise again in the eastern sea. Symbolism surrounding the dead pertained in part to their descent at death. More important, however, was their ascent from the underworld, reborn in a fashion similar to the sun or the Maize God. For this reason red, the color of dawn and the east, was painted throughout Maya mortuary spaces and *Spondylus* was placed around the bodies of the royal dead. To facilitate that ascent Maya graves were established as axes mundi, drawing on a diverse range of symbolism such as crosses, trees, navels, mirrors, and snakes. Although every death was a source of social turbulence, noble deaths were understood as particularly disruptive to the web of relations among humans and the supernatural world. The ascent of their souls came at a terrible price. Substitute sacrifices (*k'ex*) were offered to placate supernatural beings. It seems that the celestial realms could only be achieved through the exchange of flesh and blood.

Too often, however, we focus on meaning within the grave at the expense of the protracted mortuary rites that preceded these deposits. For Maya commoners, such rites may have been relatively modest and involved display and subsequent inhumation in the very houses in which they lived. But for Maya royalty these rites were quite elaborate and would have involved display and procession of the body through the ceremonial heart of the Classic period kingdoms. To that end, we must also consider how funerary objects pertain less to the grave than to the performances that occurred prior to the final deposition. Recall the dancing dead noted in chapter 2, especially the kings clad in their full performance regalia. As dancers they processed to the grave in one grand final public display that echoed similar performances that they had enacted throughout their lives.

FIGURE 4.1. The urban character of a modern Yucatec cemetery in Campeche (photograph by author).

CHAPTER 4

THE MORTUARY LANDSCAPE

This final chapter situates Classic Maya burials within the greater landscape of Maya communities and kingdoms. As is true of most lineage-organized societies, ancestors constitute a vital organizing force for the Maya. It has become accepted wisdom that the Classic Maya used their dead to map control of land and other resources. Though perhaps true in some contexts, such a position oversimplifies how the ancient Maya articulated the world of the living to the places of souls and the remains of the dead. Fundamentally, burials and bones were not ancestors but were instead the indices by which such beings were accessed. As the following pages show, possession and veneration of those remains did not legitimize claims to *resources* but to the *ancestors* that facilitated such claims.

In the limited space here, I focus primarily on the western mortuary landscapes of Piedras Negras, Yaxchilan, and Palenque, with reference to Tikal as a point of comparison from the Central Petén.[1] I revisit the symbolic importance of heat, fire, and the sun in the Maya worldview, vital to understanding the Classic Maya mortuary landscape. Although different Maya communities may have employed the same basic ritual grammar, much diversity nevertheless existed, even among neighboring kingdoms, in the specific rites and practices pertaining to the dead. Yet within these kingdoms some common patterns are evident in graves from all social strata, illustrating how kings and commoners were united in shared ritual practice.

LIVING WITH THE DEAD: NEGOTIATING WITH THE ANCESTORS

The majority of burials recovered in the Maya lowlands have been found within or adjacent to domestic structures. The practice of burying the dead in domestic complexes has deep roots. Excavations directed by Norman Hammond at Cuello, Belize, provide some of the best evidence for early mortuary behavior among the Maya. The Cuello burials demonstrate remarkable variability in body position and mortuary offerings, yet with symbolism that foreshadows mortuary practices of the Classic period. Of the 21 burials at Cuello that date to the Middle Preclassic period, 17 (81 percent) were placed below the platforms of houses.[2] The remaining graves were located adjacent to such structures. All of the bodies were placed in relatively simple earthen graves, some with minor architectural elaboration. Offerings at Cuello included a grinding stone (Burial 123), red pigment, jade, ceramics, and vessels placed over the head of the deceased. Although the diversity of mortuary practices at Cuello may simply reveal diachronic evolution in funerary rites, it may also reflect a lack of centralization in ritual practice at this early Maya village. Similar mortuary patterns were noted at K'axob, Belize, though without the same level of diversity observed at Cuello.[3]

To understand why the Maya placed their dead within residential complexes we must pay careful attention to Maya concepts of the house, the kin group, inheritance, and, most importantly, the soul. As is evident from the preceding three chapters, both ancient and modern Maya belief and practice reflect a general concern with the departure of souls from places of the living. This seems in direct conflict with the claims of archaeologists that Classic Maya funerary practices were intended to tether souls of ancestors to the house and community.[4] Domestic inhumation in particular seems counterintuitive for a society that sought to separate the souls of the dead from the world of the living.

The logic behind domestic inhumation among the ancient Maya is likely much more complex than has generally been recognized. The Classic Maya likely practiced household burial for a variety of reasons to be explored here: the practice ensured that the souls of the dead were content with the status of their former body and prevented bodies from being disturbed, especially by enemies. Domestic inhumation also aided souls in finding their way back to the dwellings of their descendants to be reincarnated in the next generation. Most importantly, however, I suggest that domestic inhumation was an act of monopolization: a means to restrict access to the bones and burials of the dead, privileging certain individuals and families within the greater lineage.

The contemporary Maya view the souls of the dead with ambivalence. Immediately after death, newly dead souls are feared as possible sources of illness or even death, especially if they linger within the community. The Mopan vacate the house for eight days while the *pixan* (soul) wanders restlessly.[5] At the end of those eight days a feast is held to calm the restless spirit; if the ceremony is deemed a success, the house is reoccupied. Robert Redfield and Alfonso Villa Rojas indicate "the rituals that attend the dying express the wish to speed the soul safely on its heavenward journey, that it may be quit of earth and not remain below to trouble the living."[6] Similarly, as described by Pitarch in regard to the souls of the dead and dying Tzeltal: "During the body's agony, the *ch'ulel* leaves it for the last time and spends a number of days (twenty, forty) wandering about what was its home. During this period the *ch'ulel* poses some danger, because since it is always emotive, it may try to induce other *ch'ulel* it is fond of to join it. A prayer may be said that gently but firmly persuades the deceased's *ch'ulel* to take leave of this world once and for all."[7] Only after these souls depart and join other ancestral souls at places away from the house and village, in otherworldly places outside of the community, is the danger abated. Eventually the souls may be invited back, but only at certain times of the year, such as Día de los Muertos. As June Nash notes for the Tzeltal, "while souls are honored relatives on this day [Día de los Muertos], if they appear at any other time there is a great fear of them."[8]

Throughout the Maya world Día de los Muertos marks a time when relatives travel to the cemetery and provide meals and other offerings to deceased family members to ensure that they are content. In the Yucatán this is when the recently dead (two to three years) are exhumed and the bones are cleaned and fed. Redfield and Villa Rojas note that only after this act is performed "the soul is safely out of the world of the living."[9] Although Día de los Muertos is fixed in the Christian calendar, it likely has a Precolumbian precursor that corresponds to the end of the agricultural cycle. For the Tzeltal, Pitarch gives this date as the first day of the month of *pom* in the traditional solar calendar.[10]

To ensure that the souls of the dead are content, great care is taken in regard to the treatment of the body at the funeral and the selection of its burial place. As Nash describes for Tzeltal souls, "If dissatisfied in its resting place, a soul, even when properly buried, cannot find peace and wanders back to the house. Fearing this, relatives try to comply with the wishes of the dead." Often the old and dying request that they be buried with the family, "between their dead parents, but if they were unhappy with their parents, they might choose a *padrino* or uncle."[11] During our recent visit to a Yucatec community in Campeche, one local informed us of a current problem with overcrowding in the cemetery. A second cemetery had been set aside, but he suggested that nobody wanted to be buried in such an isolated, lonely place. Indeed, walking through this Campeche cemetery is like exploring the streets of any vibrant Mexican town, each crypt painted brightly like a modern house, some with windows and other architectural details (fig. 4.1).

The Classic period practice of interring the dead within the domestic complex assured that the dead were content to have their bones laid to rest in a familiar place and also guaranteed that those remains could not easily be disturbed, stolen, or violated in some other way. The captive from the sixteenth-century *Rabinal Achi*, musing on his imminent demise, laments that after his decapitation the only way in which the "bone of my crown, bone of my head" would return home was as an item of "trade for five score seeds of pataxte, five score seeds of cacao," a sizable sum that his descendants would potentially need to pay for his remains.[12] As long as an ancient Maya house complex remained inhabited, the bodies of the dead were safe from desecration and theft.

But what happened when domestic complexes or even entire communities were abandoned? People probably were not much concerned for those who had died a few generations earlier, the anonymous dead. The bones of close relatives and especially focal members of the lineage, however, may have been exhumed and taken along when the community was abandoned. The removal and transport of important bodies finds precedent on Tikal Altar 5 (see fig. 2.58). Over the years my colleagues and I have documented quite a few mortuary spaces along the Usumacinta River void of human remains (or nearly so), including graves at Piedras Negras (Burial 115), Tecolote (Burial 6 and 7), and El Kinel (Burial 11), among others (see fig. 3.25). All of these burials date to the latter half of the eighth century or early ninth century, within a generation or two of the time when these sites were abandoned. In all of these cases few if any skeletal remains were present, though some graves still contained funerary objects (for example, Tecolote Burial 6 and 7, El Kinel Burial 11). At El Porvenir, a community just north of Piedras Negras, Burials 1, 2, and 3 were located adjacent to one another in the same patio group and contained remarkably little bone material despite being earthen burials, where preservation is generally good in this region (fig. 4.2). I was able to recover only forty fragments of bone combined, including a nearly complete set of teeth in Burial 1. The presence of human bone in each mortuary space indicates that they all once contained bodies.

Chapter 2 shows how bones served as conduits for communication with the souls of the dead. Domestic inhumation ensured that the souls of the dead did not linger and molest the living, yet their remains (or their graves) could be used to conjure ancestors as needed, as evident in contemporary Maya Día de los Muertos rituals. Domestic inhumation may also have facilitated the inheritance of souls. Recall that the contemporary Maya perceive children as ancestors reborn. Susan Gillespie calls attention to the importance of the inheritance of souls.[13] Although Gillespie may have erred in her suggestion that the Maya meant to keep the "spirits in place," she aptly suggests that by situating the dead

FIGURE 4.2. Nearly empty graves from El Porvenir, Guatemala (Burials 1, 2, and 3) (photograph by author). No bone was removed prior to the photograph being taken.

within the domestic complex the Maya were facilitating the process of *k'ex* (newborns as replacements for dead relatives who inherited their souls). According to Evon Vogt, souls, like names, are passed through the patriline at Zinacantán: a "new member of the patrilineage is *k'esholil* ('replacement' or 'substitute') in the system for the departed ancestor."[14] The concept was employed by Classic period kings when they chose the regnal names of auspicious ancestral kings, especially grandfathers, upon their accession.[15] The concept is also resonant in Classic period depictions of the Maize God resurrecting from a subterranean skull; new life is reborn from the old (see fig. 1.12). In this sense the bones of the dead may have served as conduits or beacons, drawing ancestral souls back to the homes of their lineage descendants.

Aside from *k'ex* and the contentment of dead souls, the most important factor relating to domestic inhumation may have been issues of inheritance and control of resources. Much of the Classic Maya literature on dead bodies, ancestors, and property emphasizes competition among lineages. This idea has been eloquently developed by Patricia McAnany in her influential volume *Living with the Ancestors*.[16] McAnany's argument is inspired by Arthur Saxe's proposition that corporate groups maintained communal disposal areas for the dead in a given society when resources were attained or legitimized through lineal descent.[17] According to McAnany, "ancestor veneration, through lineage organization, charted and legitimized resource rights, through the mechanisms of oral memory, written records, and, most importantly the continued physical presence of buried ancestors in domestic complexes which were, in effect, a type of *domestic mausolea*." In this sense, "ancestors 'slept' within the construction mass of residential compounds—to insure the chain of continuity in resources as transmitted between generations." For McAnany, the location of the ancestors was essential; they were effectively "markers of place." She situates the origins of this process in the Preclassic period, such that by "400 BC, many Maya residences were assuming the characteristics of a domestic mausoleum. In effect, ancestors come to symbolize the coalescence of lineage and locale."[18]

What is left underdeveloped in McAnany's model is the mechanism by which placing ancestors within the house ensured a chain of inheritance. The most important heritable resource for the Maya, both ancient and modern, was agricultural land. If Classic Maya bodies were strategically placed to legitimize claims to land resources against other kin groups, as suggested by McAnany, they should be situated in shrines across the landscape, much like the Chinese ancestral tombs that she cites in her study, and not located within house groups.[19] It is possible that some Classic period Maya were buried in the milpas and have been largely missed by archaeologists, who excavate primarily in and around architectural complexes. At the beginning of the eighteenth century Francisco Ximénez observed Maya burying their dead in the milpa.[20] Similarly, in a rural community in Chiapas near where I currently work, the dead are buried at the edge of town, where the cluster of houses ends and the fields begin. Nevertheless, if Classic Maya ancestral shrines were located within the milpas, surely there should be some evidence for such constructions. Instead, McAnany raises the possibility that "the conflation of the ancestral shrine with the physical remains of ancestors within the residence may be partly a result of the importance of near-residential fields and orchards in lowland tropical areas."[21] Certainly, good evidence indicates the linkage between ancestors and trees. Pakal's sarcophagus shows the ancestors as fruiting trees, as many have noted, and the Tz'utujil of Santiago Atititlán plant trees over the graves of the dead in the town cemetery.[22]

The problem, however, is that we must assume that the Maya perceived a spatial connection between the remains of ancestors and the resources that they legitimized. As Ian Morris points out in his critique of Saxe's hypothesis, "ancestral cult can be, and often is, carried out at the grave side, but equally it can be, and often is, physically distinct, and even archaeologically invisible."[23] If we look to the Maya today, this is indeed the situation. Venerative acts are essential for claims to resources, yet the graves and bones that are accessed in rites of veneration are spatially removed from the land and resources that are being negotiated. Instead the kin either travel to the cemetery to perform their rites of veneration on Día de los Muertos or bring the bones to the family house for the celebration.

Therefore domestic inhumation probably did not map access to resources but instead facilitated control over the rites of veneration. By burying the dead within the domestic complex or, more accurately, possessing the home in which the focal ancestors lay, family heads had a direct claim to the remains of the dead that kin who lived outside of that house or house group simply did not have. Competing claims to the bones of a focal ancestor would only have been relevant against other members of the same lineage, people who could also claim descent from the same ancestors and introduce competing claims to resources. Thus the Classic Maya practice of locating the dead within domestic complexes says little about competition *among* lineages. Rather, it was a mechanism to legitimize claims to resources *within* the lineage. McAnany acknowledged the likelihood of this within-lineage competition in her own model, noting that "ancestor veneration in particular is not a practice that promotes social equality; rather, it promotes and perpetuates inequality and alienation from resources *within* the household as well as the polity. The principle of first occupancy gave preferential access to land to certain lineages and *within lineages to certain families*" (emphasis added).[24]

Ethnography highlights the tensions involved in land inheritance within contemporary Maya communities. As Pitarch explains of the Tzotzil of Cancuc:

> The various domestic groups that compose a lineage (in reality its male heads) do not have to be physically close to each other or even in the same hamlet, but

they do share the same lands or group of allotments (*jun lum k'inaltik*) that have been inherited from the "grandfather." Most of the time, land is not divided up, so each domestic group works it according to its needs and capacity—an arrangement that is a *never-ending source of antagonism* between heads of family. "Uncles" with the most sons prefer that the land remain undivided, while those with fewer sons would rather it were carved up; and in the case of such a division (the definitive settlement of boundaries can give rise to feuds that can last for generations), the allotment's subsequent sale to any other Cancuquero becomes possible. There are times when a father, getting on in years and unhappy with his offspring's attitude, simply sells the land for money or a continuous supply of corn and beans for the remainder of his life [emphasis added].[25]

By locating the houses of the living in association with the graves of the dead, Maya families gained immediate monopoly over access to those remains and the rituals performed at the grave. For the Tzeltal described by Pitarch, it is during these rites of veneration at the beginning of the month of *pom* when the souls of the dead return to the earth and families gather to offer food for the dead. At this time lineage members have "an opportunity to meet and decide which plots of land are going to be farmed by each domestic group during the coming season."[26]

As a point of ethnographic comparison, today the cost of funerals in Ghana, which are often quite elaborate and expensive, is expected to be paid by the family. Precisely who is "family" is negotiable: by paying for part of a funeral, claims to property are established. As Marleen de Witte explains:

> Family membership is widely recognized as a legitimate basis for making claims to family property, but because kinship is negotiable, access to property is just as subject to negotiation and contest. It is usually accepted that those who share in the debt are entitled to part of the inheritance. Funerals are so important then, because in coming together to honor the dead and share funeral expenses, people assert claims to family membership and the right to share in the enjoyment of family property. A funeral debt binds the *abusua* [matrilineal kin group] together.[27]

Although the public and elaborate funerals in Ghana are opportunities for the matriline to display its status and wealth, the actual organization and payment of that funeral is a mechanism for members of the kin group to lay claim to property within the matriline against other members of the immediate kin group. For the Classic Maya, mortuary rites and subsequent acts of veneration may have been an opportunity for the lineages to display their solidarity. Yet these events also would have offered an occasion for the kin to gather and renegotiate claims to heritable resources possessed by the lineage.

The process was likely organic, beginning with the relatively egalitarian communities of the Preclassic period, as McAnany's model suggests. We should not conceive of these early Maya farmers as scrabbling over the bones of their ancestors; they likely hassled over houses or house groups that contained the ancestral remains. To live with the dead and care for their graves was to maintain an intimate link to those ancestors. In the United States siblings sometimes argue that greater investiture in the care of an elderly parent legitimizes the claim to inheritance once that parent dies. For the Classic Maya, raising children and grandchildren among the graves of the ancestors likely made it easier to claim that offspring inherited the souls of those ancestors and were thus appropriate inheritors of their land and resources. As my Greek wife points out, the tradition in her home country is for the firstborn grandson to be named after his paternal grandfather. This nominal connection inevitably affords that grandson an edge when it comes to inheritance from the grandfather.

William Haviland's excavations in a nonelite residence group at Tikal (Group 2G-1) are informative in this regard (fig. 4.3).[28] He found that one of the structures, 2G-59, was the oldest, most complex, and most heavily modified and contained the most burials ($n = 12$) relative to other structures in the house group. Moreover, this building was situated on the eastern side of the patio group, a direction that was often associated with burials and ancestors at Tikal (see the discussion below).[29] Haviland suggests that each episode of remodeling of 2G-59 resulted from the death of the household head and a shuffling of residences as the new family head moved into the principal house. Based on ceramic chronology, the group was occupied from the late Early Classic period (late Manik phase) through Terminal Classic times (Eznab phase), circa AD 550 to 900. Remodeling and expansion of the primary structure indicates that it remained in use throughout the history of the group and that the growth of the lineage prompted the construction of additional structures. The final tally of five buildings, however, would have been insufficient to accommodate the growth of this kin group over more than three centuries. Quite simply, only some descendants were able to remain each generation; others had to move on and establish new domestic complexes elsewhere at Tikal or beyond. Those that remained inherited the house and the ancestral graves. Notably, the oldest burial in the principal house, Problematic Deposit 64, is a secondary inhumation, a disarticulated skeleton that either was temporarily exhumed and then put back into place or was a bundle of bones moved there from another location.

Haviland suggests that these skeletal elements are the remains of the founder. Perhaps his bones were transported to the structure when the house was established by a descendant, or maybe his skeleton was exhumed periodically, as in the venerative rites performed in the Yucatán today. Acts of veneration were likely performed at this house group if not in the house itself. Monopolizing access to those remains must have guaranteed or at least favored 2G-59's residents' claims to ancestral property against other members of the kin group.

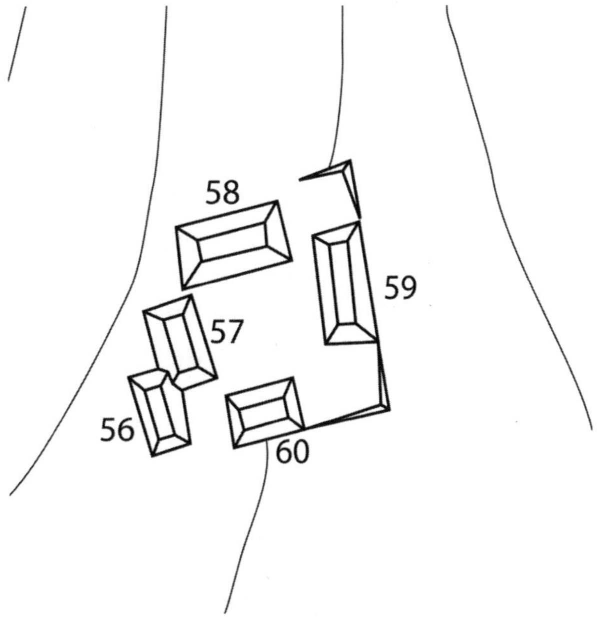

FIGURE 4.3. Tikal Group 2G-1 (drawing by author after Carr and Hazard, *Tikal Report No. 11*).

At Tikal and some other Central Petén sites some lineages constructed shrines to house the bodies of their dead. This arrangement, dubbed Tikal Plaza Plan 2, has been thoroughly investigated by Marshall Becker, who describes the placement of the focal burial as the "opening of the earth (bedrock) for the grave and the covering of it with a tall structure that is equivalent to a mountain (the enclosure of most caves). The cultural rule is that all subsequent burials need only penetrate the tall structure and not extend down into the original bedrock, since the structure now is the cognitive equivalent of a mountain."[30] The placement of shrines on the eastern side of domestic groups may relate to the concept of solar rebirth. Becker suggests that burned floors at these locales are evidence of the ceremonial use of fire. Arlen Chase and Diane Chase have suggested that as many as 60 percent of the residential groups at Caracol conform to Plaza Plan 2.[31] Only about 14 percent of the residential groups at Tikal include the eastern shrine.[32]

The domestic ancestral shrines at Tikal and Caracol reflect increasing energies directed at rites of veneration over the course of the Classic period, presumably as populations grew and competition for limited inherited resources escalated. Although the domestic landscape of the Maya Preclassic period remains poorly understood, the evidence from K'axob and Cuello noted earlier suggests a lack of formalization of funerary rites. Variability in the presence or absence of ancestor shrines at Tikal, from one domestic group to the next, may reflect different lineage customs or could relate to the temporal depth and the extent of growth of some lineages relative to others. The construction of domestic ancestor shrines may also have been triggered by increasing inequality within society, as older lineages claimed primacy of land against younger lineages.

The processes that I am suggesting here are partly informed by the tensions that I have witnessed in modern communities near the Usumacinta River in Guatemala, which are useful for thinking about how kinship and community intersect in regard to the negotiation of land and resources. When many of these modern communities were first settled in the mid- to late twentieth century, land was held collectively by the community. As these towns have grown in the past two decades, some community members have advocated that land be divided into family plots, advantaging their interests and those of their kin against the greater needs of the community. Private, family-held land can be sold to community outsiders, which is impossible if it is held collectively by the village. The dimensions of land inequality (and therefore wealth) in the region have become quite complex. Within families, elder sons generally have primacy in terms of access to land. Between families, older more established families have the choicest land to farm. Even between communities, older towns have the best territory and younger towns must settle for lesser territory. In Mexico impoverished settlers without land of their own simply work as hired help. Wealth is directly tied to the length of time of land ownership.

Although J. Eric Thompson notes that some Maya communities still buried their dead below the house as recently as the early twentieth century, most Maya communities abandoned domestic inhumation in favor of burial in Catholic churches and cemeteries not long after the conquest.[33] Nevertheless, ancestor veneration remains important to the fabric of Maya society. We could argue that the relative rapidity by which the Maya quit domestic inhumation reflects the strong influence and will of Spanish friars. Yet many contemporary Maya beliefs and practices pertaining to the dead are decidedly non-Spanish in origin. I would suggest instead that the Maya relinquished domestic inhumation because it was secondary to more important aspects of ancestor veneration. As long as family heads could utilize other means to monopolize rites pertaining to their focal ancestors, it was feasible to shift burial from below the house to underneath the church or in the cemetery.

A parallel was observed by William Douglass in his study of funeral rites in a Basque village in the mid-twentieth century.[34] The family *brasseria* (farmstead) is inherited by only one child and is indivisible. Each estate is associated with a *sepulturie*, a symbolic (though once functional) burial plot on the floor of the village's church. Today corpses are buried outside of the church and receive little subsequent attention. The focus of veneration is on the *sepulturie*, which may leave family hands if the estate is sold. Primacy is not given to the body but to a symbol of veneration that could be publicly monopolized. Among the Maya today, ancestors and other supernaturals are invoked by erecting crosses in the fields and by offering sacred substances at shrines throughout the landscape.[35] Yet these rites and locations are easily transferable and duplicable among various members of a lineage. In contrast, graves and bundles of bones, like the Basque *sepulturie*, are indivisible and inalienable and can be monopolized by a single family.

FIGURE 4.4. (*opposite*) Map of La Venta, Mexico (drawing by author after González, "La Venta," fig. 1).

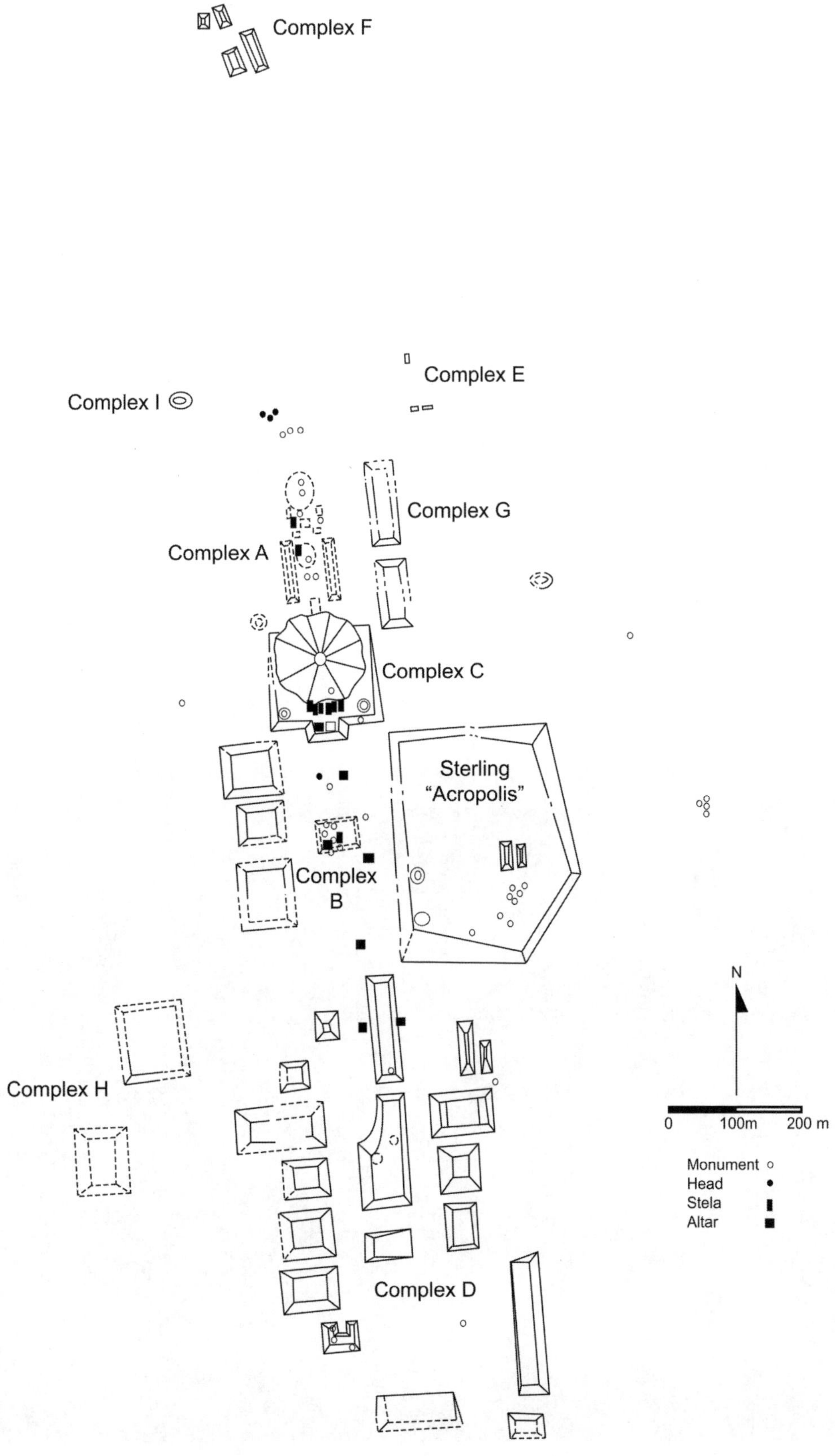

Complex F
Complex E
Complex I
Complex G
Complex A
Complex C
Sterling
"Acropolis"
Complex
B
N
0
100m
200 m
Monument
Head
Stela
Altar
Complex H
Complex D

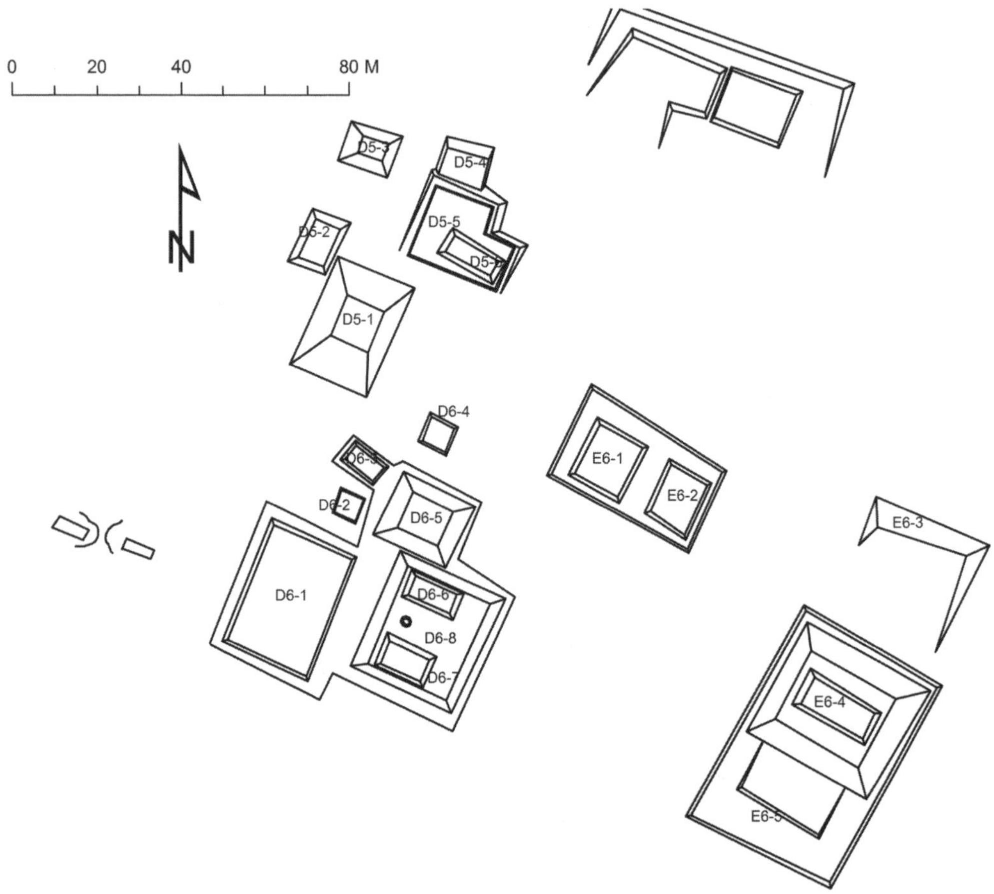
0
20
40
80 M
N
D5-3
D5-4
D5-5
D5-2
D5-1
D6-4
D6-3
D6-2
D6-5
D6-1
D6-6
D6-8
D6-7
E6-1
E6-2
E6-3
E6-4
E6-5

Domestic inhumation was not an act of place-making but a means to heighten the inalienability of ancestral remains in order to minimize contesting claims to resource and inheritance.

THE POLITICAL AFTERLIFE

The practice of royal ancestor veneration emerged from these domestic roots.[36] The bodies of dead kings and queens were vital to the negotiation of authority throughout the Classic period. Placed in monumental pyramids and commemorated with massive stone sculptures, royal bodies populated the civic-ceremonial hearts of Classic period kingdoms, referenced by the living king in rites that legitimized claims to the throne through their ancestral connections. As McAnany explains, the elite subverted "ancestral veneration from a practice that linked family and lineage to landholdings to one that validated the semidivinity of the royal lines, legitimized systems of taxation and tribute, and in general sanctioned kingly prerogative."[37] Certainly the use of funerary monuments and other memorials as political capital is not unique to the Classic Maya. If anything the practice has become more pervasive as modern nation-states populate their coins, public parks, and street signs with images and names of dead state officials.

In Mexico in the late nineteenth century state funerals and commemorative festivals were essential for the construction of a Mexican national identity during the rule of Porfirio Díaz.[38] Inversely, the dismantling of statues and the removal of the corpses of Communist leaders from public display were among the first post-socialist reforms following the demise of the Soviet Union.[39] Mortuary practice is political practice at the level of the state. We tend to focus on how the tombs of Maya kings and queens of the Classic Maya demonstrate the immense authority and power of these individuals.

FIGURE 4.5. (*opposite*) Map of Rancho Búfalo, Mexico, and Burial 1 located in Structure D5-3 (map by Jeffrey Dobereiner, Charles Golden, and Bryce Davenport; photograph by author).

More importantly (and interestingly), we must consider what these contexts reveal regarding social and political negotiation and even competition that followed in the wake of a royal death.

The Olmec site of La Venta provides some of the earliest evidence for public, centralized mortuary rites by Mesoamerican elites. High-status tombs and representations of four ancestors as colossal heads were located in Complex A to the north of the principal pyramid (Complex C) (fig. 4.4; see also fig. 1.7c). The northerly placement of the mortuary precinct presages the layout of some Maya Classic period communities.[40] At the Preclassic Maya site of Rancho Búfalo we have identified a similar pattern: a tomb located in a shrine (D5-3) north of a larger pyramidal building in the civic ceremonial center of the site (fig. 4.5). Recent excavations at Chiapa de Corzo have demonstrated that elite tombs were placed within artificial mountains that were used for inhumation at least as early as 700 BC.[41]

Classic period royal courts successfully claimed power and authority for five centuries through a careful balance of highly public ritual performances combined with secretive acts performed in exclusive, restricted contexts.[42] Royal mortuary rites were both. They were public events that united the polity around the veneration of focal figures who were memorialized by highly visible mortuary monuments. Yet royal funerals also had a more intimate dimension: proximity to the dead king was presumably indexed by those who were allowed to enter (or reenter) his tomb or to erect sculptural memorials in his honor.

A careful reading of the epigraphic record provides multiple examples of mortuary rites used to legitimize claims to the throne, good evidence of intrapolity turmoil and competition recorded in sculptural text and image. Although the Classic Maya situation was likely less chaotic, the comparative case of the Byzantine Empire reminds us that status in a particular political office may be ephemeral and perpetually at risk even though the office itself may be stable and enduring. The office of

FIGURE 4.6. Temple O-13 and Panel 3 of Piedras Negras (illustration by Heather Hurst and Mark Child, photograph by Charles Golden).

autokrator of the Roman Empire (emperor of Byzantium) persisted for over a millennium, yet roughly half of the men who held that title were overthrown and forcibly deposed.[43] It seems that much of the competition for the throne among the Maya was the result of competing claims from within the same lineage, whether from a legitimate usurper or from an outsider who married into the royal lineage. This may have been the case with the so-called *entrada* at Tikal, where an apparent lord of Teotihuacan, Spearthrower Owl, married a local Tikal woman and had their son installed as king.[44] We have also the well-known example of royal intralineage rivalry in the seventh century that escalated to full-out war between Nuun Ujol Chahk, *k'uhul ajaw* of Tikal, and Bajlaj Chan K'awiil, his possible half-brother and founder of the dynasty at Dos Pilas. Both lords claimed the same title, *mutal k'uhul ajaw*.[45]

Amid this tumultuous political landscape, strategic memorialization of dead kings and queens was vital for newly seated kings. When Ruler 7 of Piedras Negras took the throne following the apparent abdication of Ruler 6, among his earliest acts was to venerate Ruler 4 by opening his tomb (Burial 13), located in front of Temple O-13 (fig. 4.6).[46] Although Temple O-13 originated as an Early Classic period construction, it was refashioned into the mortuary shrine for Ruler 4 in the mid-eighth century, its new design mimicking the Temple of the Inscriptions at Palenque (compare with fig. 4.7). The renovation of Temple O-13 was likely commissioned by his successor, Ruler 5. Ruler 4 was apparently a successful king, though we do not understand what (if any) familial relationship existed between him and the subsequent three kings. Rulers 5, 6, and 7 all made Temple O-13 the ritual focus of their reigns, placing the majority of their known monuments (fifteen in all) on or adjacent to the building.[47] The intensity of ritual activity at this site is further evident in the fifty-seven caches that have been recovered from the structure, all postdating Ruler 4's tomb.[48] Through this intense ritual activity the subsequent kings sought to link themselves and their reign to that of Ruler 4. Ruler 7's veneration of Ruler 4 is described on Panel 3, which depicts Ruler 4 seated on a throne in one of the finest sculptural scenes known for the Classic Maya (fig. 4.6). One year after Ruler 7's accession he opened the tomb of Ruler 4 in AD 782 and performed an *el naah umukil* (house burning at the burial).[49] The very fact that Ruler 7 bothered to record this particular event in stone, while failing to reference either of his predecessors, underscores both the importance of ancestor veneration in legitimizing Maya kingship and the potentially contentious nature of such acts.

Certainly there is precedent for the use of royal mortuary rites to mitigate competing claims to the throne among other societies. As Peter Metcalf and Richard Huntington note, traditional Thai royal mortuary rites served to "refocus attention on the center . . . mortuary rites provided a major opportunity, if not the major opportunity, for ritual display, they also occurred at a dangerous moment of transition." Just as the death of a Thai royal is fundamentally a disruption, "paradoxically, the funeral could be used to express the continuity of

FIGURE 4.7. *(Left to right)* The Temple of the Inscriptions, Temple XIII, and Temple of the Skull (Temple XII) of Palenque (photograph by author).

kingship." Proximity to the mortal remains of Thai royalty was one means by which successor kings legitimized their rule in the long term. As Metcalf and Huntington explain, the cremated remains of royalty became "royal relics that reinforced the centripetal tendencies of the kingdom . . . they formed a kind of charismatic stockpile, distilled from the genius of ancient kings."[50] For most Maya kingdoms, proximity to those sacred remains was advertised through the funerary pyramid. In kingdoms like Piedras Negras burial reentry afforded an even more tangible connection to the dead kings.

Even when succession was assured, mortuary rites and ancestor veneration were employed by Maya kings to calm the turmoil and anxiety caused by the death of an important king and the accession of a new, untested ruler. The death of K'inich Janaab Pakal after a remarkable sixty-eight years of rule at Palenque is a fine example. Few if anyone within the kingdom at the time of his death would have remembered a Palenque that was not ruled by this great king. Although we might assume that K'inich Kan Bahlam's status as the son of a revered and successful king would be a great advantage, comparison to similar events in global history suggests that it is equally likely that Kan Bahlam came to the throne amid a cloud of immense expectation and uncertainty. A comparable situation is currently unfolding in Thailand, where Bhumibol Adulyadej has reigned since 1946. He is a much loved king there, and the populace has significant concern about the capacity of the crown prince, Maha Vajiralongkorn, to follow in his father's footsteps.[51] It is little surprise that Kan Bahlam remade the ceremonial center of Palenque in honor of both the patron deities and Pakal, constructing what is arguably the greatest funerary pyramid (the Temple of the Inscriptions) and ritual precinct (the Cross Group) in the Maya lowlands (see fig. 4.7).

Throughout the world monumental funerary shrines like Temple O-13 and the Temple of the Inscriptions

C-Group
Northwest Group Plaza
West Group Plaza
Burial 5
Burial 13
East Group Plaza
USUMACINTA RIVER
South Group Court
South Group Plaza
Burial 10
N
0
m
200

FIGURE 4.8. (*opposite*) Map of Piedras Negras, Guatemala, including location of the Late Classic period tombs (map courtesy of Stephen Houston).

FIGURE 4.9. (*above*) Piedras Negras Burial 10, located within the South Group Plaza (photograph by author).

operate not only to commemorate the dead but also as stages for the social and political performances of their successors and later generations. The Lincoln Memorial in Washington, D.C., one of the most hallowed spaces in the United States, has repeatedly been chosen as the site for important events in the history of the civil rights movement. It is where Martin Luther King Jr. delivered his "I Have a Dream" speech and President Barack Obama hosted his inaugural celebration in January 2009. In both cases Lincoln's memory was explicitly invoked. In a similar fashion massive plazas front Temple O-13 and the Temple of the Inscriptions in order to ensure that hundreds if not thousands of people could witness spectacles performed on their steps and summit.[52]

The location of royal tombs in large plaza spaces at Piedras Negras ensured that these sacred sites could easily be accessed for later acts of veneration, according well with the abundant evidence for such practices in the hieroglyphic record of the site and presumably facilitating a sizable audience at such events (figs. 4.8, 4.9).[53] The burial space of Ruler 3, however, was decidedly different from the one used for Ruler 4. Ruler 3's tomb, Burial 5, is located in a more secluded location at the site, below the floor of the J-5 terrace, in front of the stairs leading to Structure J-8, a range structure that seems not to have served as a memorial (fig. 4.10). No large audiences could have surrounded his grave, and no funerary pyramid was raised to mark the place. Despite the abundant textual evidence for tomb reentry at Piedras Negras, Burial 5 was never reopened; archaeological evidence does not indicate any disturbance and the historical record does not mention any postmortem acts of veneration. Ruler 4 did not reference his predecessor, much less summon his soul for advice or judgment. Instead Stela 40 shows Ruler 4 performing an act of veneration at the tomb of a

FIGURE 4.10. Piedras Negras Burial 5 within the J-5 Terrace (photograph by author).

noble woman. This act was significant enough for him to memorialize it on a massive stela (see fig. 2.39). It would seem that Ruler 4's claim to the throne derived not from Ruler 3 but from this woman. Indeed Ruler 4's patriline is not mentioned in the extant epigraphic record from the site. Ruler 3 was certainly given a funeral fit for a king, judging by the richness of his tomb. As a final telling clue, a shell cloak worn by Ruler 2 on Panel 15 was apparently retired with the death of his son, Ruler 3, and placed in the northeast corner of the tomb. It may be that Ruler 4 intentionally discarded objects connected to those two earlier kings in order to shift focus away from them and instead emphasize his own claims to the throne through his affiliation with the unknown noble woman on Stela 40.

Burial 10 presents a further enigma in regard to the veneration of kings at Piedras Negras (see fig. 4.9).[54] The burial presumably dates to the Late Classic period, as indicated by ceramics found within the chamber's niches. The tomb is located within the South Group Plaza (south of the South Group Court), in front of Structure U-3, in an especially open and accessible place. Burial 10, like Burials 5 and 13, is located below a plaza/patio floor. What is distinctive about Burial 10, however, is that the majority of its contents were removed when it was reentered. The dating of this event is unclear, though the near absence of the remains of the primary occupant, aside from a single molar, suggests that the objective was to remove this person's remains. Burning on the floor of the chamber likely reflects rites enacted at this time. It is unclear why these remains were removed. Perhaps the grave was exhumed when the site was abandoned. We do know that some royal paraphernalia left Piedras Negras at some point in its history, as evident in the jade head recovered from the cenote of Chichen Itza, carved in the likeness of Ruler 3.[55] Alternatively, the tomb may have been opened and the bones discarded in an act of desecration by enemies of Piedras Negras or by much later by visitors to the site, after the city had been abandoned and its kings long forgotten.

We should also consider Maya funerary pyramids and other monumental constructions as testimony to the leadership acumen of Maya dynasts and to the civic engagement of the greater populace. Consider, for example, the mausoleums built by the Berawan people of Borneo in the early twentieth century. These funerary monuments were commissioned when a new leader came to power but were not built for him or even for the preceding leader. Rather, the new Berawan leader had the mausoleums built to establish his authority and demonstrate his capacity to lead. As Peter Metcalf and Richard Huntington observe, "what the new leader requires to cement his position is a demonstration of community solidarity behind him personally. The building of mausoleums provides an opportunity for such demonstration." Thus the most elaborate mausoleums were not necessarily built for the most important people but instead for relatives that "happened to die at the right moment." As Metcalf and Huntington explain, by "honoring them with a mausoleum, the leader ennobled himself."[56]

The construction and dedication of the Temple of the Inscriptions was among the most important early political acts of Kan Bahlam II of Palenque. Such efforts demonstrated that recently enthroned Maya kings could effectively manage the will of the people and channel their energies into monumental constructions that served not merely as a memorial to a dead king but as testimony to the vitality of the kingdom. All too often archaeologists treat ancient monuments as evidence of aggrandizement, subordination, and inequality. Yet, as Charles Golden and I have recently argued, we must also consider how laboring on a massive public construction cultivated trust among community members and between the community and their king. Pakal was described posthumously in the texts of Palenque as "he of five pyramids," a reference not to his clever architectural abilities but to his capacity as a leader and manager, capable of organizing the labor force of the kingdom around important civic constructions.[57]

Even in recent history the pomp of state funerals and memorials was intended to create societal solidarity around the nation-state. Matthew Esposito explores this issue in regard to Mexico in the late nineteenth century and observes that "Porfirian political culture involved a high level of cross-sectional appeal to keep all citizens emotionally invested in state initiatives. During funerals and commemorations, the state encouraged citizens to remember an official version of the past, appreciate the present, and visualize the future of their national community."[58] For tyrannical rulers, such memorials may have an air of coercion. Yet not all rulers are despised by their people. After the death of a successful and long-lived king like Pakal, the populace of Palenque was likely more than willing to raise a spectacular pyramid in his honor. As Elliot Abrams has shown, the construction of Maya monumental buildings is not as labor intensive as we might imagine. Most constructions could have been accomplished in less than a year by rotating work crews laboring in shifts of a few weeks.[59] Much of the work was unskilled and involved hauling and piling uncut or roughly shaped block. Nevertheless, the construction of pyramids did require coordination and trust to ensure that the structure was properly built and that people were not injured or killed in the construction. By raising the Temple of the Inscriptions Kan Bahlam II quite literally built the foundation of his reign.

KING, COMMONER, AND THE MORTUARY COMMUNITY

Past scholarship on Maya mortuary practices has emphasized the alienation generated by royal funerary monuments. As McAnany notes, "I view the realms of kinship and kingship as an arena of conflict."[60] In terms of labor, scale, opulence, and sheer physicality of the mortuary space the interments of Maya royalty no doubt were fundamentally distinct from the graves of the rest of society. Yet a divisive mortuary system would do little to organize the community around the authority of the court. In describing Huron funerary ceremonies, Robert

Hertz famously observed the importance of mortuary rites as socially and political constructive:

> We find here, in a striking form, a phenomenon already observed among the Indonesians: the final ceremony always has a pronounced collective character and entails a concentration of society. But here it is not the family or even the village, but the nation that intervenes directly to re-integrate the dead into social communion. This action thus takes on a *political significance*: by dealing with all their dead in common the various domestic and local groups that form the higher unity become conscious of, and consequently maintain, the ties that unite them. In establishing a society of the dead, the society of the living regularly re-creates itself (emphasis added).[61]

Here Hertz is describing collective interment, something not widely practiced among the Classic Maya, though the mausolea of Caracol are a notable exception.[62]

Nevertheless, Hertz's observation has wider import when we consider how shared mortuary practices build solidarity across different segments of society. After Abraham Lincoln's assassination, the tour of his body was important to the apotheosis of the president, now one of America's most revered civic patron saints. The journey of Lincoln's corpse, coupled with the tragedy of the Civil War, helped give birth to the distinctly American practice of death: embalmment. This practice has become the norm in the United States yet is absent in most other parts of the world. Commenting on this and other aspects of Western funerals, Metcalf and Huntington observe that "variation in the form of death rites between classes within one nation (aside from sheer cost) is less noticeable than variation from country to country. This relative uniformity may indicate that, insofar as the nation is a solidary community, it needs to share a system of beliefs and practices."[63]

Solidarity through shared mortuary rites is especially salient among the Classic Maya, who placed the bodies of kings and commoners throughout the community; the living landscape and the mortuary landscape were one and the same. Funerary rites and ancestor veneration triggered the gathering of society. A dead family member brought together the kin, a dead king required the attention of the kingdom. In the western Maya lowlands, as each kingdom developed distinct material culture and practices, common mortuary rites reflect social cohesion within each polity.

This unity of practice is evident, for example, in decisions regarding the placement of ceramic vessels within the graves of western Maya polities (see chapter 3). Of a sample of 99 Late Classic period graves at Piedras Negras, only 13 (13.1 percent) contained ceramics. Even the royal tombs only contained one (Burial 10, 13, 82) or two (Burial 5) vessels (see figs. 3.31, 3.35). The lack of ceramics is evident elsewhere in the polity. Only one of the eight burials excavated at Budsilha to date contained ceramic vessels. Obviously the absence of ceramics in these graves was not due to "pottery impoverishment" but reflects the distinct beliefs, traditions, and practices of the kingdom.

A majority of burials at Palenque (39/51, or 76.5 percent) contain ceramics, usually no more than one or two vessels placed near the feet of the dead. This trend holds for royal interments. The tomb of Pakal contained only four ceramic vessels and the Red Queen's tomb had three. This situation is in marked contrast to the royal tombs at Yaxchilan. Although Tomb 2 from Structure 23 contained only five ceramic vessels, Tomb 3 had thirty-four, more ceramics than the entire mortuary sample at Piedras Negras combined. Of a sample of sixteen Late Classic period burials from the Yaxchilan subsidiary centers of El Kinel and Tecolote, eleven (68.8 percent) contained ceramic vessels (see figs. 2.45, 2.46, and 3.25). Typically, only the graves of children are devoid of ceramics. When at least one vessel is present in graves of the Yaxchilan polity, it is always a perforated dish found over the head or in the center of the grave. Additional ceramics included cylinders and bowls placed near the head. As noted in chapter 3, perforated dishes (common

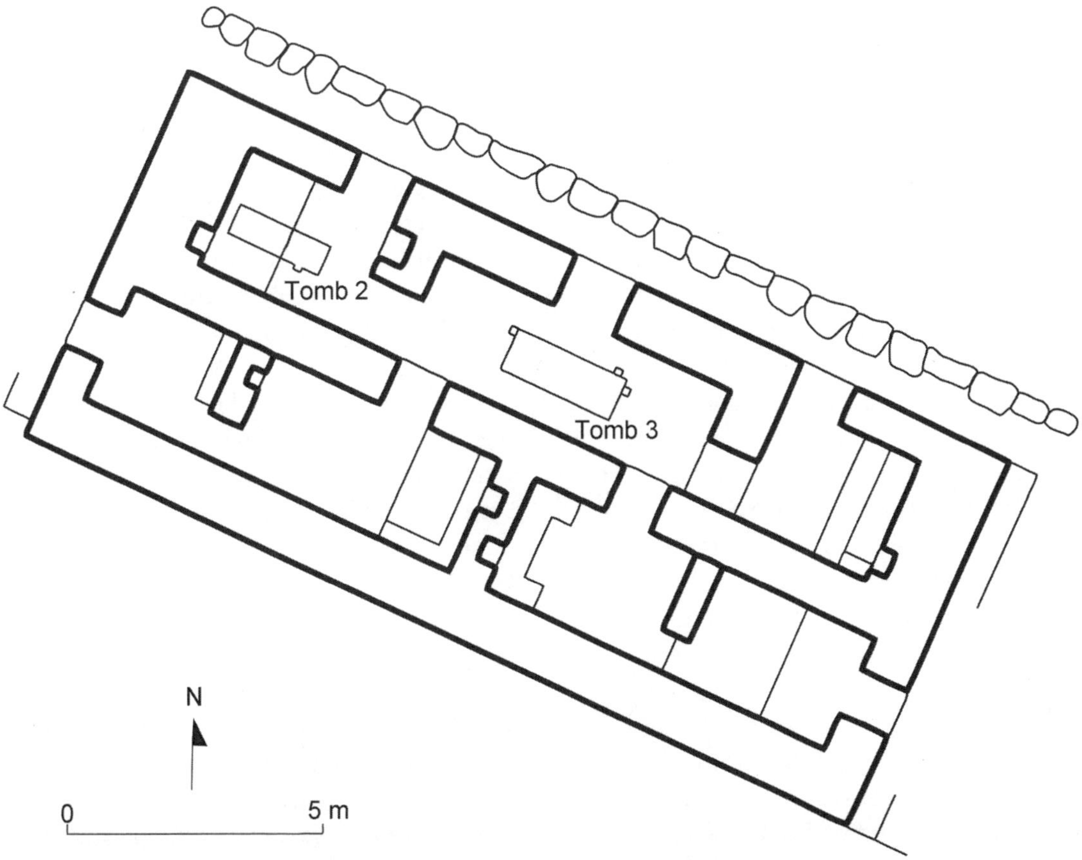

FIGURE 4.11. The location of royal tombs within Yaxchilan Structure 23 (drawing by author after García Moll, "Shield Jaguar and Structure 23 at Yaxchilan," fig. 95).

at sites in the Central Petén and in the Pasión drainage) have never been recovered at Piedras Negras or Palenque to the best of my knowledge. The placement of bowls and cylinder vessels near the head of the corpse in Yaxchilan-affiliated graves contrasts with the tendency to place them near the feet or sides of bodies in burials at Piedras Negras and Palenque.

A similar polity-specific pattern exists in regard to burial orientation. All of the reported royal tombs at Yaxchilan are oriented 120° E of N. The heads of the deceased were found to the southeast, including the bodies recovered from the tombs below the floor of Structure 23, likely those of Shield Jaguar III and his wife, Lady K'abal Xook (fig. 4.11).[64] Although data are still forthcoming, a review of the preliminary site reports indicates that the majority of Yaxchilan's burials were also oriented 120° E and N.[65] At the subsidiary centers of Tecolote and El Kinel twelve burials out of a sample of seventeen (70.6 percent) were oriented to 120° E of N. In contrast, at Piedras Negras all of the Late Classic period royal burials are oriented 30° E of N. From a sample of seventy-four nonroyal Late Classic burials, forty-eight (65 percent) followed the same northeast burial orientation. The next most frequent orientation is north ($n = 6$); only four burials were oriented to the southeast, in a fashion similar to the Yaxchilan burials. Only three out of sixty-three burials at Palenque (4.8 percent) are not on the roughly 12–18° E and N axis orientation that was dominant at that site. All of the known Palenque royal interments are oriented on this ritual axis.

Burial orientation and the placement of ceramic vessels are merely the most archaeologically salient examples of diversity in mortuary rituals among neighboring Maya kingdoms. These polity-specific patterns observed for burial orientation and the placement of ceramic vessels united kings and commoners, not only within the polity capitals but across the landscape, including communities at the very margins of the kingdom. Other practices exist, however, that are distinct for elites, such as the location of royal interments. Palenque kings and queens were interred within pyramids. At Yaxchilan nobles were placed below or in front of three-doored temples. At Piedras Negras the kings were entombed below patio and plaza floors.

PATH OF THE SUN

It would be absurd to presume that the Maya oriented their dead merely to profess their political allegiance. Rather, the orientation of bodies reveals a common ideology, belief, and practice that presumably originated at the polity capitals and permeated throughout each of the western kingdoms during the Late Classic period as populations expanded onto the landscape.[66] In this sense western Maya polities operated as communities of shared ritual experience.

At Piedras Negras the 30° E of N mortuary axis is a Late Classic period phenomenon. There was no predominant burial orientation during the Early Classic period; only five out of the eighteen burials from that era (28 percent) follow the northeasterly orientation. Although the sample size is small ($n = 3$), all known Preclassic burials from the region are in the opposing direction, 210° E of N. This axis was still favored in the early phase of the Early Classic period at Piedras Negras, as evident in Burial 110, the one royal tomb from that time (see fig. 2.43). The popularity of the 30° axis during the Late Classic period presumably reflects the adoption of new ritual ideology at that time. Considering that all of the Late Classic period royal burials use this axis, it may be that this orientation was first favored at the court and was subsequently adopted by people living throughout the site (see figs. 3.31, 3.35a). The lack of variability in mortuary alignment among royal interments is also evident at Yaxchilan (120° E of N) and Late Classic period Palenque (12–18° E and N). The courts inspired and at least some of the people followed.

For each polity capital the favored mortuary orientation is one of the two principal perpendicular axes that were employed in the overall design of the settlement. For example, at both Piedras Negras and Yaxchilan (and most Usumacinta River sites) most buildings during the Late Classic period were constructed to face roughly 30°, 120°, 210°, and 300° E of N (figs. 4.8, 4.12). In some sense the favored mortuary orientations placed the bodies of the dead in harmony with the greater built environment; mortuary structures aligned with living structures. Yet each orientation must have had specific meaning within the ritual communities of these respective Maya kingdoms. Otherwise we would expect an equal distribution of burials oriented 30°, 120°, 210°, and 300° E of N among the Usumacinta River communities.

Considering that much of Maya spatial syntax is explained by understandings of the movement of the sun, these orientations likely had solar significance. Recall that the sun's east-west elliptic is the principal axis for the Maya. This is an unfixed axis, however, and changes daily as the earth makes its annual course around the sun. To the observer, the location of sunrise, sunset, and the elliptic path in between wander north and south over the course of the year. If the Maya oriented their principal east-west axis to sunrise and sunset on the equinox, then we would see Maya sites with buildings that face true north (0°), south (180°), east (90°), and west (270°). But this is almost never the case. Rather, the principal axes of the Usumacinta River sites are 30° greater than true north, south, east, and west, as noted. The principal axes of the majority of Maya sites outside of this region are between 8° and 18° greater than the true cardinal directions.[67] Two of the principal orientations of the Usumacinta sites, 120°

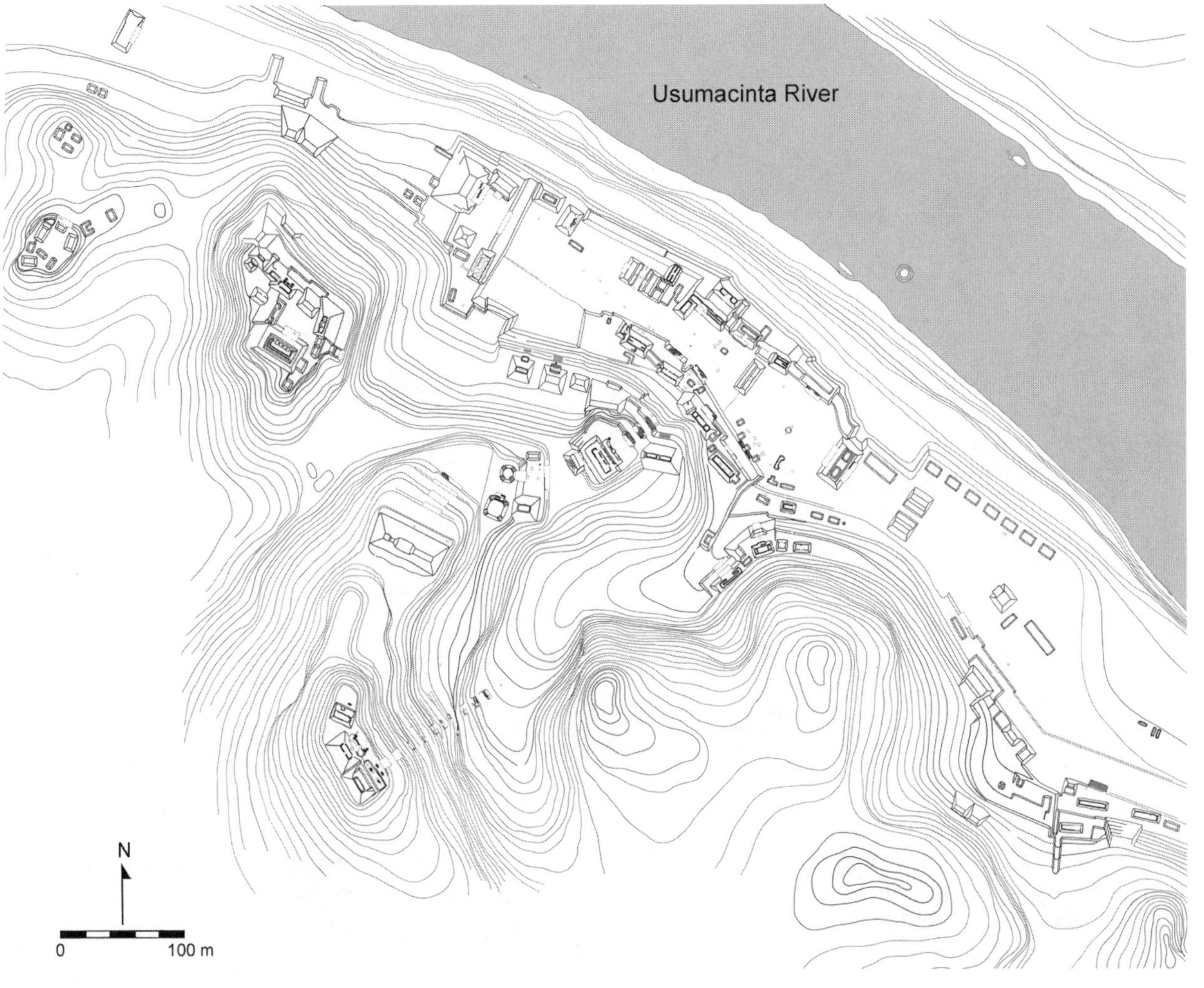

FIGURE 4.12. Map of Yaxchilan, Mexico (drawing by author after Graham and von Euw, *Corpus of Maya Hieroglyphic Inscriptions*, vol. 3, part 1, *Yaxchilan*, 6–7).

and 300° E of N, correspond roughly to the horizon location of sunrise on the winter solstice and sunset on the summer solstice. The other two directions, 30° and 210°E of N, are their perpendiculars.

Ethnographic work by Robert Carlsen among the Tz'utujil of Santiago Atitlán provides some clues as to how Maya conceptions of the horizon relate to understandings of space and time.[68] The places of sunrise on the summer solstice (June 21) and sunset on the winter solstice (December 21) mark two corners of the world. The other two corners are at sunset four days after the summer solstice (June 25) and sunrise four days after the winter solstice (December 25). The four-day lag marks a ritual period. Although this particular range of dates was likely meant to accommodate the influence of Christian theology at Santiago Atitlán (the birth of Christ, a solar being, likened to the birth of the new sun), at least as far back as the Classic period the Maya maintained a ritual period of five days at the end of the solar calendar. A line drawn from the horizon point of the sunrise of the spring equinox and the sunset of the autumnal equinox is used to center Santiago Atitlán and divides space (the world) and time (the year) in half.

Although this particular configuration of sunrises and sunsets is specific to Santiago Atitlán, the tracking of solstices, equinoxes, and other celestial events has deep antiquity and likely explains much of the spatial and temporal ordering of the Classic Maya world. At Santiago

FIGURE 4.13. Main plaza and church at Santiago Atitlán in July 2006. The view is toward the southeast: note the cross at the center of the plaza (photograph by author).

Atitlán horizon locations are specifically linked to places on the landscape that surround the town, reminding us that the Classic Maya likely made local accommodations in their cosmology.

The main church at Santiago Atitlán is oriented to 117° E of N (fig. 4.13). A visitor enters the church from the southwest, facing the direction of the rising sun on the winter solstice. At the opposite end of the church is the famous altarpiece, a representation of a sacred mountain crested by a cross that is likened to both sacred maize and a world tree.[69] The church was completed in 1582 and likely retains the original Precolumbian orientation of the town's civic ceremonial center (fig. 4.14).[70] In a similar fashion the church in Zinacantán is oriented 120° E of N. Although Vogt did not notice the possible solstice orientation there he nevertheless understood the importance of the basic symbolism, which mirrors that at Santiago Atitlán: when "people enter and approach the main altar to pray to the patron saint, San Lorenzo, they are facing in the direction of both the rising sun and the principal sacred mountain, *bankilal muk'ta vits*."[71] The main plaza at Santiago Atitlán no longer retains right angles to its corners (if it ever had them) for unknown reasons. But the main plaza at Zinacantán is still a square with faces oriented to 30°, 120°, 210°, and 300° E of N: the same axes as at the Classic period sites of the Usumacinta River. These orientations contrast with those employed at other colonial settlements (for example, Antigua, Guatemala, or San Cristóbal de las Casas, Chiapas), where Spanish orientations based on true north and a gridded settlement were imposed (fig. 4.14).

I point out these parallels not to claim a descendant-ancestor relationship between the contemporary communities of Zinacantán and Santiago Atitlán and the ancient Maya settlements along the Usumacinta River but instead to suggest that the same solar logic informed the spatial design of all of these places (compare fig. 4.12 to fig. 4.14). These orientations were especially important in the arrangement of plazas and structures in both the Classic period and early colonial communities.

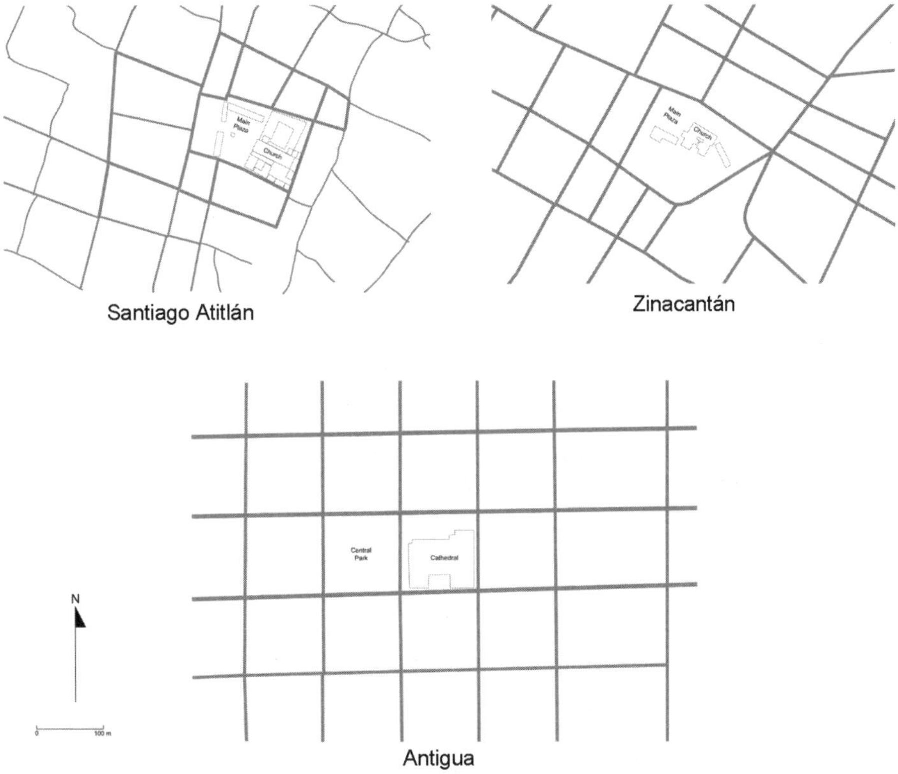

FIGURE 4.14. Street map, principal churches, and main plazas of the epicenters of Santiago Atitlán, Guatemala; Zinacantán, Mexico; and Antigua, Guatemala (plans by author based on Google Earth Imagery).

Counterclockwise ritual circuits are performed in contemporary Maya plazas and are also described in textual sources dating to the time of the conquest.[72] These circuits replicate the path of the sun and generally start in the east (sunrise), move to the north (midday), west (sunset), and south (nighttime), and end again in the east. Although we can never know the specifics of ritual circuits in antiquity, good evidence indicates that a similar logic was followed in the Classic period.[73]

The principal axis at Yaxchilan is 120° E of N, the same as its mortuary axis, as evident by the layout of its main plaza (fig. 4.12). This is also the course of the Usumacinta River at Yaxchilan. We could argue that the design of the community simply mimicked the flow of the river, a parallel that likely did not go unnoticed by the ancient Maya. But this orientation was also followed at Yaxchilan communities located off of the river, as at Tecolote.[74] Carolyn Tate first observed that the 120° orientation (or more accurately 114°) at Yaxchilan marks the position at which the sun rises on the eastern horizon on the winter solstice and suggests that the "winter solstice was the time and direction for the commemoration of the deceased" at Yaxchilan. Surprisingly, Tate generated her hypothesis without any reference to the positioning of the burials at the site, instead relying on the orientation of buildings and sculpture that pertained to the veneration of ancestors. In the iconography of the Yaxchilan kingdom, ancestors appear as celestial beings (the ancestor cartouches) on Yaxchilan sculpture and in the murals at Bonampak (see fig. 3.32).[75] The winter solstice, the shortest day of the year, marked an especially auspicious day to celebrate such transformations; each day thereafter the sun would spend increasingly less time in the underworld and more time in the celestial paradise above.

Few sites outside of the Usumacinta River region incorporate orientations that reference the solstices. Instead they have principal axes that are oriented between 8 and 18° east of the true cardinal directions.

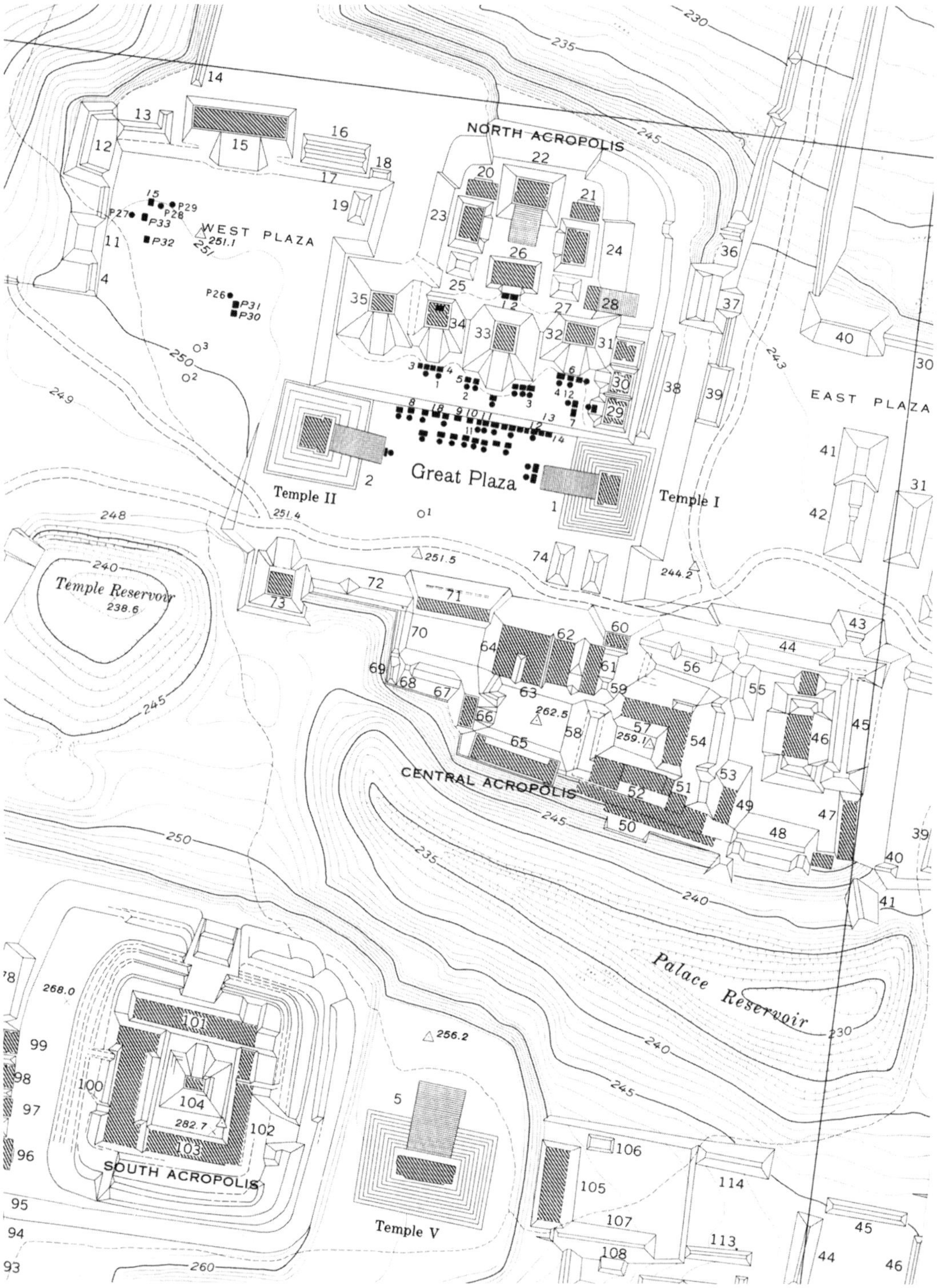

FIGURE 4.15. Map of the epicenter of Tikal (5D quadrant) (originally published as Carr and Hazard, *Tikal Report No. 11*; image courtesy of the University of Pennsylvania Museum of Archaeology and Anthropology). The Tikal map is usually shown oriented to magnetic north, but I have instead presented it here oriented to true north, as is typical of most other maps of Classic Maya sites.

If we assume that the basal platform of the North Acropolis at Tikal corresponds to its principal east-west axis, we find an east-west axis of 98° E of N; its perpendicular is 8° E of N on the northern horizon (fig. 4.15). The sun rises at 98° E of N twice a year, on February 27 and October 14. The sun sets on this axis, at 188° E of N, on April 11 and September 2.[76] To my knowledge these dates have no particular importance in terms of celestial events. They could correspond to important mythic or historical events in the early days of Tikal's history. If this is correct, the variation in the ritual axes from site to site may reflect very local explanations. But I suspect that these orientations and their corresponding sunrise and sunset dates may be better explained by the agricultural calendar.

Most architectural orientations at Palenque range between 12° and 18° E of N (or their perpendiculars). The general variability suggests that particular buildings each have their own alignments to different celestial events (fig. 4.16).[77] Nevertheless, considering that the burials at Palenque are overwhelmingly oriented north-northeast (12–18° E and N), this must have been the principal ritual axis. The platform edge leading up to the main plaza and the palace is oriented 102° E of N and is perpendicular to the ballcourt's playing alley (12° E of N), providing further confirmation that this is the primary axis of the site. Sunrise on the eastern horizon at 12° E of N occurs on both February 17 and October 24. The sun sets on the western horizon on this axis on April 22 and August 22. These dates are roughly ten days different from those noted for the Tikal horizon lines.[78] Considering the internal variability in mortuary axes that can occur within a site (they vary as much as 6° at Palenque), however, the particular sunrise and sunset dates may not have been as important as a more general span of days (a week or so) within each community.

Although the precise meaning and function of these axes is elusive, Raphael Girard's ethnographic work on the Cho'rti' ritual and agricultural calendar provides some possible explanation.[79] New Year rites are celebrated in February, corresponding to the visible northward shift of the sun's elliptic from the southern horizon, signaling that it is time to prepare the milpa for the first planting. Late April and early May mark the sun's first passage through zenith and are a time of intense ritual activity accompanying the sowing of maize and the onset of the rainy season.[80] The first maize crops are bent in preparation for harvest in mid-August and a second crop is planted. This corresponds to the sun's second passage through zenith and is also accompanied by ritual activity. The rains end in late October, when the second crop of maize is bent in preparation for harvest, marking the beginning of the end of the agricultural cycle.[81] In sum, the ritual calendar is closely linked to the agricultural cycle. As Pitarch observes, it is at the end of the agricultural cycle, "from September to December, that the Tzeltal pay most attention to their souls."[82]

As is evident in the Cho'rti' agricultural calendar, the solstice horizon dates suggested by the Usumacinta River architectural orientations do not correspond well with actual dates in the lowland Maya agricultural calendar. Horizon positions for sunrise and sunset in the agricultural calendar are better accommodated by Maya sites whose primary east-west axis falls in the range 98° to 108° E of N, which includes Palenque and Tikal. The sun rises on the horizon at this declination around the time when fields were being prepared in early to mid-February. The sunset at the western declination of this axis corresponds to mid- to late April, the time of planting in preparation for the onset of the rains. As noted, it is possible that the agricultural cycle was acknowledged in the myths of the Classic Maya. For example, the Chahk that appears in the myth of the Baby Jaguar is named Yax Ha'al Chahk (Green/First Water Chahk), an apparent reference to the onset of the rainy season.[83] The subtle variability in some of these site orientations (for example, Tikal versus Palenque) may reflect microclimatic difference in the onset and cessation of the rainy season or simply different festival dates on this basic agricultural calendar.[84]

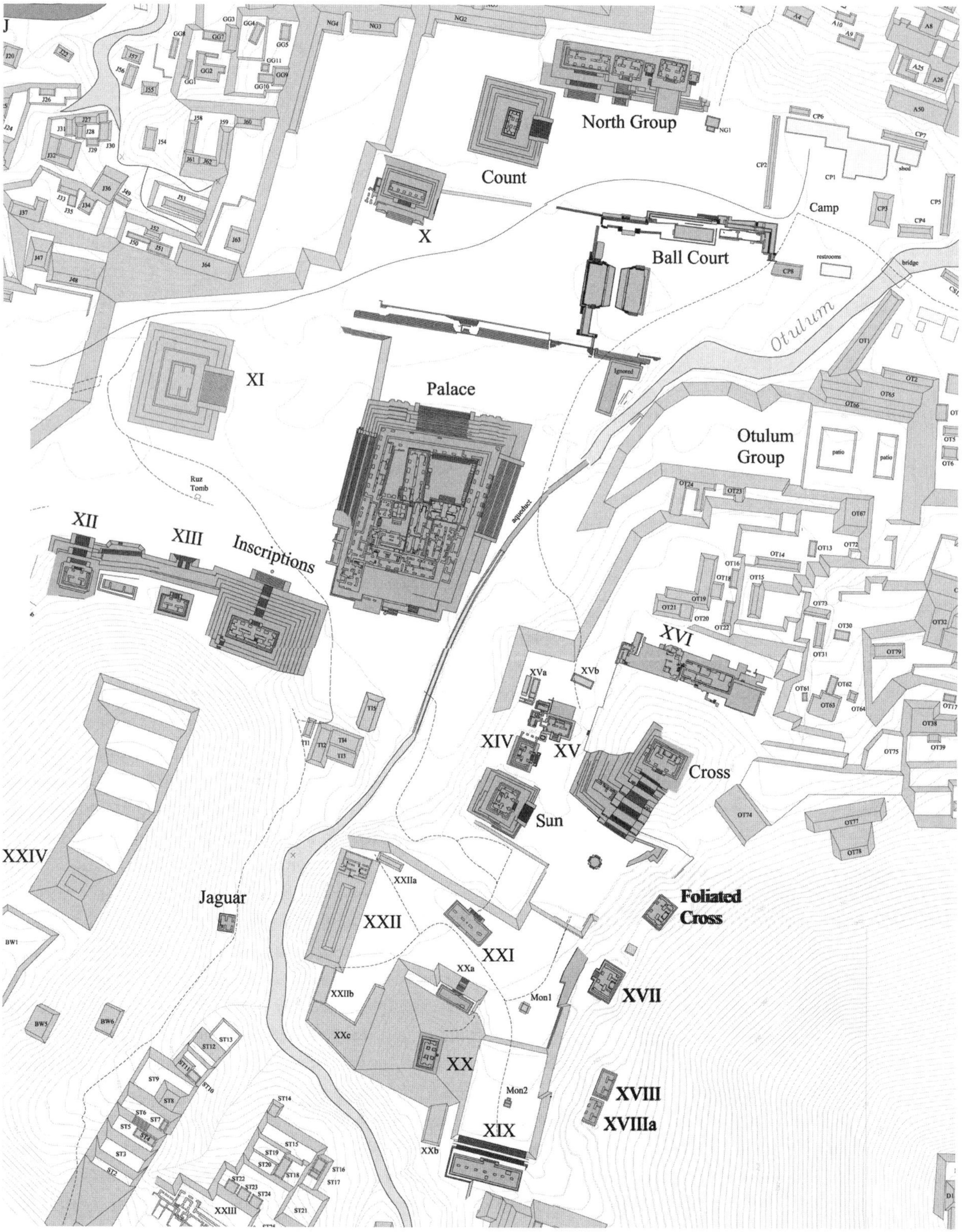

FIGURE 4.16. Palenque site map core (map courtesy of Edward Barnhart).

THE MEANING OF NORTH

At Late Classic period Piedras Negras, Palenque, Tikal, and most other Classic Maya sites bodies were not aligned along the east-west path of the sun as they were at Yaxchilan. Instead the long axes of the grave were most often oriented perpendicular to the solar path, usually with the head of the deceased pointing in a northerly direction. In her study of settlement design at Tikal and Copan, Wendy Ashmore emphasizes the site's north-south axis and suggests that "this axial emphasis maps the cosmos in such a way as to situate the king and his arenas for public action in metaphorically powerful, supernaturally potent positions, especially in the heavens to the north or the underworld to the south."[85] Ashmore sees the Twin Pyramid groups and the North Acropolis at Tikal (which contains the graves of many of Tikal's kings) as illustrative of a connection between north, the celestial heavens, and the place of ancestors.

Although north-south was unlikely to have been the primary axis of Tikal, it is the perpendicular of the solar elliptic and was likely important for conceptualizing the celestial heavens and the underworld. For the Tzotzil of Chamula, north is the "horizontal equivalent of the point of 'maximum heat' of the sun at noon at the zenith of his orbit. . . . as a result, the cardinal direction north shares with the east the sign of good omen and positive orientation, while west and south are generally negative in the cosmological system."[86] As Karl Taube shows, the path of the sun is the route to the celestial paradise, "Flower Mountain," the home of ancestors and other supernatural beings.[87] Thus north is conflated with ascension, the sun at its zenith, and celestial paradise. This may explain the preference at Tikal and many other Maya sites for burying the body with the head at the northern end of the burial, particularly if we recall that souls enter and exit primarily through the orifices of the head.

If this is correct, the 30° E of N Late Classic period orientation at Piedras Negras may simply be the northern perpendicular of the primary east-west axis of the site, in the same fashion that 8° was perpendicular to the principal solar axis at Tikal. At Piedras Negras a natural ravine leads from the East Group down to the Usumacinta River and also happens to be aligned 30° E of N (see fig. 4.8). The Maya may have also been referencing descent into the river at Piedras Negras by orienting the burials in this fashion. Certainly there is precedent for the movement of souls into and out of watery places at death, including the spirals carved on the lid of Piedras Negras Burial 103 (see fig. 3.71). This burial was located in a structure on a ridge that overlooked similar carved spirals at the river's edge below.

The northerly burial orientation is even more regular at Palenque; all but two burials contained primary occupants with their heads oriented to the north, between 12 and 18° E of N. Two Early Classic tombs are oriented north-northeast, indicating that the axis was in place at least by the turn of the seventh century. Focusing on the burial of Pakal, Wendy Ashmore and Pamela Geller suggest a connection between the orientation of the dead king's body and the sacred tree inscribed on the sarcophagus lid: Pakal's feet, like the roots of the tree, were in the underworld (south) and his head, like the treetop, was directed toward the heavens (north).[88] This orientation also put the deceased's feet toward the mountains to the south, which may have related to local concepts of the underworld or the dwelling places of ancestors. As at Yaxchilan, burial (and architectural) orientation at Palenque may also have been influenced by the local river, the Otulum, which flows in a north-northeasterly direction through the site epicenter.

As the perpendicular to the daytime path of the sun, north-south may also have been linked to the nighttime sky.[89] As we saw in chapter 3, in the Early Classic period paintings from Río Azul Tomb 12, the northern eagle is marked by *uh* (moon) and the southern eagle by *ek'* (star) (not the true cardinal directions, as Río Azul is oriented about 10° E of N) (see fig. 3.1). Of the possible celestial bodies in the night sky, the Milky Way is perhaps the best candidate for a perpendicular axis to the

FIGURE 4.17. Palenque Temple of the Sun (*left*) and Temple of the Cross (*right*) (photograph by author). The Temple of the Cross superstructure is named as a Six Sky place and the *naah* and *otoot* of GI.

elliptic of the sun. David Freidel, Linda Schele, and Joy Parker have proposed that the Milky Way was the *sacbe* (white road) of the ancient Maya and was a path taken by the dead, as suggested in the *och bih* (road entering) euphemism.[90] Their hypothesis was influenced by Girard's observations of Ch'orti' cosmology: "The elders attentively observe the changes in position of the Milky Way in relation to the sun. The latter runs from East to West and the Milky Way approximately from north to south, and that is how the elders see it at midnight on 'the day of Santiago.' By that time the solar trajectory and the Milky Way form a gigantic cross in the sky."[91] The "day of Santiago" is July 25, the feast of the patron saint of the Spanish conquistadors. For the Cho'rti', this astronomical event marks the beginning of the *canícula*, the temporary break in the summer rains. Undoubtedly, the Cho'rti' were influenced by Spanish perceptions of the Milky Way. Even today Spaniards refer to the Milky Way as "el camino de Santiago," in reference to an important pilgrimage route in northwestern Spain. Yet Cho'rti' perceptions of the Milky Way also show indigenous theological concepts: as Santiago rides the Milky Way in July he uses his sword to control the rains, lightning, and thunder. Such imagery is presumably drawn from Precolumbian understandings of Chahk, the axe-wielding patron of storms. The linkage of the north-south axis, the Milky Way, and concepts of the afterlife is attractive but cannot be established definitively, in part because no evidence confirms that the Classic Maya understood the Milky Way as a road.

NAAH AND *WITZ*

The ethnographer William R. Holland, inspired by his ethnographic fieldwork among the Tzotzil, proposed long ago that the Classic Maya understood their pyramids to be representations of *witz* (hills or mountains).[92] For the Tzotzil, natural mountains are the places of spirits and a

FIGURE 4.18. Cleared, partially collapsed tomb from Rancho Búfalo (Burial 1) (photograph by author).

destination for ancestor veneration. Holland proposed that Classic Maya pyramids were functionally the equivalent, constructed for the veneration of Maya kings and queens. Pyramid glyphs appear at Palenque and a few other sites, though their reading remains elusive. Many are marked by the **NAAH** sign, perhaps in reference to the structures on top that were conceived of as houses for supernatural beings.[93] These superstructures served as places to store effigies of supernatural patrons, other sacred objects, and perhaps even spoils of war.[94] Some *naah* are named as *wahybil* (literally places of sleep and dreams). A possible double meaning is that these are also places of *wahy*, in reference to the spirit travel that occurs when a person sleeps.

Evon Vogt and David Stuart draw a parallel between the pyramidal superstructures of the Classic Maya and the mountain-top home of San Juan, the patron saint of Chamula. His dwelling is visible as a cave, San Juan's Window, on nearby Tzontevitz (Moss Mountain).[95] The animal co-essences of the living Chamula dwell within this mountain. Vogt and Stuart liken San Juan's window to a sky cave and see parallels in the superstructures of the Palenque Cross Group, which are replete with cave and mountain imagery, further substantiating that pyramids are images of *witz* (fig. 4.17). The concept of sky caves seems to have deep antiquity in Mesoamerica. For example, a being emerges from a monstrous cave whose mouth is marked by a crossed-band sky sign on Altar 4 of La Venta (see fig. 3.24).

Tombs, defined here as mortuary spaces with an internal height greater than one meter, have long been recognized as conceptual houses for the dead (fig. 4.18).[96] Most Maya tombs are chambers built within the architectural cores of pyramids, though exceptions include tombs placed in patios or hewn out of bedrock. Although royal tombs are most often described as *muk* (literally "burials"), others are named as *naah* (sacred buildings to house the remains of ancestors).[97] Even today resting places for bones can be conceptualized as houses, as evident in the symbolism employed in crypts in Yucatán (see fig. 4.1).

Throughout the Maya lowlands pyramids and other earthly constructions were adorned with stucco masks that symbolized the animate spirits that dwelled within. In some instances the masks show celestial beings, such as Chahk, the Maize God, or the Sun God and his many variants (fig. 4.19). In other instances the masks are the names of ancestors entombed within. For example, the Copan's founder's name, K'inich Yax Kuk Mo', adorned the façade of the Margarita Structure.[98] The so-called Cauac masks that decorate buildings of the northern lowlands are representations of the animate spirit of the *witz*. In the Chenes region of the Yucatan *witz* masks frame doorways, their mouths serving as symbolic caves and portals (fig. 4.20). The *witz* are also evident in the mosaic masks of the Puuc region, often mistaken for Chahk (fig. 4.21). Andrea Stone and Marc Zender call attention to the stepped or curling nose and foreheads of the *witz*, which are suggestive of the steps of pyramids (figs. 4.21, 4.22).[99]

Iconographically, the opening in the *witz* head seems to be suggestive of a cave and is rendered as a stepped-fret design. Beings emerge from the step-fret cleft in the *witz* head (fig. 4.22a) in some depictions. In other instances the entrance to the *witz* is only abstractly suggested by the step-fret sign, as on the façade of the Labna arch building and a cache vessel excavated at Tecolote, a secondary center in the Yaxchilan kingdom (fig. 4.23). A nearly identical vase is shown on Yaxchilan Lintel 24, used to collect the blood that spilled from the mouth of Lady K'abal Xook. It is possible that these two vases were produced in the same workshop, perhaps for the purpose of use in ritual bloodletting. In any case the step-fret symbol itself is widespread in Mesoamerican iconography, where it may symbolize sacred mountains, caves, and other related concepts.

FIGURE 4.19. Solar masks on the Kohunlich Temple of the Masks (photograph by Holly Scherer).

FIGURE 4.20. *Witz* mask doorway on Structure II at Chicanna (photograph by Holly Scherer).

FIGURE 4.21. *Witz* masks covering the façade of the Codz Pop at Kabah (photograph by the author).

(a)

(b)

(c)

FIGURE 4.22. *Witz* in Classic period iconography: (a) Bonampak Stela 1 (drawing by Linda Schele, © David Schele, courtesy Foundation for the Advancement of Mesoamerican Studies, Inc.); (b) Palenque Tablet of the Foliated Cross (drawing by Linda Schele, © David Schele, courtesy Foundation for the Advancement of Mesoamerican Studies, Inc.); (c) an Early Classic period vase (drawing by David Stuart).

FIGURE 4.23. *Witz* cave entrance in Classic period iconography: (a) on the façade of the Labna arch building; and (b) marked with jaguar spots on a cache vessel from Tecolote (compare to the bloodletting vessel on Yaxchilan Lintel 24) (photographs by author).

As noted in chapter 3, the earflares of the *witz* and other stucco masks are often shown to be the glyph for road, **BIH**. In some contexts, serpents pass through the earflares and can be read *och bih* (the "road-entering" death euphemism). An Early Classic depiction of the *witz* shows the **BIH** earflares marked by the **KUH** signs, perhaps showing mountains as places of supernatural emergence and departure (see fig. 2.31). Similarly, some *witz* masks, such as the Chicanna façade, show curling breath emanating from the sides of the mouth, indicating that they are both alive and presumably the source of wind (see fig. 4.20).[100] Iconographically, *witz* are often shown with three circles infixed with three smaller circles (fig. 4.22). This may mark them as three stone places, representing the axis mundi. The *witz* on the Early Classic vase has three shiny stones (jade celts) in its mouth, possibly its teeth (fig. 4.22c). In other instances a series of three animate *witz* is depicted, perhaps in reference to Maya concepts of centering. The *witz* on the Palenque Tablet of the Foliated Cross is shown with three faces: a forward gazing face and two other faces looking laterally, as evidenced by the three noses and snouts (fig. 4.22b).

As a basic architectural form, the pyramid allowed ancient Mesoamericans to construct buildings of massive bulk and dizzying heights with relatively simple building techniques and materials. Pyramids operated as monumental and highly visible proclamations of exclusivity. Their massiveness was intended to awe viewers, to make them feel small, and to establish the heart of Classic Maya kingdoms as unique places on

FIGURE 4.24. Calakmul Structure I (photograph by Jeremiah Scherer).

the landscape (fig. 4.24). Hundreds if not thousands of people could gather in plazas to witness performances enacted on their flanks and summits, yet only a handful of people were allowed to the top, much less into their inner sanctuaries. Much like the imperial box of the Roman Coliseum, pyramidal summits were highly visible and utterly exclusive. Pyramids also manifest the Classic Maya belief in verticality as a metaphor for authority, as evident in the composition of Late Classic period scenes of the court. On the Bonampak murals, Piedras Negras Stela 12, and Piedras Negras Panel 3, the king is positioned on top, and the lower registers are occupied by visiting dignitaries, subordinate lords, and captives (see fig. 4.6).[101]

Pyramids also brought Maya lords in proximity to the power of the sun, a principle that the contemporary Maya continue to value. As Gossen observed at Chamula, the community residents perceive themselves as living on the highest place on earth, closest to the sun, and thus their community and the highlands in general is "less dangerous and less asocial than the lowlands." Anyone who has ever climbed the pyramids in the lowlands—especially in the remarkably flat Yucatán peninsula—can appreciate how reaching the summit of these structures conveys

FIGURE 4.25. View from atop Yaxha Structure 216 (photograph by author).

a sense of being above the world (fig. 4.25). Gossen notes that the Tzotzil concern with distance from the sun explains why saints habitually "ride on litters and sit on platforms that raise them above the level of men. Cargoholders who serve them thereby achieve the elevation of goodness, virtue, and prestige."[102] Indeed many Maya pyramids were adorned with images of the Sun God in his various guises and were generally painted red to evoke their solar power (see fig. 4.19).

Mountains and the sky are closely linked conceptually in Maya cosmology. As Holland observed for the Tzotzil: "The sacred mountains symbolize the sky, and to ascend one is tantamount to rising into the heavens. Each level is presided over by distinct ancestor gods. The strata of the sky are presumed connected by the sacred cottonwood tree; while those of the sacred mountain are imagined as linked from bottom to top by a huge stairway, in many ways suggesting the ancient Maya pyramids."[103]

Holland also notes that the tops of the sacred mountains of Zinacantán and Larraínzar are each marked with Christian crosses.[104] Although he suggests that this reflects Christian replacement of effigies of ancestor gods, it may be that these crosses instead marked these mountains as axes mundi, conceptually parallel to the cottonwoods that connected the levels of the sky. The altar piece at Santiago Atitlán, another symbolic mountain, is also topped by a cross that is explicitly understood as an axis mundi by the Tz'utujil of the community.[105] The depiction of a *witz* on the Early Classic vessel shows its forehead-entrance marked by a sky-band with a pole emerging from it (see fig. 4.22c).

Mountains, especially the "Flower Mountain," were also understood to be places of abundance. The Maize God dancer vessels bore texts that described their ascent to a place that corresponds with the emblem of different polities (likely mythical places of origin) or described them as "inside/before the *ch'e'n* [cave]" (see chapter 1).[106] The texts may hint at an association between these maize beings and either a mountain or a celestial place above. There certainly is abundant iconographic evidence of the Maize God descending from the sky bearing a tether or umbilicus that links the earth and heavens (see chapter 2). A diving or hanging being is shown on a mask on Tikal Platform 5D-4, the basal platform for the upper part of the North Acropolis (fig. 4.26). Here the symbolism suggests the pyramid; the greater North Acropolis complexes were an axis mundi, a liminal space between the earth and the place of the ancestors. It may be that the pyramids of Tikal and elsewhere should be understood as mountain-sky places, akin to the celestial paradise, "Flower Mountain," of the Maya afterworld.

FIGURE 4.26. Diving or hanging being on Tikal Platform 5D-4 (originally published as Coe, *Excavations in the Great Plaza, North Terrace, and North Acropolis of Tikal*, fig. 164a, image courtesy of the University of Pennsylvania Museum of Archaeology and Anthropology).

PLACING THE SPIRIT

Much contemporary Maya ritual practice is vested in the movement of souls and other supernatural beings between the earth and otherworldly places with the aid of crosses, sacrifice, and other rites. Iconographic and epigraphic data suggest some parallels in Classic Maya rites of conjuring. Although the reconstruction of Classic period ritual process is immensely difficult, ethnography offers a useful analogy for considering how such supernatural movements may have been enacted. William Hanks describes the movement of spirit in a contemporary Yucatec house:

> Spirits, who are located in cardinal places throughout the vertical and horizontal universe, are "lowered" to the place of performance, actually "moved" from their appointed places and brought down to an encounter with the shaman and those he presents in the performance space. The space of the altar, referred to in performance as "there on the earth of sin" and "at the base of X's altar (where X is a spirit name)," is constituted in an absolutely orderly process, whereby spirits are lowered in the order of their originative locations. . . . The first phase is called "opening the road," or "binding (the) altar" and the second is "untie (the) altar."

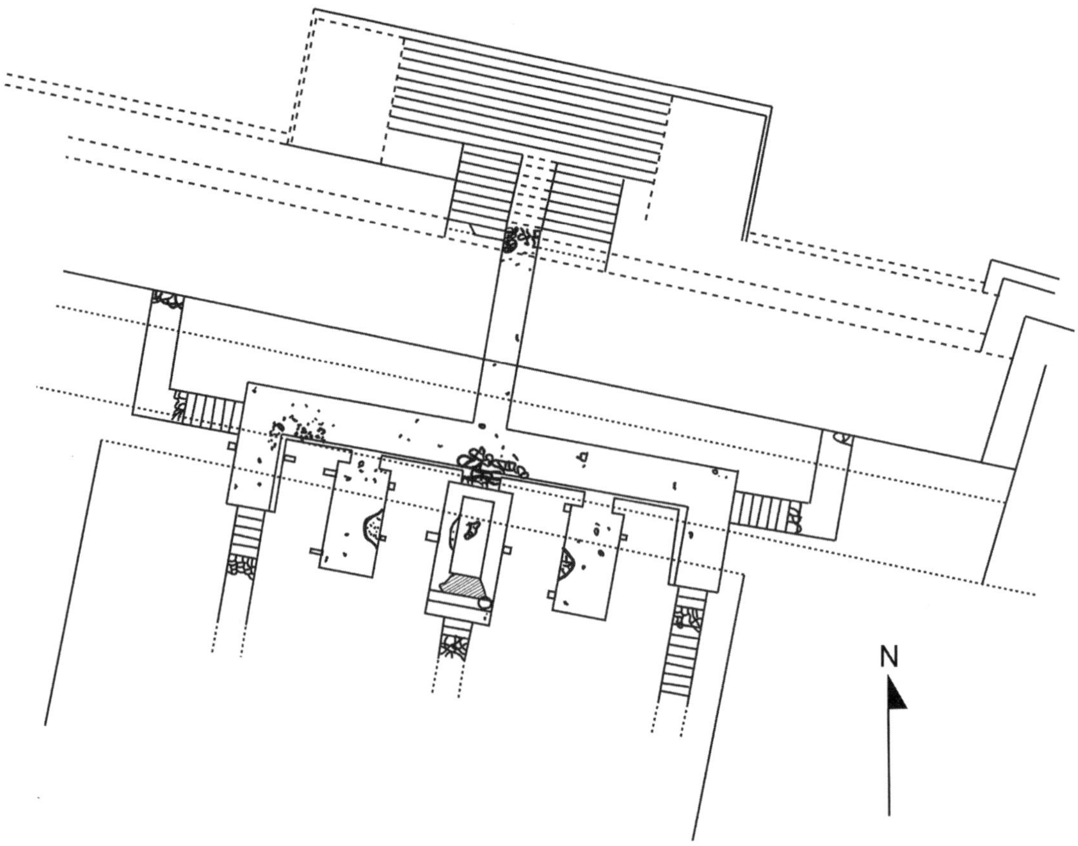

FIGURE 4.27. Plan view of Palenque Temple XIII showing access ways into the substructure and the tomb of the Red Queen (*center*) (drawing by author after González Cruz, *La reina roja*, 125).

These Yucatec rituals proceed in clockwise and counterclockwise motions, acknowledging the four directions and the center. By centering ritual spaces Yucatec shamans open a conduit or road "between earth and absolute cosmological locations."[107]

Concerns with cardinal orientations, ritual circuits, and supernatural conduits suggest that ancient ceremonies may have been vaguely similar to these Yucatecan rites. In the Classic period ritual landscape such ritual invocations must have occurred in a range of liminal spaces: inside tombs, on altar tops, within caves, and at the summit of pyramids. Moreover, such rites employed a range of important images of ancestors and supernaturals: stone monuments, figurines, masks, and human remains that manifest the spirit when invoked and summoned with offerings of smoke, flame, and sacrifice.

The iconographic evidence for the use of human remains in spirit conjuring (as explored in chapter 2) points to the likelihood that graves were reentered to bring the living in close proximity to the bones and bodies of the dead that were to be invoked.[108] At Palenque extensive measures were taken to keep tombs accessible for postinterment rituals. The Red Queen's tomb was entered by way of a passage in the front of Temple XIII and was also accessible via stairways that led to the substructure from later constructions above (fig. 4.27). The most elaborate tomb access was the massive stairway that descended into the heart of the Temple of the Inscriptions to reach Pakal's tomb. Other Palenque pyramid-tombs were accessible via internal stairwells, including the tombs in the Temple of the Jaguar, Temple XV, and Temple XVIIIA. Frans Blom reports passageways into two corbel-vaulted tombs in Group A (Burials S-2 and S-5) and a vaulted tomb in Group B (Burial S-6).[109] The lid of the Red Queen's sarcophagus was perforated by a small circular hole that was likely used to "feed" her bones with copal and other offerings, similar to the scene depicted on Piedras Negras Stela 40 (see fig. 2.39).[110] A

number of crypts at the site, including a slab-crypt from Temple XVIII and two crypts from the Temple of the Foliated Cross, also had perforated lids for such activities.

FIGURE 4.28. Psychoduct that runs along the stairway that leads to Pakal's tomb (photograph by Linda Schele, © David Schele, courtesy Foundation for the Advancement of Mesoamerican Studies, Inc.).

Despite the extensive efforts to make the Palenque tombs accessible after their initial construction, most were eventually sealed away. The most extensive closure was the filling of the Temple of the Inscriptions stairway with rubble, which took Alberto Ruz Lhuillier three years to clear. The eventual sealing of these passageways may have been necessary in part to protect the remains of the king from disturbance during periods of turmoil. Perhaps later Palenque lords feared desecration of the king's tomb as Palenque's political and military fortunes waned over the course of the eighth century. Such elaborate closure may also have been mandated by later kings who sought to distance themselves from earlier rulers. To that end the sealing of the king's tomb may have been an active act of "forgetting": to move the king from the realm of active protagonist to the more distant pool of ancestors. Obviously, people have good reason to want to forget a bad ruler. But it can also be strategic for later rulers to distance themselves from great kings if they themselves cannot live up to the memory. By the final decades of the eighth century to recall the great king Pakal in acts of veneration may have served as a bitter reminder that Palenque's golden age was fading. The ease at which the Temple of the Jaguar, Temple XV, and Temple XVIIIA tombs were looted in the nineteenth century or earlier suggests that the internal stairwells in these pyramids were never sealed with rubble.[111]

Even when tomb chambers were finally sealed, mechanisms were put in place that allowed the living to communicate directly with the remains of the dead. Such devices are commonly referred to as "psychoducts" and have been found throughout the Maya area. Among the most impressive of these conduits was the one that wound through the stairwell of the Temple of the Inscriptions, constructed of fine limestone blocks

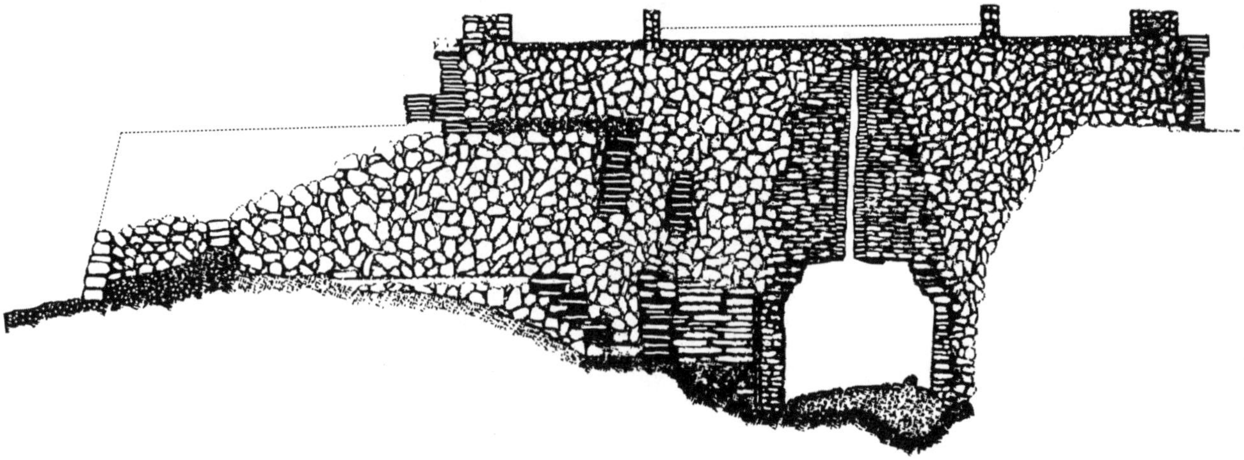

FIGURE 4.29. Psychoduct in the north-northeast to south-southwest cross-section of Palenque Temple XVII (drawing by author after Ruz Lhuillier, "Exploraciones arqueológicas en Palenque: 1957," fig. 5).

(fig. 4.28).[112] The conduit starts on the fifth step of the staircase and descends until it ultimately passes into the tomb chamber and finally connects to the sarcophagus itself. Inside the tomb artisans covered the conduit with stucco that was modeled to resemble a serpent, perhaps the fire serpent, the solar essence of royalty, shown rising from the tomb on Piedras Negras Stela 40 (see fig. 2.39). Another such conduit was extending directly up from Tomb III of Temple XVIIIA, 2.7 m in length and terminating 0.5 m below the floor (fig. 4.29). A third was found leading from the Temple XX tomb, ending below a buried floor that had traces of burning.[113]

Although the Palenque psychoducts came close, these tubes did not reach the outside world but instead terminated before they exited the surface of the structure. It may be that these architectural features were used for one final mortuary rite, perhaps evidenced by the burning in Temple XX, before any connection to the mortuary chamber was ultimately sealed. Alternatively, they may have been constructed in this fashion so that they could be uncovered on occasion to facilitate movement of the dead's soul but remain otherwise sealed to prevent spirit travel. Similar conduits have been observed at Calakmul, and small conduits were noted exiting tombs at El Zotz and nearby Bejucal.[114] The conduits from El Zotz and Bejucal were also sealed.

These psychoducts resonate with modern practices that encourage the departure of souls. Redfield and Villa Rojas describe how a round opening is made in the thatch roof of houses in Chan Kom to allow souls to leave. Even in modern masonry constructions a small window is placed near the ceiling for the soul to use during its departure.[115] Kerry Hull and Michael Carrasco suggest that the painted capstones at Becan, Caracol, and Ek Balam are symbolic portals.[116] The spiral painted on the crypt at Piedras Negras may be another example of a mortuary portal (see fig. 3.71).

The inscriptions of Piedras Negras contain numerous accounts of both burial reentry and the activities of dead souls. For example, Stela 8 records that Ruler 2 danced on the three-*k'atun* anniversary of his son's birth, even though he had been dead for almost forty years.[117] But masonry work at Piedras Negras was simply too poor to facilitate the sort of internal stairwells and pyramidal passages that allowed burial reentry at Palenque. Instead

FIGURE 4.30. Tecolote Structure D3-1 with looted crypt beneath bench in Room 3 (photographs by author).

royal tombs were placed below the patio floors and were accessed years if not decades after the death of the king (see figs. 4.9, 4.10).

Unlike in most other Maya kingdoms, the kings, queens, and nobility of Yaxchilan were buried directly beneath the floors and stairway of buildings that were not only used for ceremony but may have served as reception halls and for other quotidian functions. Structure 23 contained the remains of what may be Shield Jaguar III and Lady K'abal Xook (see fig. 4.11). Yet the building also had benches, architectural features usually not found in pyramid superstructures with the primary function of ritual performance and, presumably, the storage of ritual paraphernalia. This arrangement is also evident at Yaxchilan's eighth-century subordinate centers. Crypts were located below benches in Yaxchilan-style structures at La Pasadita, Tecolote, and Bonampak (fig. 4.30). In the recently excavated burial at Bonampak the cranium had been removed, indicating that these crypts were reentered on occasion. In this fashion the royal mortuary landscape of Yaxchilan was distinct from that of most other Classic period kingdoms in that the dead were not spatially removed from the day-to-day affairs of the living.[118] It may be that Yaxchilan lords sat and governed above the very remains of their ancestors.

PLACES OF VIOLENCE, DEATH, AND ASCENSION

The ancient Maya ball game was many things, depending on context and situation: a spectacle to be watched and enjoyed, a tense political competition involving players from different kingdoms, a ritual reenactment of mythic foundations of a kingdom, or, presumably, a casual game played for practice and fun.[119] Scholars have long recognized that paired structures with sloping walls and a central playing alley served as ballcourts,

FIGURE 4.31. Yaxchilan ballcourt (Structure 14) (photograph by author). Note the monuments that line the alley.

where the ball game was played (fig. 4.31). Mary Miller and Michael Coe have both pointed out, however, that images of ball playing on Classic period vases and on sculpture at Yaxchilan demonstrate that the game was also played against the steps of pyramids.[120] Thus it seems unlikely that all ball games were played in such sacred ballcourts, much as modern soccer games can be played just about anywhere on the landscape with sufficient space. Moreover, Karl Taube and Marc Zender have shown that ballcourts were arenas for more than just the ball game: instead a whole spectrum of violent contests, including combat between dueling rivals, was waged within this space.[121] We should not simply think of the Classic Maya ballcourt as a place where the ball game was played; its function was more broad: it served as a ceremonial stage and especially a place to frame interpersonal violence and competition as sacred ritual acts.

Material evidence for the violent competitions that were staged in Classic Maya ballcourts remains markedly thin. Osteological signatures of sacrifice and other forms of interpersonal violence generally are not as robust for the Maya as they are in other parts of the world. Recent excavations in the Grupo Cheje of Yaxha by Laura Gámez and Adriana Segura offer an important exception. The group is located about 250 meters west of the Blom Causeway. Gámez and Segura uncovered the remains of a middle-aged man (thirty-five to fifty years old) that not only show signs of traumatic violence but were deposited in such a way as to suggest that he was a victim of sacrifice (fig. 4.32). Based on the position of the skeletal remains, the body was interred prone, with the right arm and both ankles trussed at the pelvis. The man was decapitated, and the head was placed over the lower back. The left arm was missing. The first cervical vertebra was fractured in multiple places as a result of the blow that removed the head. Aside from the decapitation and the missing limb, the arrangement of the body was exactly like that of a trussed captive on Tikal Altar 8 and other monuments from that site (fig. 4.33).

What may connect this deposit to the bloody competitions identified by Zender and Taube is the assemblage

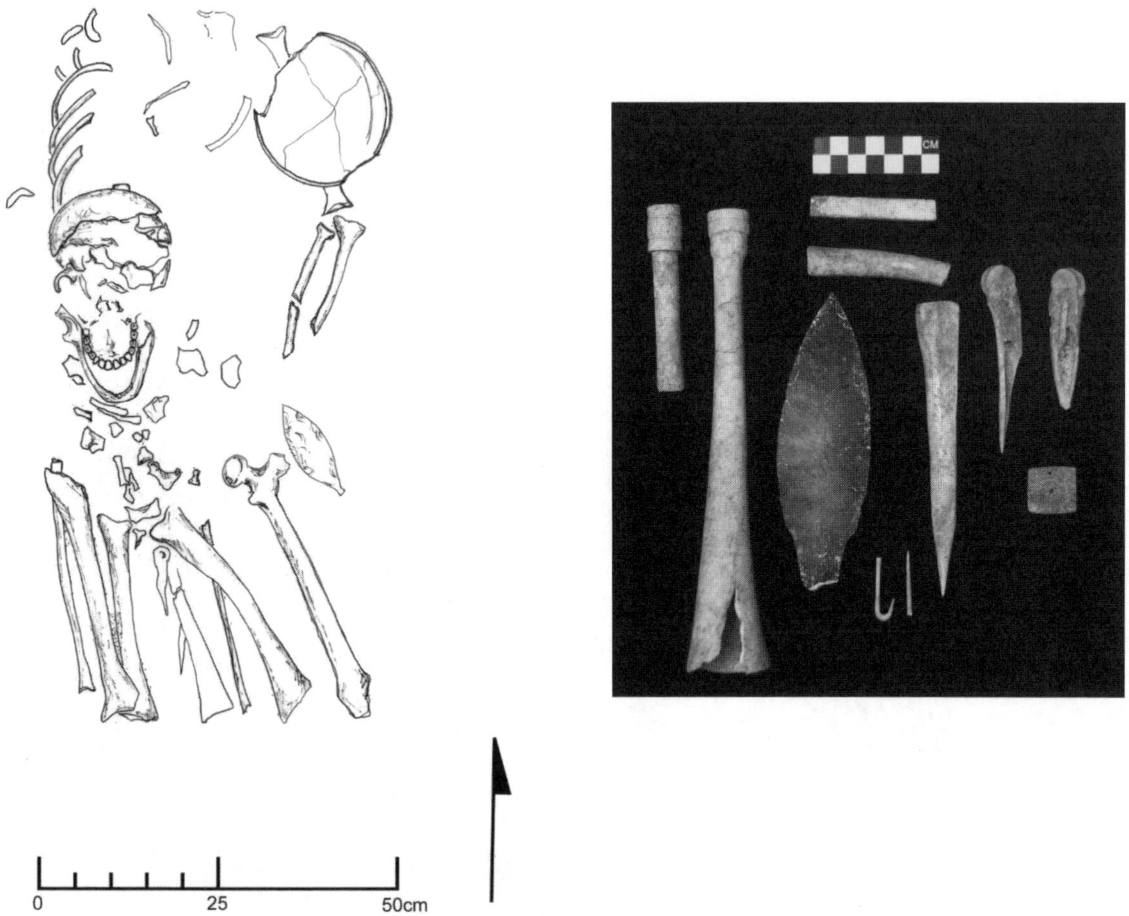

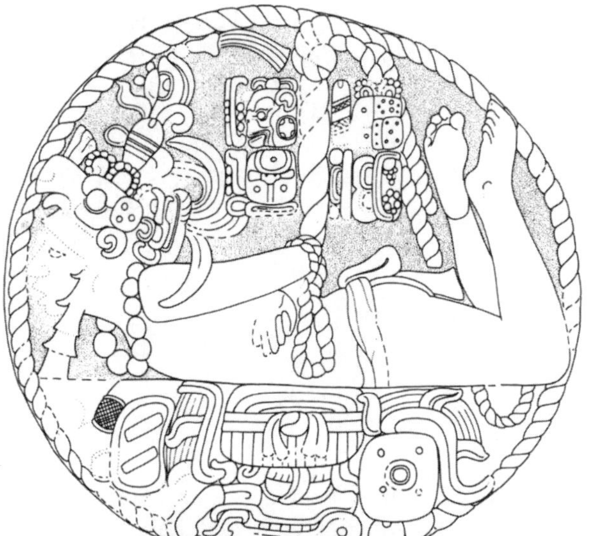

FIGURE 4.32. (*above*) Yaxha Burial 14, including photograph of associated material culture (excluding ceramic vessel) (drawing by Laura Gámez, photograph by author).

FIGURE 4.33. (*left*) Tikal Altar 8 (drawing by William Coe from Jones and Satterthwaite, *The Monuments and Inscriptions of Tikal*, fig. 30, image courtesy of the University of Pennsylvania Museum of Archaeology and Anthropology).

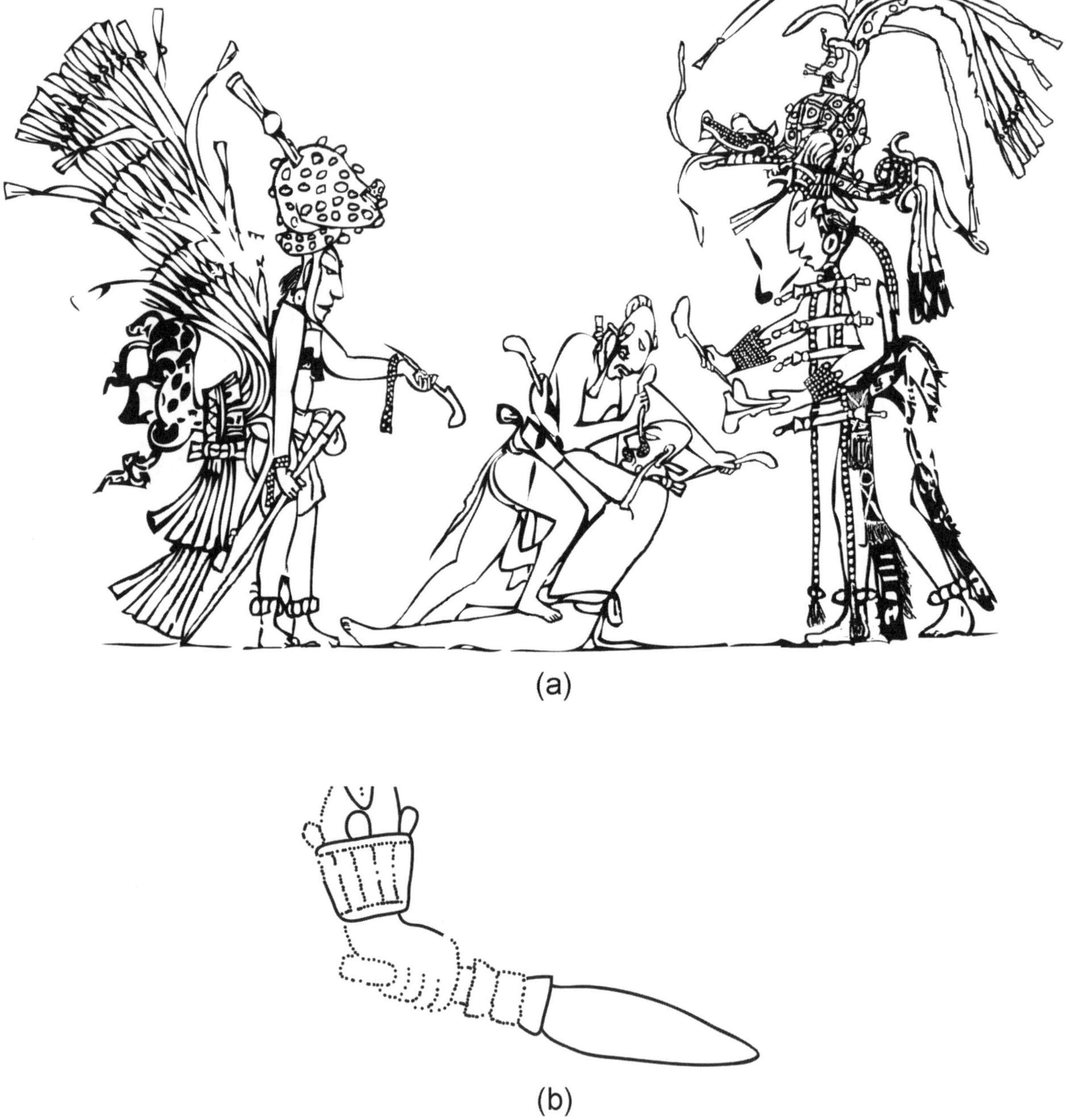

FIGURE 4.34. Classic Maya close combat weapons: (a) combatants on an unprovenanced vase (drawing by author after K7749); and (b) knife from Yaxchilan Lintel 26 (drawing by author).

of bone implements located between the man's legs, including a collection of awls that are similar to those used by a group of ritual combatants on an unprovenanced ceramic vessel (fig. 4.34a). Near the body was a large bifacial blade that may have been hafted into one of the bone handles that was found with the body, the largest of which was crafted from the femur of a child. Stone knives mounted in bone handles are evident in the iconography of Yaxchilan, including Lintels 4 and 26 (fig. 4.34b).[122] Healed antemortem fractures of an unidentified rib and the fifth metacarpal, a so-called boxer's fracture, suggest that this individual experienced a fair share of violence in his life. We can only wonder if these are the remains of someone killed in ritual combat or as a captive sacrifice, perhaps deposited near the home of the person who slew him.

In design ballcourts were the conceptual inverse of the pyramid. To climb the stairs of a pyramid was to move closer to the celestial realm, while to enter the ballcourt was to descend toward the underworld. This is likely why most Classic period ballcourts are low, squat structures, to contrast with the soaring height of pyramids. The ballcourts at some highland and western lowland sites were sunken into the earth (fig. 4.35). Conceptually, Classic Maya ballcourts were rifts in the earth, openings to underworld spaces within the civic ceremonial heart of the kingdom. It may be that the cleft that appears in the *witz* heads could be understood as a ballcourt (or vice versa) (see fig. 4.22).[123] As underworld entrances, the blood spilt in these arenas may have been perceived as an offering to the supernatural domain below.

FIGURE 4.35. Tonina ballcourt. The author is standing in the playing alley; an adjacent plaza is located to the north (*foreground*) (photograph by Charles Golden).

Such concepts are certainly evident in the sacred landscape of the sixteenth-century K'iche', as described in the *Popol Vuh*.[124] A ballcourt was located along the path to the underworld, and the relentless ball playing of Hun Hunahpu and his brother angered the underworld lords. The twins were summoned to the underworld and were sacrificed and buried within the underworld ballcourt.

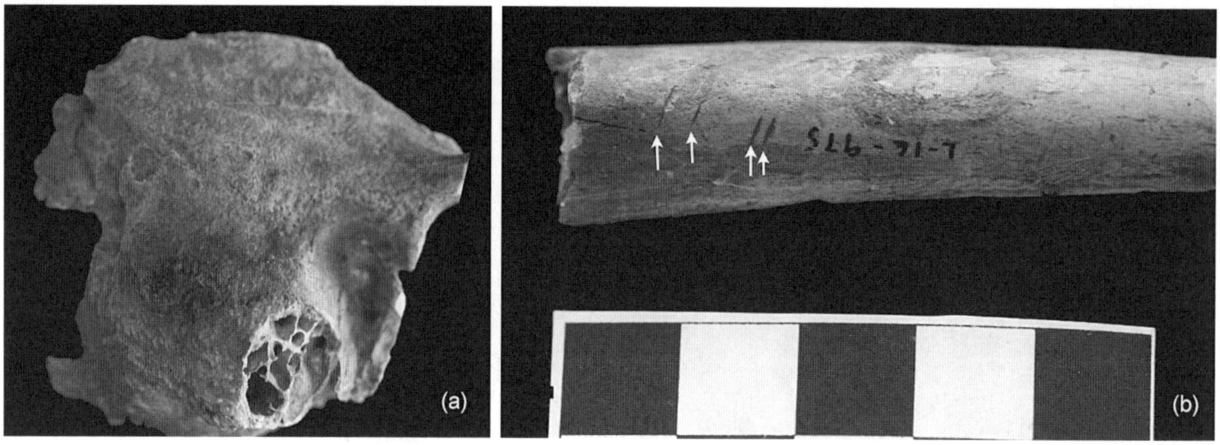

FIGURE 4.36. Human remains from Piedras Negras Burial 4: (a) thermal blackening of the right temporal bone (damage to the mastoid process from excavation); and (b) cut marks on a radius fragment (photographs by author).

The Hero Twins ventured into the underworld and managed to outsmart the lords of the underworld after playing a series of ball games.

At Maya sites with significant topographic relief, such as Piedras Negras, Tonina, Yaxchilan, and Plan de Ayutla, the ballcourts are situated in the plazas at the lowest levels of the ceremonial center (see figs. 4.8, 4.12, 4.35). At Yaxchilan and Tonina dead kings were buried in more elevated levels of the ritual precinct, closer to the sun and away from the ballcourt below. Aside from replicating the vertical cosmology, this spatial arrangement also allowed spectators to witness the ball games without passing through restricted areas of the royal precinct. In the Yaxchilan kingdom ballcourts are only located at Yaxchilan itself; none have been identified at any of its secondary centers. Presumably the ball game and other violent contests were played throughout the kingdom. When such events were staged as a performance of sacrifice, however, subjects must have been required to travel to the polity capital to witness such spectacles.

The location of ballcourts at some Mesoamerican sites suggests that they played an important role in mortuary rites and postmortem acts of veneration. The elite tombs at Preclassic La Venta were separated from the main pyramid by a monumental ballcourt (see fig. 4.4). Stephen Houston notes that the ballcourts at some Classic Maya sites seem to have been intentionally aligned to important tombs.[125] A ballcourt at Tikal is directed toward the tomb of Jasaw Chan K'awiil in Temple I, and the ballcourts at Piedras Negras are aligned with important funerary pyramids (see figs. 4.15, 4.8). Ballcourts are generally oriented along the same ritual axes as the burials within each Maya kingdom. At Palenque the ballcourt is oriented 12° E of N, and at Piedras Negras the courts are oriented 31–32°. One court at Yaxchilan is oriented to 120° (the mortuary axis) and the other is oriented to 27° (the perpendicular of the mortuary axis).[126]

Recall that in the K'iche' *Popol Vuh* Hun Hunahpu, the rough equivalent of the Maize God, is sacrificed and buried in the ballcourt. The University of Pennsylvania project excavated a skeleton (Burial 4) from the center of the playing alley of the K-6 ballcourt at Piedras Negras, not far from the location of the sculpture of ritual combatants.[127] The largely incomplete skeleton is that of an adult male, consisting of fragments of the cranium, mandible, both humeri, and a radius and ulna fragment. It is unclear if the incompleteness reflects J. Alden Mason's inadequate excavations or the true nature of the deposit. During a cursory review of the skeletal materials, I noted thermal blackening of the right temporal bone and cut marks on the radius, ulna, and mandible (fig. 4.36). These cuts most likely reflect flaying or defleshing of the body, though they may also have been injuries sustained shortly before death. In any case they provide the best evidence for adult human sacrifice at Piedras Negras, a death that involved fire either before or shortly after the individual was slain. The ballcourt's playing alley is aligned with the K-5 pyramid, the probable funerary monument of Lady Hix

FIGURE 4.37. View of the K-5 pyramid from the playing alley of the K-6 ballcourt at Piedras Negras during the 1930s excavations and in 2010 (earlier photograph courtesy of the University of Pennsylvania Museum of Archaeology and Anthropology, recent photograph by author).

Witz Ajaw, the principal consort of Ruler 2 (fig. 4.37).[128] The temple stairway is flanked by bearded zoomorphic masks that may be representations of the "Bearded Dragon," a serpent being that seems to be related to celestial ascent.[129] The northerly position of the K-5 pyramid relative to the K-6 ballcourt further suggests ascent from the underworld to the celestial sanctuary atop the pyramid. The traumatized skeletal elements within the ballcourt not only identify it as a place of sacrifice but are perhaps the remains of a *k'ex* offering to underworld supernaturals in exchange for the queen's solar ascension.

The concept of celestial ascent from the ballcourt is further suggested at Yaxchilan and Tonina. The playing alleys of the ballcourts at both sites were lined with three circular altars, incised with images of dead royalty (see figs. 2.9, 4.31, 4.35). The text of the center sculpture at Tonina, Monument 69, describes the lord's period of 260 days inside the earth, after which "fire entered his tomb," perhaps some sort of catalyst for the ascension of the dead king.[130] The sides of the Tonina playing alley were adorned with sculptures of captives, presumably offerings to underworld beings similar to the remains found in the ballcourt alley at Piedras Negras (fig. 4.38).[131]

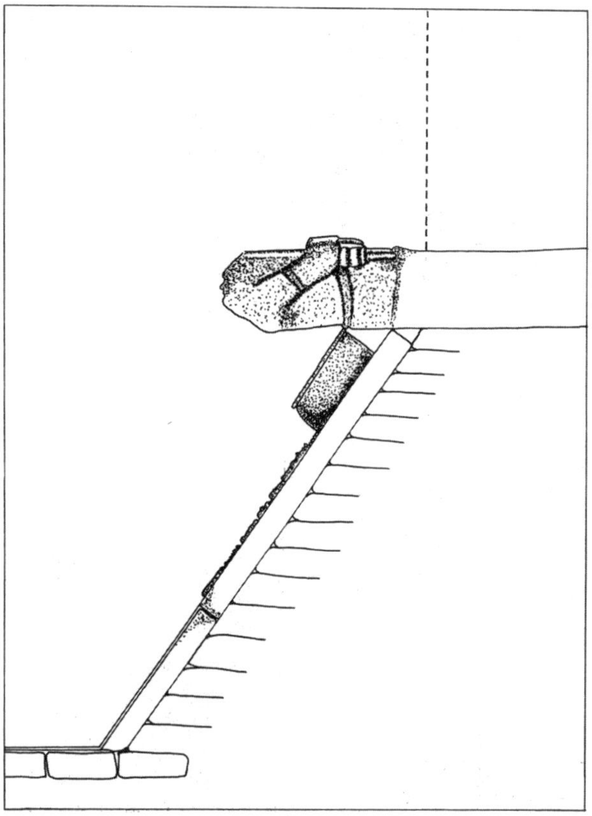

FIGURE 4.38. Sculpture of a captive at the Tonina ballcourt (illustration originally published as Becquelin and Baudez, *Tonina, une cité maya du Chiapas* vol. 1, fig. 91, courtesy of Claude Baudez).

WATERY PLACES AND UNDERWORLD SPACES

Like ballcourts, caves were understood to be entrances to the underworld (fig. 4.39). In hilly and mountainous regions caves are thought to be the source of rain and clouds and in Classic times were perceived as the home of Chahk.[132] In the karst environment of the lowlands, water may percolate or pool in subterranean caverns. Sufficient dissolution of caves will cause their collapse, creating the cenotes that abound in the region. Although some caves and cenotes are found in the heart of Classic Maya communities, others are located in uninhabited parts of the landscape and have been destinations for ritual pilgrimage since Preclassic times. The contemporary Maya visit caves to make offerings and requests of the supernatural beings that dwell within. Presumably the Classic Maya performed similar rites.

Cave inhumation was not widespread among the ancient Maya.[133] In part this is because caves, particularly caves located outside of communities, could not easily be monitored and protected. Bodies placed inside caves would have been vulnerable to disturbance and violation. Instead the most important individuals were interred underneath tons of stone rubble in the very heart of Classic Maya kingdoms. Relative to other parts of the Maya world, however, bodies seem to have been more frequently placed in caves and rock shelters in Belize, perhaps reflecting a unique social and political landscape in the eastern lowlands.[134] If caves could be monopolized by a community or simply were not at risk of disturbance, the Belizean Maya were presumably less concerned about the desecration of remains placed inside of them. In postconquest times the Lacandon of Chiapas used caves as ossuaries to house the remains of their dead.[135] This practice likely reflects the particularly demographic situation for the Lacandon Maya, who have maintained low population levels and have been semimigratory, at least until very recent times. As Didier Boremanse notes, the Lacandon moved the household as soon as the agricultural land was exhausted and, with the forest to themselves, simply selected a new place to settle with good land. Thus the veneration of the dead does not seem to relate to access to territory and other resources for the Lacandon as it might for other Maya groups. What little is inherited by the Lacandon consists primarily of incense burners and other personal items, which go to "the sons who stay with their father and treat him well. . . . the sons who leave their father and live in neolocal or uxorilocal residence do not inherit from him."[136] In this respect it is interesting that residential location still influenced inheritance among the Lacandon.

FIGURE 4.39. Interior and entrance to a cave at La Mar (photographs by author and Charles Golden).

Despite the rarity of formal cave burial in other parts of the Maya lowlands, burials and human remains are occasionally found in such contexts. Burial 56 at Piedras Negras was excavated in a cave north of the Northwest Group Plaza by Pierre "Robbie" Colas. The deposit contained the remains of at least three individuals: a more or less complete nine-year-old, a fetus, and another subadult individual, along with a dart point and a ceramic figurine depicting the emergence of God N from a shell. The nine-year-old child was placed face down in the deposit. Overall this collection of human remains is completely unlike the burials of Piedras Negras and most likely represents a *k'ex* offering to the supernatural beings that dwelled within. Similar to the God N figurine from the deposit, a rock shelter south of the epicenter bears an incised turtle shell with God N and K'awiil emerging, suggesting that such locations were perceived as openings or portals from which supernaturals emerged (fig. 4.40).[137]

Other caves in the area contain scatters of disarticulated human remains, though such deposits have not been systematically studied. In a hilltop cave located near Oso Negro, a Yaxchilan secondary center, I observed an abundance of human remains and broken ceramic vessels on the floor, much of which was burned (fig. 4.41).

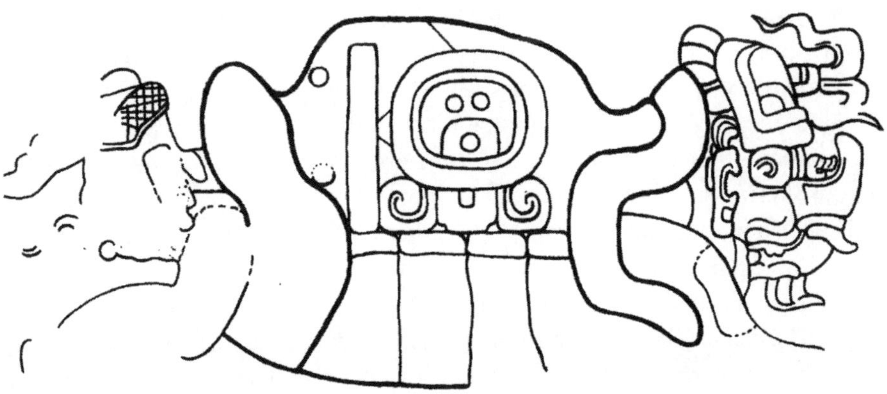

FIGURE 4.40. God N and K'awiil emerging from a turtle shell on a rock shelter on the southern edge of Piedras Negras (drawing by Zachary Hruby).

FIGURE 4.41. Burned scatter of human bone and ceramic sherds from Cueva de las Tres Entradas (photograph by author).

I performed a basic in situ analysis of the remains, which included multiple children and adults. These may be skeletal elements taken from burials elsewhere in the kingdom or perhaps the remains of people offered in sacrifice, burned as underworld offerings.

Noting the disproportionate amount of child remains in caves relative to noncave mortuary contexts, Ann Scott and James Brady have suggested that at least some of the human bones that are found in caves are the remains of human sacrifice.[138] Indeed I suspect that many of the human remains found in the caves of the western lowlands are offerings to the temperamental supernaturals that dwell beneath the earth. Numerous Classic period vessels show underworld *wahy* eagerly receiving and serving dishes of human limbs and other body parts (see fig. 3.57). The contemporary Maya are concerned about and even fear subterranean supernaturals. The Tzeltal and Tzotzil observe that one of the worst things that can happen is for a person's soul to be possessed by the "Earth Lord" or "Mountain Lord," a being likened to a greedy Ladino landowner throughout the Maya area.[139] Vogt and Stuart describe the ambivalence with which the Tzotzil approach the Earth Lord (Yahval Balamil) as "a combination of hopeful enticement and sheer terror."[140] For the Tzotzil of Zinacantán any cave is a potential place of communication with this capricious being, who may shower people with a wealth of money and livestock or may enslave their souls. In a similar example of fear of cave supernaturals, John McGee describes a Lacandon cave shrine as dedicated to the god Ah Kˀakˀ, "who was feared for the sickness he caused. He received many offerings to placate his wrath."[141]

The preferential selection of children as sacrificial victims was especially evident in Vera Tiesler's analysis of cenote remains from the northern Yucatán.[142] She notes a near absence of infants yet a preponderance of children and adolescents, the inverse of what is expected in a normal demographic mortality profile. Tiesler also identified evidence for heavy postmortem processing (burning, flaying, defleshing) of some remains from the cenote at Chichen Itza, which leads her to believe that people were not necessarily killed by being thrown into the cenote but that the remains of sacrificial victims were deposited after extensive ritual manipulation of the corpse (flaying, defleshing, disarticulation). This may also partly explain the preponderance of disarticulated remains found at some Maya caves.[143] We can only wonder how many rivers, lakes, and other watery places received human bodies (and body parts) over the course of the Classic period.

FIERY PLACES AND SOLAR SPACES

Mayanists have too long conflated the K'iche' concept of Xibalba (place of fright) with the destination of souls in the afterlife.[144] Such a dismal afterlife would make the Classic Maya a terribly pessimistic people. Quite the contrary: much of the Classic Maya imagery of death pertains to themes of resurrection, renewal, and the potential for new life. It is unlikely that the Maya envisioned their ancestors as dwelling permanently in a cold, dark, wet place. Unfortunately, the concept of an afterlife is only lightly addressed in Maya ethnography and remains inadequately understood for the Classic period. For the Tzeltal, Pitarch suggests that this is because "the origin and destiny of the *ch'ulel* [personal soul] are completely irrelevant to the circumstances of one's personal life, and vice versa." The Tzeltal have a notion that the *ch'ulel* originates from the sky and is introduced into the fetus during early pregnancy. Yet when the body dies the soul apparently descends to *k'atinbak* (bone burned to ashes), where bones are pulled from graves and used as kindling, because this underground place is both cold and dark (see the Classic Maya Death God warming himself with human bone in fig. 1.36).[145] The concept of the afterlife is given a bit more clarity by Vogt's Tzotzil informants. As he explains of the journey of the soul:

> Here the soul must cross a large river on the back of a black dog and follow a trail to a junction marked by a cross shrine. It learns whether it will continue on by the broad road to *k'atinbak* or, by a narrow, crooked path, to *winahel*, somewhere in the sky. Unless a person has seriously deviated from the moral code of Zinacantán, his soul may take the path to *winahel*, which is much like an earthly Zinacantán with similar social responsibilities. If someone has sinned and is condemned by his kinsmen and neighbors as a witch or a murderer . . . , the soul of the deceased is banished to *k'atinbak*.[146]

It is possible that the distinction between heavenly reward and underworld punishment reflects the influence of Christian eschatology. Nevertheless, Taube provides abundant evidence for the existence of a Classic period celestial paradise, "Flower Mountain," as discussed throughout this book. This place is the domain of the Sun God and other celestial beings and is the idealized destination of the souls of the dead. It may very well be that the Classic Maya understood death to be an underworld descent followed by a celestial ascent in a circuit that mimics the daily path of the sun (depicted in imagery pertaining to solar cartouches and ballcourts noted in earlier chapters).

Evidence from Palenque, more than any other site, demonstrates the conflation of death with solar rebirth and a final destination for the soul in the sky. Mortuary spaces were not cold and dark places. Rather, they were conceived as fiery, solar places. On a limited scale this is apparent in internal funerary space, such as the blazing red painted walls of the Temple XX Early Classic tomb.[147] But this concept was also integrated into the very design of Palenque by the construction of the three temples of the Cross Group, the ceremonial heart of Palenque during the eighth century and the center of both ancestor and supernatural veneration (see figs. 4.16, 4.17). The principal structures are the Temple of the Cross, Temple of the Foliated Cross, and Temple of the Sun. Each of these structures is linked to a patron of the kingdom, two of which, GI and GIII, are aspects of the Sun God.[148] The pyramids are arranged around a center in an obvious parallel to the three hearthstones of creation, emblematic of both primordial fire and the axis mundi. A low radial platform marks the heart of the Cross Group (see fig. 4.17). David Stuart and George Stuart hypothesize that this central structure may have served as the foundation for a great wooden pole, similar to those discussed earlier and emblematic of the sacred trees shown on the sarcophagus lid and on the Temple of the Cross tablet. Although we do not know when the structures of the Cross Group were originally built, we do know that the final versions were dedicated in AD 692 as part of Kan Bahlam II's construction program and his efforts to establish himself as the great venerator of Pakal and Palenque's patron supernaturals.[149] Employing the spatial logic explored earlier in this chapter, the Temple of the Foliated Cross in the east is associated with dawn. This is the shrine of GII, Baby K'awiil, the embodiment of lightning and agricultural fertility in its newborn form. The Tablet of the Foliated Cross shows the sun rising along the axis of a jeweled maize plant, the base of which is in the primordial sea. The Temple of the Sun is located in the west, the direction of the setting sun, and is the shrine of GIII, a variant of the Sun God linked to war and the underworld. The central icon of the temple's panel is the Jaguar God of the Underworld. Located in the north is the Temple of the Cross, emblematic of solar ascent and the zenith. Appropriately, it is the shrine of GI, the aquatic Sun God that rises from the eastern sea. In summation, the Cross Group operated as a world center, a map of the circuit of the sun, and a sacred place to honor and remember the patron deities and the ancestors whose remains were interred throughout the complex. The principal east-west axis of the Cross Group is not the more common 102–108° axis of Palenque. Instead the Temple of the Sun and the Temple of the Foliated Cross are aligned at 123° E of N. The sun rises over the Temple of the Foliated Cross on the winter solstice as viewed from the Temple of the Sun. The sun sets behind the Temple of the Sun on the summer solstice as viewed from the Temple of the Foliated Cross.

Inscriptions from the Temple of the Cross provide its name, which has been partially deciphered as a "Six Sky place."[150] Much of the imagery associated with the building is celestial. This Six Sky place seems to be a specific reference to the heavens in Classic period cosmology. Text from the Tablet of the Cross describes the ascent of GI: he "goes up (to) Six ?-Sky, (to) the northern Eight-GI Edifice [*naah*], (which is) the name of the house [*otoot*]."[151] The glyph after *otoot* is usually read as *xaman* and translated as north, though this decipherment is not secure. The stairway of the Temple of the Cross ascends six levels before reaching the superstructure (see figs. 4.16, 4.17). The pyramid symbolizes the levels of mountains and the sky; the temple superstructure was quite literally a Six Sky place.

"Six Sky" appears in other texts from the Maya world, as both a place and part of a name. Stela C at Quirigua describes how three supernaturals planted stones at the "first three stone place" under the supervision of a Six Sky *ajaw*, apparently the lord of the Six Sky place.[152] The inscriptions from Tomb 12 of Río Azul end with a passage that David Stuart originally translated as "was buried Six Sky" followed by two other undeciphered glyphs. Stuart thought that "Six Sky" was a reference to the name of the deceased.[153] At Naranjo Six Sky was the name of an important Late Classic period lady, and at least one of the famous Holmul dancer vases (produced in the vicinity of Naranjo and contemporary with her reign) names a Six Sky place as linked to the dynasty of Calakmul (K3400). A throne back in the Museo de Amparo from an unprovenanced locale (but likely somewhere in the western lowlands) has a passage that describes the Patron of Pax's descent from the "Six Sky place" as the messenger of God D (Itzamnaaj).[154] There are enough references to Six Sky to suggest a pan-Classic Maya concept. It is perhaps a primordial celestial place

FIGURE 4.42. Two excavated crypts on the second terrace of the Palenque Temple of the Cross (photograph by author).

occupied by God D and other supernaturals, the specifics of which varied with local mythology. At Palenque this Six Sky place seems to have been materialized as the Temple of the Cross.

The Cross Group and especially the Temple of the Cross were the focus for elite interment throughout the Late Classic period. Early reports by Edward Thompson and later excavations by Mexican archaeologists have revealed dozens of crypts and tombs inserted into the terraces of the Temple of the Cross (fig. 4.42).[155] Close to one hundred censer stands have been excavated from the Cross Group. If even a fraction of these objects were put into operation at the same time the Cross Group would have been enveloped in an impressive cloud of smoke and flame. Martha Cuevas notes an interesting pattern; censer stands associated with the Temple of the Cross are primarily supernatural beings (especially GI and the Jaguar God of the Underworld), whereas those associated with nearby Temples XIV and XV are humans.[156] Arguably, the dead buried in these latter temples were significant enough to be identified as singular ancestors, whereas those buried in the smaller burial chambers on the flanks of the Temple of the Cross and Temple of the Foliated Cross were not.

The triadic arrangement of the Cross Group at Palenque dates back to at least the Late Preclassic period, as evident from the triadic groups at El Mirador.[157] Across

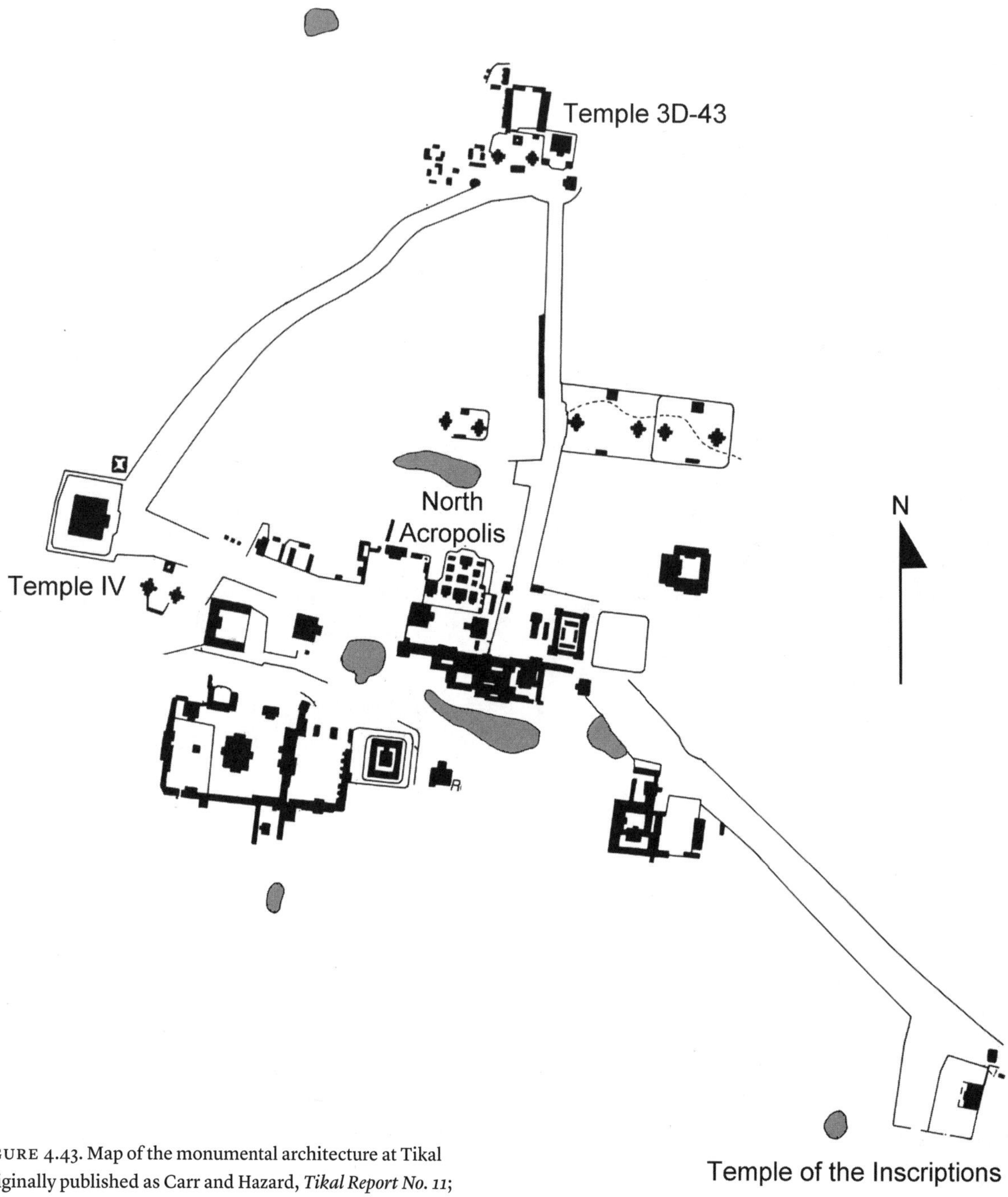

FIGURE 4.43. Map of the monumental architecture at Tikal (originally published as Carr and Hazard, *Tikal Report No. 11*; image courtesy of the University of Pennsylvania Museum of Archaeology and Anthropology). The map is corrected to show true north.

the Central Petén triadic groups were used as centers of place, as at Tikal and Calakmul.[158] Although the addition of later structures often blurs the definition of "triadic," these groups are nevertheless recognizable by the arrangement of three pyramidal structures facing into a common space. Tikal evidences a series of triadic groups of increasing spatial scale. Within the North Acropolis, 5D-21, 5D-22, and 5D-23 form a small triadic group at the summit of the North Acropolis (see fig. 4.15). A larger triadic arrangement was formed by the construction of 5D-33, 5D-1 (Temple 1), and 5D-2 (Temple 2), establishing the Great Plaza as ritual center.[159] Finally, 3D-43, Temple 4, and the Temple of the Inscriptions establish the civic-ceremonial center of Tikal as an axis mundi, each temple connected to the center by massive causeways (fig. 4.43). Although the pattern does not hold for all triadic complexes, the Tikal triadic groups are like the Cross Group at Palenque and map the daily route of the sun, with temples in the east, north, and west. Also like the Palenque Cross Group, in the largest triadic configuration Temple IV is aligned to the Temple of the Inscriptions at roughly 120° E of N.[160]

Triadic groups are just one of a range of architectural formations that map solar movement. For example, at Tikal the Twin Pyramid complexes were constructed in reference to local cosmology.[161] E-Groups are among the oldest ceremonial configurations that reference the movement of the sun (fig. 4.44).[162] These complexes consist of a radial pyramid associated with a plaza space, bounded on the east side by a long low structure, topped by an additional three buildings. Scholars have long recognized that the alignment of structures within E-Groups references sunrise at different times during the year.[163] At many E-Groups an observer on the radial pyramid will see sunrise over the northeastern structure on the winter solstice, the southeastern structure on the summer solstice, and the central-eastern structure on the equinox. As a place of solar birth, Taube suggests that these structures represent the three-stone hearth at the center of creation. Like the small radial structure

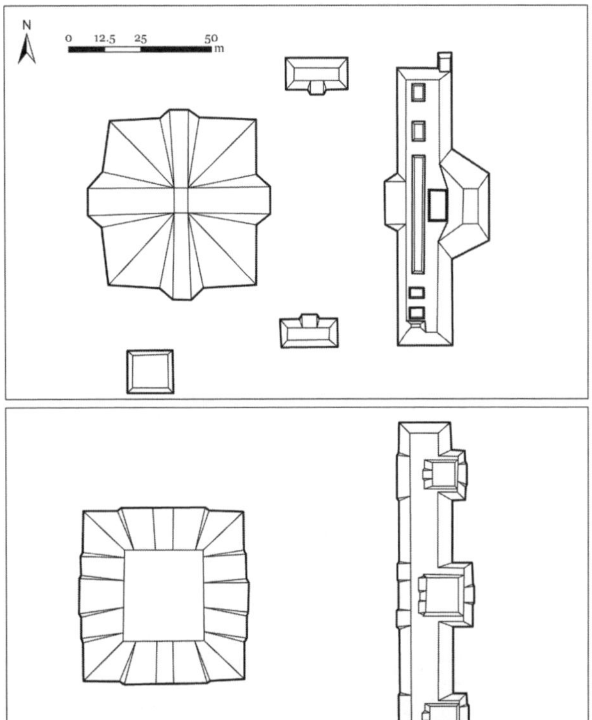

FIGURE 4.44. E-group complexes at El Palmar (*top*) and Tikal Mundo Perdido (*bottom*) (drawings by James Doyle).

in the Palenque Cross Group, the E-Group radial pyramid marks the ritual center, and the three temples of the eastern structure are the three stones laid out in a line.[164] Building on similar symbolic motifs, E-Groups track the changing path of the sun over the course of a solar year, whereas triadic groups map the daily path of the sun. Notably, despite extensive tunneling in the Mundo Perdido radial pyramid at Tikal, no burials have been found.[165] These radial structures are not funerary pyramids but instead served ritual functions pertaining to the centering of the community.

Inscriptions at numerous Maya sites reference a different sort of fiery space in royal mortuary ritual. For example, monuments at Piedras Negras record a ritual,

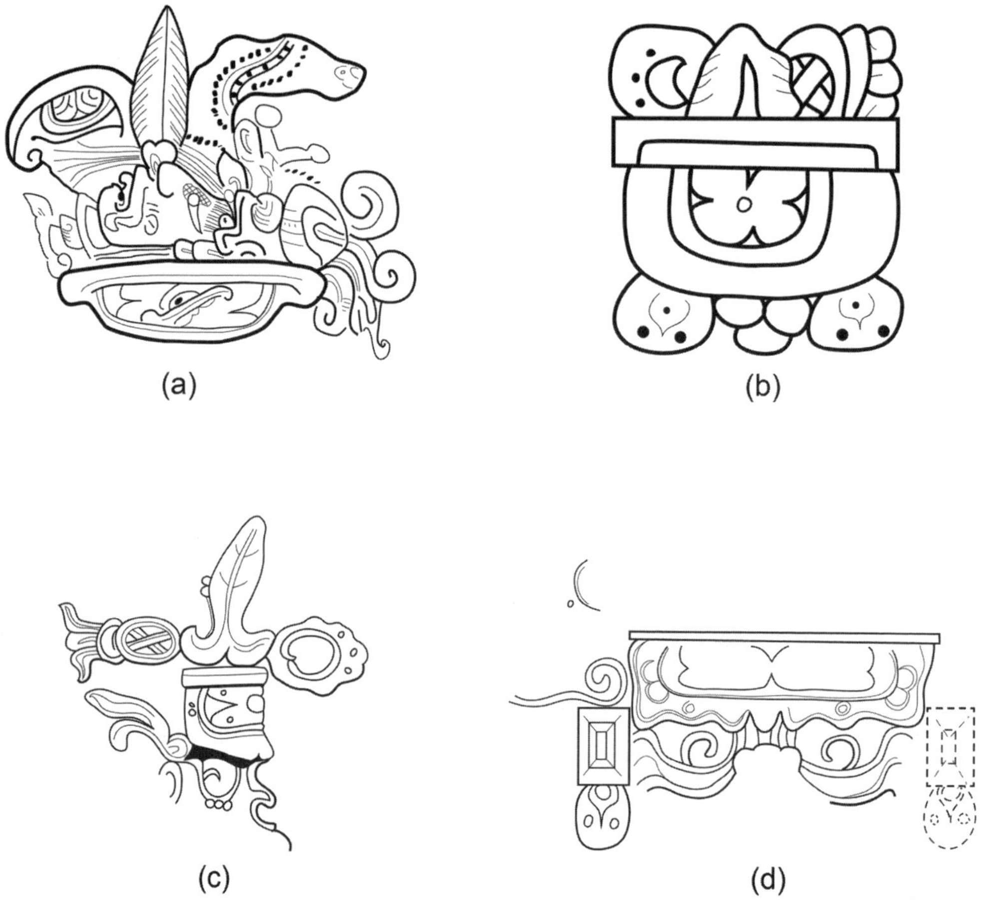

FIGURE 4.45. Solar receptacles and the Quadripartite Badge: (a) as a bowl depicted on the lid of a cache vessel from the Tikal region; (b) as a cache vessel shown in a hieroglyph from the Temple of the Inscriptions; (c) inset into the forehead of a being from House D of the Palenque Palace; and (d) as a solar receptacle on the forehead of a supernatural on Pier E of the Temple of the Inscriptions (drawings by author).

el naah umukil (house burning at the burial), performed at the tombs of Ruler 1 and Ruler 4.[166] More ambiguously, there is a reference to *pul-uy u-ts-itil* ([it] burns his long object) performed on the one *k'atun* anniversary of Ruler 2's *och bih* (road entering).[167] The tombs of Rulers 1 and 2 have not been located. Ruler 4's tomb has been excavated. As discussed in chapter 3, the human remains in the tomb were exposed to heat and flame long after the bodies had decomposed (see fig. 3.28). At the most general level the burning of tombs likely operated to renew the mortuary space and perhaps even suggests solar ascension. The Maize God plate from the Boston Museum of Fine Arts shows this supernatural being resurrecting from a torch within a skull (see fig. 1.12). Conceptually, the Piedras Negras tombs could have been understood as parallels to the earth oven (*pib*) and the sweatbath, *pib naah* (oven building). With eight monumental sweatbaths known from the site core, the lords of Piedras Negras certainly were concerned with the curative and spiritual values of such structures.[168] Moreover, both sweatbaths and tombs may have been conceived as points of entry (birth) and departure (death), respectively. Today in the Yucatán peninsula the crypts used for the decomposition of the body are likened to the *pib*.[169] Nevertheless, without textual confirmation, the link between tombs and sweatbaths at Piedras Negras remains purely speculative.

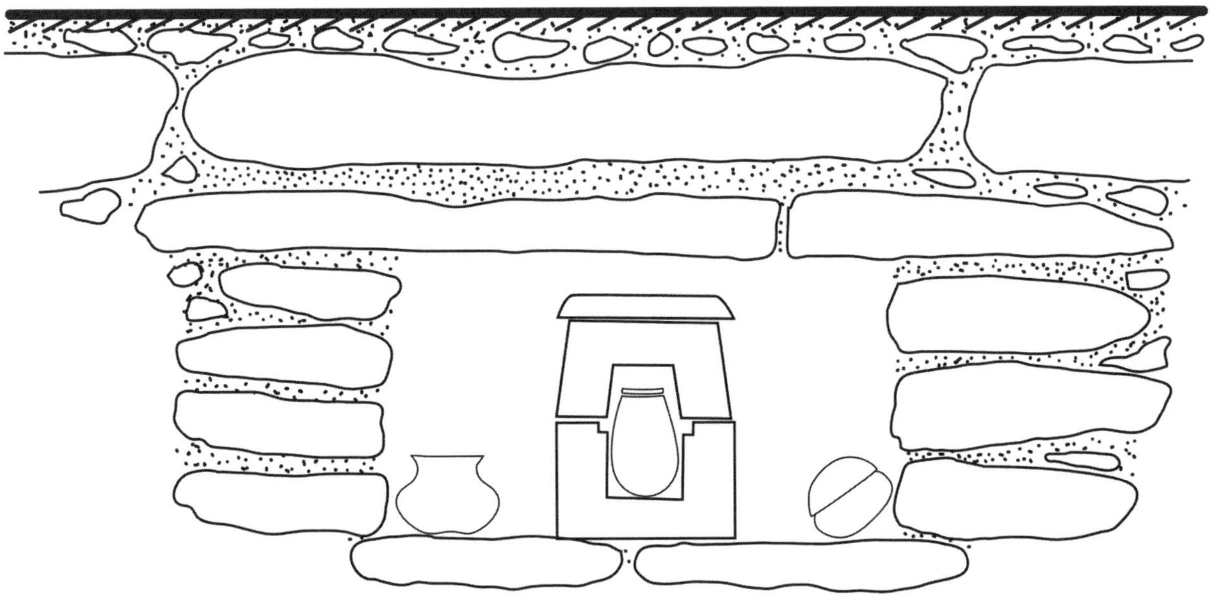

SOLAR RECEPTACLES

As spiritually potent objects, royal bodies were interred with the same consideration that was accorded to other charged objects. Inhumation, especially of royalty, was very much like caching (or vice versa), as first suggested Marshall Becker.[170] In the past hundred years an impressive diversity of caches has been excavated across the Maya area. These deposits contain a variety of ritual objects, including stingray spines, *Spondylus* shells, stone eccentrics, and other effigy objects. These deposits were not simply "offerings," however, but represent the remains of spiritually potent materials that, following their use, were carefully contained, sealed away, and interred in ritually significant places.[171]

In Classic period imagery caches are shown as the so-called Quadripartite Badge, sometimes mounted within the forehead of G1 or some other supernatural being (fig. 4.45; see also 1.20b).[172] The badge includes a *k'in*-marked object that indicates its hot, fiery, solar nature. Piled above the *k'in* object are a stingray spine, a *Spondylus* shell, and a third, less easily interpreted object. The emblem is especially common in Palenque imagery and is associated with the supernatural patron G1.[173] Although most scholars assume the *k'in*-marked object is a bowl, I suspect that in some contexts the subtle yet intentional differences in various depictions of the Quadripartite Badge refer to a particular type of container related to Maya sacrifice and rituals of renewal. Classic Maya depictions of ceramics are quite specific, and bowls are easily distinguishable from other ceramic types in images of courts, feasting, and bloodletting (for example, see fig. 3.57). Bowls are shown as wide-lipped, relatively shallow, and with restricted base circumferences.[174] Only in a few cases, however, is the *k'in*-marked object in the Quadripartite Badge clearly a bowl (for example, fig. 4.45a). Rather, the majority of *k'in*-marked objects are depicted as having vertical, slightly in-curving walls and are lidded (fig. 4.45b and c). This is the form of Maya cache vessels—objects rarely seen in Classic period imagery but found in excavation.[175] Depictions of censer stands are also marked by the *k'in* symbol, further underscoring the specificity of the vessel depicted within the Quadripartite Badge.[176]

FIGURE 4.46. Lidded cache deposit from the Temple of the Cross, Palenque, Mexico (drawing by author after Ruz Lhuillier, "Exploraciones arqueológicas en Palenque: 1953," figs. 3 and 4).

A cache deposit from the Temple of the Cross at Palenque wonderfully demonstrates how such deposits can manifest the Quadripartite Badge (fig. 4.46). This cache consisted of a masonry box containing two ceramic jars, one of which was lidded, and an unusual two-piece stone container with a false lid. Inside the stone container was a third ceramic vessel, also lidded. Thus the cache's contents were "contained" by three layers: lidded vessel, stone container, masonry box. Each container within the cache held organic and inorganic residue. The interior of the stone cylinder was painted red, identifying it as a *k'in* object.

With their stingray spines and *Spondylus* shells, we could argue that the royal tombs of Piedras Negras and other Classic Maya sites were like-in-kind to the Quadripartite Badge and were places of sacrifice and solar rebirth. The symbolism is even more overt in the rare cases of sarcophagi, two of which have been identified at Palenque and another two at Tonina.[177] The most well-known sarcophagus is Pakal's, from Palenque.[178] The internal stairway of the Temple of the Inscriptions facilitated access to Pakal's tomb long after he was interred. The sarcophagus lid, showing Pakal's ascent from the underworld along the axis of an iconic world tree, was a powerful statement of rebirth, vitality, and continuity (see fig. 2.5).[179] Beyond the normal tropes of Maya fertility and rebirth through death, this icon was especially vital, considering that Pakal ruled for almost sixty-eight years; few if anyone at Palenque would have remembered a time without him. It is no surprise that the Palenqueños prepared a funerary monument for their revered king that permitted interaction long after he was dead. Visitors to his tomb stood on a stone landing at the foot of the sarcophagus and would have looked directly at the representation of the king (presumably the focus of veneration after his death), an image that linked his celestial ascent with the vitality of the polity. The area around the depiction of the king and the Quadripartite Badge are blackened, perhaps from censers placed on the lid of the sarcophagus.

As a focus of veneration Pakal's sarcophagus operated much like the preserved corpses of political leaders in modern times. As the poet Vladimir Mayakovsky famously proclaimed following Vladimir Lenin's embalmment, "Lenin, even now, is more alive than the living."[180] Prior to the advent of embalming, a putrefying corpse was hardly ideal for social, political, and ritual action. The bundled corpses discussed in chapter 2 were the only option available to the Maya for the public display and procession of corpses. In this sense the sarcophagus was an important ritual and even political innovation, not unlike the Peruvian mummy bundles. Gordon Rakita and Jane Buikstra interpret Andean mummification as an attempt to keep the souls bound to the world of the living, where they could continue as a powerful social force. Mummies were beings that "never entirely leave this world but are nevertheless not of this world."[181] For the Maya, the sarcophagus facilitated near-direct interaction with the body of the dead king without the need to confront his putrefying remains.

The second Palenque sarcophagus was encountered in the neighboring pyramid, Temple XIII, by Arnoldo González Cruz and Fanny López (see fig. 4.27).[182] Typical of Palenque tombs, the funerary chamber was converted from an earlier three-roomed vaulted building that was then covered by the final version of Temple XIII. Like Pakal's sarcophagus, the Temple XIII sarcophagus was carved of a single gigantic piece of limestone but is smaller and bears no inscribed imagery. The inside surfaces of both Palenque sarcophagi were coated in thick layers of cinnabar, transforming them into receptacles of solar rebirth and ascent. Archaeologists also found a censer still sitting atop the lid of the Red Queen's sarcophagus when they entered the tomb chamber.

Some of the especially long Quadripartite Badges in Palenque's iconography specifically suggest the shape of the sarcophagus. The long form is shown on Pakal's sarcophagus lid, on Pier E of the Temple of the Inscriptions, and on the Tablet of the Cross (see fig. 4.45d; see also figs. 2.5 and 2.6). In these contexts the sarcophagus was substituted for the cache vessel as the receptacle of sacrifice and may be a specific reference to the tomb of Pakal as the source of the "Shiny Jeweled Tree" that is venerated by Kan Bahlam on the Tablet of the Cross.[183]

Two other probable royal tombs have been identified at Palenque, one in a substructure underneath Temple XVIIIa and the other underneath Temple XX. Both of these predate the tombs of Pakal and the Red Queen and

FIGURE 4.47. Slab-crypts within the Temple XV substructure, Palenque (originally published in Thompson, "Ancient Tombs of Palenque").

lack sarcophagi. It seems that the sarcophagi were first made at Palenque sometime in the late seventh century and were manufactured for the royal dynasty. Around the same time, however, a new style of "slab-crypt" burial came into use at Palenque.[184] These receptacles were constructed of massive flat limestone slabs arranged to form what is effectively a composite sarcophagus. A single slab was used for each crypt side, and a fifth was used as the lid. The slab-crypts differ from the more common masonry crypts not only in their construction but in their placement *above* the floor of sealed chambers, in a manner similar to the sarcophagi. To the best of my knowledge these slab-crypts are found exclusively within Palenque's site center and are the graves of nobility and other elites of eighth-century Palenque.

Many of these slab-crypts are located within the terraces, below the floors, and within buried structures of buildings in and around the Cross Group at Palenque (fig. 4.47).[185] One of the most important is a three-roomed building that was converted into a series of mortuary chambers and then buried below Temple XV. Alfred Maudslay described this chamber in his report of the site:

> The entrance to these chambers was originally from the top of the mound by a flight of steps descending to the south-eastern chamber. This entrance has been purposely closed with a large stone slab, and access to the vaults is now gained by a hole broken through the masonry. There is a descent of two steps from the outer to the inner or northern gallery, which is divided into three chambers. The doorways of the middle and western chambers have been walled across. In the east end wall of the southern gallery is a doorway now blocked up by fallen masonry and rubbish. On the floor of the middle chamber is a coffin, made of thin slabs of stone, coated on the outside with stucco and smeared on the inside with red powder. The coffin had been rifled of its contents.[186]

Slab-crypts that were placed inside tombs offered a more efficient and convenient approach to sarcophagus-style inhumation. They served as receptacles for the remains of significant personages that could be revisited and venerated without the worry of disturbing the putrefying corpse. The other advantage is that sarcophagi and slab-crypts provided a relatively sealed burial environment in comparison to tombs and masonry crypts, particularly in terms of the invasive activities of rodents and other animals. Other slab-crypts have been reported in Group A, Group H, along the ridge south of the Temple of the Jaguar, within the superstructure of Temple XVIII, and below the floor of the Temple Olvidado.[187]

The other two known Maya sarcophagi are from Tonina, the seat of the archrivals of Palenque's lords. The Tonina Acropolis is vertically organized: palatial residences and administrative buildings are located in the lowest terraces, whereas the upper terraces are dominated by ritual pyramids (fig. 4.48).[188] The massive plaza in front of the Acropolis contains the large, sunken ballcourt noted earlier. Tonina's civic-ceremonial center

FIGURE 4.48. (*top*) The Acropolis of Tonina (photograph by author).

FIGURE 4.49. (*bottom*) Sarcophagus from Tonina Burial IV-1 (photograph courtesy of Claude Baudez).

is a cosmogram that places the underworld at its lowest levels, the domain of the living in the middle, and the place of ancestors and supernaturals in the highest reaches, closest to the celestial realm. A number of tombs and crypts have been identified within Terrace 5, an upper terrace that served as the burial ground of Tonina's royal family and other nobility.

The Tonina sarcophagi were found in vaulted tombs below the open plaza floor of Terrace 5 (fig. 4.49). The Tonina sarcophagi, both Late Classic in date, are far more modest than those from Palenque. The first, Burial IV-1, was identified during the excavations of Pierre Becquelin and Claude Baudez at Tonina.[189] The sarcophagus occupies much of the tomb and consists of a single monolithic container and a lid (later broken into multiple pieces), neither of which are carved. The second Tonina sarcophagus was recently identified as part of ongoing operations at the site under the direction of Juan Yadeun and is quite similar to Burial IV-1 in design, context, and use. The narrow sarcophagi of Tonina resemble the small wooden boxes that have been recovered at Alavaro Obregón, Mexico (near Piedras Negras), Tortuguero, and Actun Polbilche, Belize. These boxes were used as containers for bloodletters, another example of the careful sealing and bundling of spiritually charged objects.[190]

COMMUNITY AND KINGDOM

This chapter offers a modest attempt at situating Maya mortuary contexts in a spatial dimension. Even the simplest burial is enmeshed in many layers of meaning: some that are easy for the anthropologist to discern, others that are hopelessly entangled and opaque. Ancient Maya burials could be houses, watery places, and solar spaces. The most important dead were placed in or near structures that symbolized sacred mountains that rose to the celestial paradise, while victims of sacrifice were cast into real (or symbolic) clefts in the earth to appease temperamental beings below.

Maya burials were places of departure, the point at which souls left on their journey to the otherworld. The travels of the soul were likened to the movement of the sun, with solar symbolism important not only for the placement of burials but for the very spatial organization of Maya communities. Mortuary spaces were also places of return. The bones and graves of the dead were used to conjure ancestors that remained vital to the livelihood of the community. For this reason Maya graves were carefully guarded, placed below the floors of the living or encased within the mass of monumental pyramids. Yet, paradoxically, many burials also needed to be accessible so that lineages and even entire kingdoms could conduct appropriate rites of veneration.

Death and burial is inevitably political. Maya lords venerated royal ancestors in acts that legitimized their authority much as the rest of the populace may have referenced their ancestors in claims to land and resources. Scholars have been especially concerned with how mortuary rites reflect competition among lineages, with less attention paid to the possibility that the greatest tension was more likely among members of the same lineage. Yet all was not divisive. Decades of preoccupation with Maya mortuary spaces as measures of wealth and power have blinded us to the essential commonalities that united king and commoner, community and kingdom, into domains of shared practice and belief.

EPILOGUE

I have had the opportunity to study the royal burials from quite a few Maya sites. Yet most of my time has been spent working with the inhumations and remains of the common people. As I hope is evident in the preceding four chapters, all graves have the potential to tell an important story and to contribute to our broader understanding of the ancient Maya. Each mortuary deposit has the potential to surprise, to contain an object or practice that we have never seen before or to suggest that something we once thought was idiosyncratic was in fact patterned behavior with important significance. Such was the case, for example, with the dancing dead of El Kinel. Similarly, it was only after we encountered a series of empty crypts at El Porvenir that we understood that such deposits could not be dismissed as poor preservation but were instead evidence of reentry and removal of bone. Both of these cases—dancing dead and empty crypts—remind us of the diversity and complexity of Maya mortuary practices and hint at the importance of the dead to the community of the living.

Archaeologists who write about burials often focus on topics such as status, power, gender, or identity. These are certainly important and are touched upon at various points in this book. Yet these are unlikely to have been the primary concerns of the Maya themselves when dealing with the uncertainty, turbulence, and social rupture that is death. The living must cope with grief and grapple with the vacuum created in the newfound absence of a relative, friend, or fellow community member. Quite simply, we can learn many more interesting things from Maya burials than such mundane matters as inequality or hierarchy. Too much tabulation of pots and counting of jade ignores the nuances and significance of Classic Maya ritual traditions.

For the Classic Maya, two very basic problems needed to be addressed at death: bodies were disposed of and souls were managed. Most corpses were likely washed, a few even painted in vibrant reds. They were carefully wrapped, some in cloth sealed with resins. Others were clad more simply, perhaps dressed in the clothing of life and placed in their graves. Containers to hold the bodies were prepared, ranging from simple holes dug near the house to spacious tomb chambers, encased within gigantic pyramids. Then the real work began: souls were encouraged to depart, content with the life and living they left behind. Rather than descending into the dangers of the underworld, they ideally rose as celestial bodies to the paradise above.

The study of ancient burials allows us to access scales of time not easily discerned from other archaeological contexts. On one end of the temporal scale burials can be understood as "moments in time," death and burial events that represent mere days or weeks. Rarely do archaeological deposits offer glimpses at such finite episodes in the lives of ancient people. But burials also reflect deep time. In the grave we find bodies that we read as "osteobiographies," stories that reveal bits and pieces of the life and death of people who lived over a millennium ago. We also can access ritual traditions that reflect decades if not centuries of shared practice. Mortuary contexts also speak to the future, at least the future envisioned by those who interred the dead. Monumental constructions such as the Temple of the Inscriptions at Palenque are testimony to perseverance in the wake of loss. Similarly, the removal of ancestral bones during the abandonment of a town speaks to the strength of the lineage despite the failings of a particular place or community.

In writing this book I have sought to shed light on these dimensions of the past, present, and future, as they were understood and enacted by the Classic Maya. The first chapter explores the skeleton as biography, the lived body. Scholars have long recognized the value of bones for teaching us about topics such as ancient health, diet, migration, or violence. What is perhaps less fully appreciated is that the remains of the dead, when properly situated in their social context, can also illuminate other dimensions of the past, including belief, ritual practice, and even notions of the self. Acts like body modification may speak less to issues of identity than to ancient Maya understandings of personhood and the soul. Moreover, these same issues extend to the dead body, its treatment reflecting the management of the soul. The consideration of beliefs pertaining to the body and soul is especially important for making sense of such diverse situations as the killing and discard of sacrificial victims and the removal of bone from ancestral graves. Such understandings are also relevant when we consider meaning in Maya mortuary contexts. When those traditions are mapped onto the landscape, new insight emerges into seemingly unrelated topics such as settlement design, political practice, or modes of inheritance.

Archaeology requires patience and rewards perseverance. It takes time to begin to understand a place, its people, and their practices. This is especially true of a society as rich, diverse, and varied as the Maya of the Classic period. For the archaeologist, insight is unlikely to come from a single field season of excavation or by reading one or two books. Rather, it takes years of excavation or laboratory analysis before we can begin to gain real perspective into the lives of the ancient Maya. Ethical obligation also requires us to pay close attention to the efforts of our colleagues and predecessors. Quite a bit of new information can be learned by simply paging through old monographs or opening the dusty boxes that line the shelves of our museums. Moreover, we must justify our future archaeological endeavors. Excavation is destruction, and new projects further burden hopelessly crowded storehouses.

Yet insight can be gained from new (judicious) excavation, especially when armed with knowledge wrestled from past scholarship. For example, our understanding of mortuary processes in the kingdom of Piedras Negras is dependent on the groundbreaking work of the University of Pennsylvania project of the 1930s and the later efforts of the Brigham Young University project of the late 1990s. Yet only in more recent years have we really begun to understand death rituals as they were practiced throughout the kingdom. Moreover, there is value in comparative work. We often treat the Classic Maya as homogeneous through time and space. Yet when we explore the data on mortuary practices even among neighboring contemporaneous kingdoms of the western lowlands, it is evident that each respective kingdom lived and died in distinct communities of shared ritual practice. Such a finding points to the fundamental importance of spiritual matters to the very fabric of Classic period kingdoms.

This book was not the first and will not be the last written about the skeletons and burials of the ancient Maya. It is an important step forward, highlighting both the commonalities and differences in life and death in the Classic period, from burial to burial, site to site, and kingdom to kingdom. Although the syntax of death rites varied in each polity, much of the basic spiritual vocabulary was shared across the Maya area. The Classic Maya were especially concerned with the departure of souls and their solar ascent to a celestial paradise. In the Classic period, as today, the dead were fundamental to the well-being of family, lineage, and community.

NOTES

INTRODUCTION

1. McAnany, *Living with the Ancestors*, 4, 7.

2. Fitzsimmons, *Death and the Classic Maya Kings*, 16. This volume is especially valuable as a synthesis of epigraphic data on Classic period mortuary rites and beliefs pertaining to life, death, and the soul. See also Eberl, *Muerte, entierro y ascensión*.

3. My colleague Stephen Houston graciously allowed me to attend his undergraduate class on Classic Maya epigraphy.

4. Spinden, *Ancient Civilizations of Mexico and Central America*, 177.

5. For a review, see Webster, *The Fall of the Ancient Maya*.

6. Estrada-Belli, *The First Maya Civilization*.

7. Saturno, Taube, and Stuart, *The Murals of San Bartolo, El Petén, Guatemala, Part 1*.

8. Restall, *The Maya World*.

9. This estimate is based on combined data from the CIA World Factbook and the Comisión Nacional Para el Desarrollo de los Pueblos Indígenas.

10. Important prior works on Precolumbian Maya souls include Houston, Stuart, and Taube, *The Memory of Bones*; Fitzsimmons, *Death and the Classic Maya Kings*; Grube and Nahm, "A Census of Xibalba"; Calvin, "Where the Wayob Live"; Houston and Stuart, "The *Way* Glyph"; Gillespie, "Body and Soul among the Maya"; Eberl, *Muerte, entierro y ascensión*.

11. Pitarch, *The Jaguar and the Priest*.

12. Vogt, "Human Souls and Animal Spirits in Zinacantan," 1154.

13. Pitarch, *The Jaguar and the Priest*, 12.

14. Watanabe, "Unimagining the Maya"; Warren, *Indigenous Movements and Their Critics*.

15. I have noticed that one of the long-term effects of working with archaeological projects is that the Classic Maya are demystified. Once our collaborators are convinced that the ancients were not superhuman and could build pyramids with basic technologies, conversation invariably turns to food. Some affinity is gained with the knowledge that maize was also the staple crop of the Classic Maya, yet our collaborators are incredulous when we explain that the Maya had no chickens or mangos.

16. Smith, *Excavations at Altar de Sacrificios*, 212. See also Welsh, *An Analysis of Classic Lowland Maya Burials*.

17. Little and Kennedy, *Histories of American Physical Anthropology in the Twentieth Century*.

18. For a recent overview, see Mascia-Lees, *A Companion to the Anthropology of the Body and Embodiment*.

19. In the past few decades much of this work has been subsumed by the field of bioarchaeology. For overviews of bioarchaeology, see Larsen, *Bioarchaeology*; Buikstra and Beck, *Bioarchaeology: The Contextual Analysis of Human Remains*; Martin, Harrod, and Pérez, *Bioarchaeology*.

20. Houston, Stuart, and Taube, *The Memory of Bones*.

21. The key foundational texts include Foucault, *Discipline and Punish*; idem, *The History of Sexuality*; Butler, *Gender Trouble*; and idem, *Bodies That Matter*. For an archaeological perspective on the notion of the materiality of the body, see, for example, Joyce, *Gender and Power in Prehispanic Mesoamerica*; idem, "Girling the Girl and Boying the Boy"; idem, *Ancient Bodies, Ancient Lives*; Meskell and Joyce, *Embodied Lives*. For a refutation of constructivist approaches to the body in Mesoamerica, see Houston and McAnany, "Bodies and Blood."

22. See especially Sofaer, *The Body as Material Culture*.

23. This might suggest that something is amiss with "materiality" if it cannot be defined. See, for example, Miller, "Materiality." The author avoids defining the topic, despite having a section entitled "What Is Materiality?" in a chapter called "Materiality: An Introduction" (4). For a critique on the study of "materiality," see Ingold, "Materials against Materiality."

24. Sofaer, *The Body as Material Culture*, 76. Again, Sofaer never defines what she means by materiality in her book.

25. Ingold, "Technology, Language, and Intelligence," 470.

26. Bahn, *Written in Bone*; Maples, *Dead Men Do Tell Tales*.

27. Saul, "Osteobiography," 373.

28. Robb, "Time and Biography," 160.

29. Bell, *Ritual Theory, Ritual Practice*, 220.

30. Bell, "Response," 278.

31. Hertz, *Death and the Right Hand*; van Gennep, *The Rites of Passage*.

32. Bell, *Ritual Theory, Ritual Practice*, 220.

33. This is often referred to as the direct historical approach. For early applications, see Wedel, "The Direct-Historical Approach in Pawnee Archaeology." See also Strong, "From History to Prehistory in the Northern Great Plains." For Mesoamerica, see Marcus and Flannery, "Ancient Zapotec Ritual and Religion."

34. Kertzer, *Ritual, Politics, and Power*, 104.

CHAPTER 1: LIVED BODIES

1. For a review of the divide between mortuary archaeology and bioarchaeology, see Goldstein, "Mortuary Analysis and Bioarchaeology."

2. Research reports are posted online at http://www.famsi.org/research/piedras_negras/pn_project/piedras_negras.htm; http://usumacinta-archaeology.blogspot.com/; and http://www.mesoweb.com/zotz/.

3. Harris, "Concepts of Individual, Self, and Person in Description and Analysis."

4. The definition presented here is largely a simplification of George Herbert Mead's definition of self: Mead, *Mind, Self, and Society*.

5. Butler, *Bodies That Matter: On the Discursive Limits of "Sex"*; Foucault, *The History of Sexuality*; Joyce, "Archaeology of the Body"; idem, *Ancient Bodies, Ancient Lives*; Sofaer, *The Body as Material Culture*.

6. Calvin, "Where the Wayob Live," 869.

7. Pitarch, *The Jaguar and the Priest*. For a report of Tzotzil souls that parallels Pitarch's work in many regards, see Guiteras-Holmes, *Perils of the Soul*.

8. Pitarch, *The Jaguar and the Priest*, 5.

9. Ibid., 22.

10. Ibid., 24.

11. Ethnographic discussions of Maya souls can be found in Carlsen, "Footpath of the Dawn, Footpath of the Sun"; Carlsen and Prechtel, "The Flowering of the Dead"; Watanabe, *Maya Saints and Souls in a Changing World*; Wilson, *Maya Resurgence in Guatemala*; Groark, "Discourses of the Soul"; Guiteras-Holmes, *Perils of the Soul*; Holland, "Contemporary Tzotzil Cosmological Concepts as a Basis for Interpreting Prehistoric Maya Civilization"; Gossen, *Chamulas in the World of the Sun*; idem, "Animal Souls and Human Destiny in Chamula"; Bunzel, *Chichicastenango*; Nash, *In the Eyes of the Ancestors*; Boremanse, *Hach Winik*; Wisdom, *The Chorti Indians of Guatemala*; Gossen, *Four Creations*; Redfield and Villa Rojas, *Chan Kom*.

12. Vogt, "Human Souls and Animal Spirits in Zinacantan," 1155.

13. Houston, Stuart, and Taube, *The Memory of Bones*, 125.

14. Love, "Yucatec Sacred Breads through Time."

15. Guiteras-Holmes, *Perils of the Soul*; Vogt, *Tortillas for the Gods*, 22–23; Boremanse, *Hach Winik*, 91; Kintz, *Life under the Tropical Canopy*, 36–41; Wilson, *Maya Resurgence in Guatemala*, 143; Watanabe, *Maya Saints and Souls in a Changing World*, 87; Fink, "Shadow and Substance," 399.

16. Pitarch, *The Jaguar and the Priest*, 175; Vogt, *Tortillas for the Gods*, 62; Boremanse, *Hach Winik*, 91; Wilson, *Maya Resurgence in Guatemala*, 147; Watanabe, *Maya Saints and Souls in a Changing World*, 88.

17. Fink, "Shadow and Substance," 407.

18. Stuart, "Kings of Stone," 164.

19. Taube, *The Major Gods of Ancient Yucatan*, 31.

20. Although it is perhaps just a coincidence, the *k'uh* glyph is a ringer for contemporary Maya monkey masks. For depictions of masks as fighting helmets, see Taube and Zender, "American Gladiators."

21. Tokovinine, *Place and Identity in Classic Maya Narratives*, 57–86.

22. *Wahy* were first identified by Houston and Stuart, "The *Way* Glyph." See also Grube and Nahm, "A Census of Xibalba"; Helmke and Nielsen, "Hidden Identity and Power in Ancient Mesoamerica."

23. Pitarch, "Almas y cuerpo en una tradición indígena tzeltal," note 6.

24. For a recent linguistic review of *wahy*, see Helmke, "Mesoamerican Lexical Calques in Ancient Maya Writing and Imagery," 7.

25. Houston and Stuart, "The *Way* Glyph," 13.

26. Stuart, "Maya Spooks."

27. Vogt, "Human Souls and Animal Spirits in Zinacantan," 1156.

28. Grube and Nahm, "A Census of Xibalba."

29. Schellhas, *Representations of Deities of the Maya Manuscripts*. For more recent synthetic reviews, see especially

Taube, *The Major Gods of Ancient Yucatan*; Miller and Taube, *The Gods and Symbols of Ancient Mexico and the Maya*.

30. Taube, *The Major Gods of Ancient Yucatan*, 50. Taube builds on earlier work by J. Eric Thompson.

31. Stuart, "The Name of Paper."

32. Houston, *The Life Within*.

33. For example, the Tzeltal would understand many of these forces as types of *lab*: see Pitarch, *The Jaguar and the Priest*, 40–50.

34. Grube and Nahm, "A Census of Xibalba."

35. The mythology shown on Classic period vases has been explored in numerous publications. Important works include Coe, *Lords of the Underworld*; idem, *The Art of the Maya Scribe*; idem, *Old Gods and Young Heroes*; Robicsek and Hales, *The Maya Book of the Dead, the Ceramic Codex*.

36. My brief discussion of the Holmul dancer vessels follows a more detailed interpretation of these objects presented in Tokovinine, *Place and Identity in Classic Maya Narratives*, 115–122.

37. Taschek and Ball, "Lord Smoke-Squirrel's Cacao Cup."

38. Tokovinine, *Place and Identity in Classic Maya Narratives*, 117–118.

39. Gossen, "Animal Souls and Human Destiny in Chamula," 451.

40. Vogt and Stuart, "Some Notes on Ritual Caves among the Ancient and Modern Maya."

41. For an excellent overview of Chahk, see Gárcia Barrios, "Chaahk, el dios de la lluvia, en el periodo clásico maya."

42. For an example of paired Chahks, see Taube, *Aztec and Maya Myths*, 66.

43. Tozzer, *Landa's Relación de las Cosas de Yucatan*, 137–138.

44. For a compilation of the many appellatives of Chahks, see Gárcia Barrios, "Chaahk, el dios de la lluvia, en el periodo clásico maya," 136.

45. Stuart, "Report."

46. Pitarch, *The Jaguar and the Priest*, 3.

47. Gossen, "From Olmecs to Zapatistas," 566.

48. Groark, "Social Opacity and the Dynamics of Empathic In-Sight among the Tzotzil Maya of Chiapas, Mexico," 428. See also idem, "Toward a Cultural Phenomenology of Intersubjectivity"; Danziger, "On Trying and Lying"; Haviland and Haviland, "Privacy in a Mexican Indian Village."

49. Groark, "Discourses of the Soul."

50. Pitarch, *The Jaguar and the Priest*, 89.

51. Groark, "Pathogenic Emotions," 137.

52. Guiteras-Holmes, *Perils of the Soul*, 304.

53. Tedlock, *Rabinal Achi*, 78.

54. The concepts were first explored in a series of articles: Houston and Stuart, "Of Gods, Glyphs, and Kings"; idem, "The Ancient Maya Self"; Stuart, "Kings of Stone."

55. Houston, Stuart, and Taube, *The Memory of Bones*, 60.

56. Stone and Zender, *Reading Maya Art*, 21–22.

57. Tozzer, *Landa's Relación de las Cosas de Yucatan*, 131.

58. Stuart, "Kings of Stone."

59. See, for example, K'ahk' Tiliw Chan Chahk on Naranjo Stela 22, illustrated in Houston, Stuart, and Taube, *The Memory of Bones*, fig. 2.7. The king's name is both worn in his headdress and located in the glyph panel directly in front of his gaze.

60. Houston and Taube, "An Archaeology of the Senses," 283.

61. Carlsen and Prechtel, "The Flowering of the Dead," 54; Christenson, *Popol Vuh*, 126.

62. Groark, *Pathogenic Emotions*, 183.

63. Pitarch, *The Jaguar and the Priest*, 89.

64. Christenson, *Popol Vuh*, 126.

65. Ibid.; Tedlock, *Rabinal Achi*.

66. Saturno, Taube, and Stuart, *The Murals of San Bartolo, El Petén, Guatemala, Part 1*, fig. 11.

67. Christenson, "Maize Was Their Flesh," 581.

68. Fitzsimmons, *Death and the Classic Maya Kings*, 22; Martin, "Cacao in Ancient Maya Religion."

69. Taube, "The Classic Maya Maize God"; Robicsek and Hales, *The Maya Book of the Dead, the Ceramic Codex*; Bassie-Sweet, *Maya Sacred Geography and the Creator Deities*; Quenon and Le Fort, "Rebirth and Resurrection in Maize God Iconography"; Martin, "Cacao in Ancient Maya Religion"; Taube, "The Birth Vase"; Saturno, Taube, and Stuart, *The Murals of San Bartolo, El Petén, Guatemala, Part 1*.

70. Christenson, *Popol Vuh*.

71. Taube, "The Maize Tamale in Classic Maya Diet, Epigraphy, and Art"; Martin, "Hieroglyphs from the Painted Pyramid."

72. See figure 5 in Carrasco Vargas, Vázquez López, and Martin, "Daily Life of the Ancient Maya Recorded on Murals at Calakmul, Mexico."

73. Ambrose, "Isotopic Analysis of Paleodiets."

74. These data are compiled from the following sources: Wright, *Diet, Health, and Status among the Pasión Maya*; Scherer, Wright, and Yoder, "Bioarchaeological Evidence for Social and Temporal Differences in Subsistence at Piedras Negras, Guatemala"; White et al., "Social Complexity and Food Systems at Altun Ha, Belize"; Montero López et al., "Diet and Health at Chinikihá, Chiapas, Mexico"; Williams, White, and Longstaffe, "Maya Marine Subistence"; Whittington, "Analysis of Human Skeletal Material Excavated by Guillemin"; Wright, "La muerte y el estatus económico"; Chase and Chase, "Secular, sagrado, y 'revisitado'"; Tykot, "Contribution of Stable Isotope Analysis to Understanding Dietary Variation among the Maya"; Gerry and Krueger, "Regional Diversity in Classic Maya Diets." The data from El Zotz-Bejucal ($n = 7$), Yaxha ($n = 13$), Tecolote-El Kinel ($n = 10$), and El Porvenir ($n = 3$) are from my own research and are unpublished.

75. Prowse et al., "Isotopic Paleodiet Studies of Skeletons from the Imperial Roman-Age Cemetery of Isola Sacra, Rome, Italy."

76. Redfield and Villa Rojas, *Chan Kom*, 42.

77. Ambrose, "Isotopic Analysis of Paleodiets."

78. For a recent synthesis of Maya head shaping, see Tiesler, *Tranformarse en Maya*; idem, *The Bioarchaeology of Artificial Cranial Modifications*. For other recent interpretations of ancient Maya head shaping, see Duncan and Hofling, "Why the Head?"

79. Schellhas, *Representations of Deities of the Maya Manuscripts*, 247.

80. Tiesler, "'Olmec' Head Shapes among the Preclassic Period Maya and Cultural Meanings."

81. Tiesler, "Studying Cranial Vault Modifications in Ancient Mesoamerica"; idem, *La costumbre de la deformación cefálica entre los antiguos mayas*.

82. Miller, "Extreme Makeover."

83. Tiesler, "Studying Cranial Vault Modifications in Ancient Mesoamerica"; Tiesler and Cucina, "La deformación craneana como emblema de identidad, etnicidad y reproducción cultural entre los mayas del clásico."

84. Four Late Classic period skulls at Piedras Negras are unmodified, and the fifth exhibits tabular oblique deformation.

85. Tiesler, "Studying Cranial Vault Modifications in Ancient Mesoamerica," 13.

86. Boot, "What Happened on the Date 7 Manik' 5 Woh?"

87. Tozzer, *Landa's Relación de las Cosas de Yucatan*, 125.

88. Stuart, "The Name of Paper."

89. Stuart, "Kings of Stone."

90. Reilly, "Middle Formative Origins of the Mesoamerican Ritual Act of Bundling."

91. See the various contributions in Guernsey and Reilly, eds., *Sacred Bundles*.

92. Hendrickson, *Weaving Identities*, 102.

93. See Saturno, Taube, and Stuart, *The Murals of San Bartolo, El Petén, Guatemala, Part 1*.

94. Tozzer, *Landa's Relación de las Cosas de Yucatan*, 125, 29.

95. For Landa's summary of the ceremony, see ibid., 102–106. For a recent interpretation of this ritual, see Solari, *Maya Ideologies of the Sacred*, 47–49.

96. Pedro Guzmán López, personal communication, 2012.

97. Romero Molina, "Dental Mutilation, Trephination, and Cranial Deformation"; idem, *Catálogo de la colección de dientes mutilados prehispánicos*.

98. Fitzsimmons et al., "Guardian of the Acropolis."

99. Geller, "Altering Identities."

100. Cummings, "Apache Puberty Ceremony for Girls."

101. For a photograph of the monument, see Martin and Grube, *Chronicle of the Maya Kings and Queens*, 61. For a recent interpretation of the scene, see Houston, "A Splendid Predicament," 165.

102. Though it is not an unbiased source, Landa's *Relación* reports that cranial shaping caused much pain for the infant: Tozzer, *Landa's Relación de las Cosas de Yucatan*, 125.

103. For rites of passage among the Lacandon, see Boremanse, *Hach Winik*, 81–88.

104. Tiesler Blos, Ramírez Salomón, and Oliva Arias, "Técnicas de decoración dental en México," 2.

105. Tozzer, *Landa's Relación de las Cosas de Yucatan*, 125–126.

106. Vera Tiesler, personal communication, 2014.

107. Headland, "Teeth Mutilation among the Casiguran Dumagat."

108. Fischer and Estiti Andarawati, "Tooth-Filing in Bali," 40.

109. "Body Perfect," *Taboo*, National Geographic Television and Film Production, 2003.

110. Tiesler Blos, *Decoraciones dentales entre los antiguos mayas*.

111. Blom, *A Maya Skull from the Uloa Valley, Republic of Honduras*.

112. Houston, Stuart, and Taube, *The Memory of Bones*, 143.

113. Ibid., 190.

114. See figure 20 in Taube, *The Major Gods of Ancient Yucatan*.

115. Ibid., 48.

116. Vera Tiesler first pointed out the lack of cranial modification among the Death Gods in a conversation in 2012.

117. Rodolfo Cid Beziez and Torres Sanders, "Patrones de mutilación dental en el sector oeste de Teotihuacan," fig. 3; Serrano S., Pimienta Merlin, and Gallardo Velázquez, "Mutilación dentaria y filiación étnica en los entierros del templo de Quetzalcoatl, Teotihuacan," fig. 4; Spence and Pereira, "The Human Skeletal Remains of the Moon Pyramid, Teotihuacan," fig. 2.

118. Compare with Wright, "Identifying Immigrants to Tikal, Guatemala," fig. 3; Buikstra et al., "Tombs and Burials in the Early Classic Acropolis at Copan," fig. 10.1.

119. Stuart, "The Arrival of Strangers"; idem, "The Beginnings of the Copan Dynasty."

120. Stone, "Disconnection, Foreign Insignia, and Political Expansion."

121. Berlo, "Artistic Specialization at Teotihuacan."

122. See figure 65 in Sugiyama, *Human Sacrifice, Militarism, and Rulership*.

123. There is a bit of circular logic here: jade is treated by archaeologists as one of the defining markers for identifying Maya elite.

124. Christenson, *Popol Vuh*, 99–100. Note that Maya languages do not distinguish between blue and green.

125. Modern experimentation confirms that copal resin would have been sufficient to keep the inlays in place. See Neiburger, "Jaded Smiles."

126. Taube, "The Symbolism of Jade in Classic Maya Religion."

127. Christenson, *Popol Vuh*, 100.

128. For a recent synthesis, see Joyce, *Ancient Bodies, Ancient Lives*.

129. Geller, "Skeletal Analysis and Theoretical Complications," 603–604.

130. Walker and Cook, "Brief Communication."

131. Hanks, *Referential Practice*, 110–112.

132. Advocatory positions on Classic Maya gender-bending include Geller, "Altering Identities"; Joyce, "The Construction of Gender in Classic Maya Monuments"; Looper, "Women-Men (and Men-Women)." For a critical position, see Houston and McAnany, "Bodies and Blood," 33.

133. Christenson, *Popol Vuh*, 136–138.

134. Tozzer, *Landa's Relación de las Cosas de Yucatan*, 96.

135. Miller and Martin, *Courtly Art of the Ancient Maya*, plate 47.

136. The scene itself is impossible, as Pakal was already dead by the time these particular grandchildren reached adulthood.

137. The lack of infectious disease is in part due to the diminished presence of domesticated animals relative to the Old World.

138. Smith, "Dental Attrition in Hunter-Gatherers and Agriculturalists"; idem, "Patterns of Molar Wear in Hunter-Gatherers and Agriculturalists."

139. For a comparison of dental caries across the region, see Scherer, Wright, and Yoder, "Bioarchaeological Evidence for Social and Temporal Differences in Subsistence at Piedras Negras, Guatemala." For other contributing factors to dental caries, see Cucina et al., "Carious Lesions and Maize Consumption among the Prehispanic Maya."

140. For a recent comparative discussion of antemortem tooth loss, see Lukacs, "Dental Trauma and Antemortem Tooth Loss in Prehistoric Canary Islanders." A much higher incidence of antemortem tooth loss, which may reflect dietary differences among the samples or could be due to different research methodologies, is reported in Cucina and Tiesler, "Dental Caries and Antemortem Tooth Loss in the Northern Peten Area, Mexico."

141. Houston, Stuart, and Taube, *The Memory of Bones*, 143–156.

142. Sahagún, *General History of the Things of New Spain: Florentine Codex*, book 10, 147.

143. Ibid., 89.

144. Mickenautsch et al., "Sugar-Free Chewing Gum and Dental Caries."

145. Houston and Escobedo, "Nuevas intervenciones en la estructura de Piedras Negras."

146. Buikstra et al., "Tombs and Burials in the Early Classic Acropolis at Copan."

147. Tiesler, "Life and Death of the Ruler: Recent Bioarchaeological Findings," 39.

148. Ibid.

149. Taube, "The Birth Vase," 657.

150. Vogt and Stuart, "Some Notes on Ritual Caves among the Ancient and Modern Maya," 159.

151. Casagrande, "Ecology, Cognition, and Cultural Transmission of Tzeltal Maya Medicinal Plant Knowledge," 77; Luber, "The Biocultural Epidemiology of 'Second Hair' Illness in Two Mesoamerican Societies," 73.

152. One position is that the hot-cold balance reflects diffusion of the doctrine of humors of Hippocrates. See, for example, Foster, *Hippocrates' Latin American Legacy*. I do not accept this hypothesis. For a refutation, see Chevalier and Sánchez Bain, *The Hot and the Cold*.

153. Gossen, "Animal Souls and Human Destiny in Chamula," 449.

154. Berlin and Berlin, *Medical Ethnobiology of the Highland Maya of Chiapas, Mexico*, 61; Casagrande, "Ecology, Cognition, and Cultural Transmission of Tzeltal Maya Medicinal Plant Knowledge," 122.

155. Ibid., 77; Christenson, *Popol Vuh*, 99.

156. Casagrande, "Ecology, Cognition, and Cultural Transmission of Tzeltal Maya Medicinal Plant Knowledge," 77.

157. On the use of dreams and sacred objects in bone setting, see Paul, "The Maya Bonesetter as Sacred Specialist."

158. Ibid., 79.

159. Christenson, *Popol Vuh*, 99.

160. Stephen Houston, personal communication, 2012.

161. Helmke and Nielsen, "Hidden Identity and Power in Ancient Mesoamerica."

162. Maffi, "A Linguistic Analysis of Tzeltal Maya Ethnosymptomatology," 213.

163. Ibid., 214, 21.

164. The Tzeltal generally classify disease into one of three categories: "illnesses of the belly," "illnesses of the airways," and "illnesses of the skin." For more information, see ibid., chapter 5.

165. Edmonson, *The Ancient Future of the Itza*, 24, 37, 60, 106, 87.

166. Stephen Houston, personal communication, 2012.

167. Stuart, "Maya Spooks."

168. Vogt, *Zinacantan*, 407.

169. Helmke and Nielsen, "Hidden Identity and Power in Ancient Mesoamerica."

170. Pitarch, *The Jaguar and the Priest*, 46–47.

171. Guiteras-Holmes, *Perils of the Soul*, 302.

172. Silver and Fábrega, *Illness and Shamanistic Curing in Zinacantan*, 24.

173. Pitarch, *The Jaguar and the Priest*, 147.

174. Edwards, "The Funeral Ceremony in Zinacantan," 27.

175. Grube and Nahm, "A Census of Xibalba," 707.

176. Only tibiae represented by at least 60 percent or more of the diaphysis were included. If both tibiae from the same individual were present, the individual was only counted once for presence or absence.

177. Wright and White, "Human Biology in the Classic Maya Collapse: Evidence from Paleopathology and Paleodiet."

178. Bogin et al., "Rapid Change in Height and Body Proportions of Maya American Children."

179. Danforth, "Coming Up Short," 108, 103.

180. Bogin et al., "Rapid Change in Height and Body Proportions of Maya American Children."

CHAPTER 2: DEAD BODIES

1. Seminal studies include Hertz, *Death and the Right Hand*; van Gennep, *The Rites of Passage*.

2. On the meaning of ancestors, see Houston et al., "Classic Maya Death at Piedras Negras, Guatemala"; Lau, "Feasting and Ancestor Veneration at Chinchawas North Highlands of Ancash, Peru."

3. Watanabe, "From Saints to Shibboleths."

4. Houston and Inomata, *The Classic Maya*, 211.

5. Wilson, *Maya Resurgence in Guatemala*, 80–87.

6. Redfield and Villa Rojas, *Chan Kom*, 203–204.

7. McAnany, *Living with the Ancestors*; Gillespie, "Body and Soul among the Maya." For nonelite ancestor veneration in the mortuary shrines associated with Tikal Plaza Plan 2, see Becker, *Excavations in Residential Areas of Tikal*.

8. See, for example, Christenson, "Maize Was Their Flesh"; idem, *Art and Society in a Highland Maya Community*; Carlsen and Prechtel, "The Flowering of the Dead."

9. Vogt, *Tortillas for the Gods*, 18.

10. Redfield and Villa Rojas, *Chan Kom*, 199.

11. Christenson, *Popol Vuh*, 126–127.

12. Ximénez, *Historia de la provincia de San Vicente Chiapa y Guatemala*, 1:100.

13. Carlsen and Prechtel, "The Flowering of the Dead," 55.

14. McAnany, *Living with the Ancestors*; Gillespie, "Body and Soul among the Maya."

15. Houston and Stuart, "Of Gods, Glyphs, and Kings."

16. Taube, "The Classic Maya Maize God"; idem, "The Birth Vase"; idem, "The Maya Maize God and the Mythic Origins of Dance"; Quenon and Le Fort, "Rebirth and Resurrection in Maize God Iconography."

17. Martin, "Cacao in Ancient Maya Religion."

18. Saturno, Taube, and Stuart, *The Murals of San Bartolo, El Petén, Guatemala, Part 1*, 6–7.

19. Christenson, *Popol Vuh*.

20. Vogt, *Zinacantan*, 218.

21. Ibid., 222.

22. Moholy-Nagy, *The Artifacts of Tikal*, figs. 189–192.

23. Coe and Van Stone, *Reading the Maya Glyphs*, 62; Taube, "Flower Mountain," 81.

24. Vogt, "Zinacanteco Astronomy."

25. de León, *Diccionario Quiche-Español*.

26. Girard, *People of the Chan*, 301–302.

27. Hull, "Journey from the Ancient Maya Tomb."

28. Girard, *People of the Chan*, 280; Astor-Aguilera, *The Maya World of Communicating Objects*.

29. Guiteras-Holmes, *Perils of the Soul*, 294; Pitarch, *The Jaguar and the Priest*, 147–151.

30. Pitarch, *The Jaguar and the Priest*, 151.

31. Vogt, *The Zinacantecos of Mexico*, 219.

32. Estrada-Belli, "Lightning Sky, Rain, and the Maize God."

33. Stuart and Stuart, *Palenque*, 176.

34. Gossen, *Chamulas in the World of the Sun*, 31.

35. Pitarch, *The Jaguar and the Priest*, 148, 75.

36. Astor-Aguilera, *The Maya World of Communicating Objects*.

37. Taube, "The Symbolism of Jade in Classic Maya Religion," 28.

38. Stuart and Stuart, *Palenque*, 198.

39. Taube et al., *The Murals of San Bartolo, El Petén, Guatemala, Part 2*, 43.

40. Schele and Miller, *Blood of Kings*, 72; Martin, "Cacao in Ancient Maya Religion."

41. Taube, "The Symbolism of Jade in Classic Maya Religion," 25.

42. The flowery nature of Maya essences is discussed in Houston and Taube, "An Archaeology of the Senses."

43. Taube, "The Symbolism of Jade in Classic Maya Religion," 30–32.

44. Taube, "Flower Mountain."

45. Grube and Nahm were among the first to suggest that the "*ajaw*-face" sign outside of a day cartouche be read as flower (*nik*); as cited in Schele, *Notebook for the XVIth Maya Hieroglyphic Workshop at Texas*.

46. Stuart, "'The Fire Enters His House,'" 397.

47. Vogt, *Zinacantan*, 45–51.

48. The 260-day span and reference to the rebirth of the Tonina king were first mentioned to me by Steve Houston, personal communication, 2012.

49. Pohl, Pope, and von Nagy, "Olmec Origins of Mesoamerican Writing."

50. Stuart, "The Name of Paper."

51. For foundational work on the Jester God, see Taube, "The Jade Hearth." For evidence that three distinct entities have been conflated as the Jester God in the literature, see Stuart, "The Name of Paper," 128–129.

52. For a discussion, see Stuart, "The Name of Paper," 135. For an illustration of the Palenque sarcophagus ancestors, see Robertson, *The Sculpture of Palenque*, figs. 174–201.

53. There is an interesting pattern to their use. The ancestors with the "*ajaw*-face" lived and reigned during the sixth century. The one exception is Kan Bahlam I, who wears the avian diadem and reigned briefly in the latter part of the century. Pakal's parents, who lived most recently of all the ancestors shown, are twice depicted with the avian variant of the headdress. The distinction may therefore be temporal. The Palenque artisans may have used the "*ajaw*-face" to designate the ancestors more distantly removed from Pakal in time, consistent with the sign's prominence in the headdress iconography of the Early Classic period. Alternatively, this pattern may say something about the relations of these individuals. How Pakal's parents relate to the earlier Palenque royals is unknown. Perhaps one of them was a descendant of Kan Bahlam I—or it may be coincidence.

54. Stuart, "The Name of Paper," 140.

55. For a recent exploration of the **YAX** sign, see Stone and Zender, *Reading Maya Art*, 47.

56. For evidence of these three objects as the three sacred burning hearthstones, see Taube, "The Jade Hearth," 432–435.

57. Taube et al., *The Murals of San Bartolo, El Petén, Guatemala, Part 2*, 23. As discussed in chapter 3, good evidence indicates that copal was also placed inside the abdomens of sacrificial victims, as seen, for example, on stelae from Piedras Negras.

58. Tate, *Yaxchilan*, 59.

59. Stephen Houston, personal communication, 2013.

60. For an excellent review of the Maya royal body, see Houston, Stuart, and Taube, *The Memory of Bones*.

61. For similar discussions, see Fitzsimmons, *Death and the Classic Maya Kings*; idem, "Classic Maya Mortuary Anniversaries at Piedras Negras"; Stuart, "'The Fire Enters His House.'"

62. Quoted in Miles, "The Sixteenth-Century Pokom-Maya," 749–750.

63. Quoted in ibid., 749.

64. Quenon and Le Fort, "Rebirth and Resurrection in Maize God Iconography," 892.

65. González Cruz, *La reina roja*.

66. Nash, "The Reassertion of Indigenous Identity," 15.

67. Golden et al., "Polities, Boundaries, and Trade in the Classic Period Usumacinta River Basin."

68. Kovacevich, "Ritual, Crafting, and Agency at the Classic Maya Kingdom of Cancuen," 82.

69. Ruz Lhuillier, *El Templo de las Inscripciones*.

70. The lack of queens likely also reflects the misidentification of some queens as kings. Generally, the most reliable skeletal element for sex identification, the os coxae (especially the pubis), is poorly preserved. Without this essential element many archaeologists are overly eager to assume that most royal tombs contained the bodies of men.

71. Information on the Red Queen's tomb can be found in González Cruz, *La reina roja*. The greenstone mask that covered her face was made of malachite, a copper carbonate mineral.

72. Miller and Martin, *Courtly Art of the Ancient Maya*, plate 102.

73. García Moll, "Shield Jaguar and Structure 23 at Yaxchilan," 269.

74. González Cruz, *La reina roja*.

75. Bell et al., "Tombs and Burials in the Early Classic Acropolis at Copan"; Bell, "Engendering a Dynasty."

76. Redfield and Villa Rojas, *Chan Kom*, 201.

77. Stuart, "Kings of Stone"; idem, "The Name of Paper."

78. Whittington, "Description of Human Remains and Burial Structures."

79. Stuart, "The Name of Paper."

80. Ruz Lhuillier, *El Templo de las Inscripciones*; Moholy-Nagy, *The Artifacts of Tikal*, figs. 115 and 21; Coe, *Piedras Negras Archaeology*.

81. Stuart, *The Inscription from Temple XIX at Palenque*.

82. Stuart, "The Name of Paper."

83. Taube, "The Jade Hearth," 454–463; idem, "The Symbolism of Jade in Classic Maya Religion," 28–30.

84. Stuart, "The Name of Paper," 129–131.

85. Ibid., 129.

86. For relevant depictions of the Olmec Maize God, see Taube, "The Olmec Maize God," figs. 1–3, 5f, 6, 13.

87. Beaubien, "Ceramic Laminates."

88. We must also entertain the possibility that this object was a backrack worn during dance performances.

89. Hall, "Realm of Death," 63.

90. Beaubien, "Ceramic Laminates."

91. Díaz del Castillo, *The History of the Conquest of New Spain*, 31.

92. Coe, *Excavations in the Great Plaza, North Terrace, and North Acropolis of Tikal.*

93. Similar mother-of-pearl ornaments were recovered at Tikal: Moholy-Nagy, *The Artifacts of Tikal*, fig. 116.

94. Kidder, Jennings, and Shook, *Excavations at Kaminaljuyu, Guatemala*, 110–111, fig. 47h.

95. For a complete exploration of these concepts, see Taube, "The Symbolism of Jade in Classic Maya Religion," 32–42; Houston and Taube, "An Archaeology of the Senses."

96. Houston, Stuart, and Taube, *The Memory of Bones*, 154.

97. Taube, "Gateways to Another World."

98. Redfield and Villa Rojas, *Chan Kom*, 199.

99. Nash, *In the Eyes of the Ancestors*, 131.

100. Gossen, "Animal Souls and Human Destiny in Chamula," 450.

101. Cited in Miles, "The Sixteenth-Century Pokom-Maya," 749.

102. Coe, "The Ideology of the Maya Tomb." There are too many jade-in-mouth interments to list all of them. Some examples include Pakal's sarcophagus at Palenque, Burial 9 at El Zotz, and Burial 82 at Piedras Negras.

103. Roach, "Pyramid Tomb Found."

104. For a full exploration of the meaning and symbolism of jade, see Taube, "The Symbolism of Jade in Classic Maya Religion."

105. Ibid., 30.

106. Sahagún, *General History of the Things of New Spain: Florentine Codex*, book 11, 222.

107. Tozzer, *Landa's Relación de las Cosas de Yucatan*, 130.

108. Houston and Taube, "An Archaeology of the Senses," 270.

109. Moholy-Nagy, *The Artifacts of Tikal*, figs. 89–91; Martínez, *Rostros de la divinidad*.

110. The use of funerary masks at Dzibanche and Calakmul was common mortuary practice at both sites, perhaps reflecting royal mortuary tradition in the Kaan dynasty.

111. Roach, "Pyramid Tomb Found."

112. Beekman, "Recent Research in Western Mexican Archaeology," 71; Brooks, "Industrial Use of Mercury in the Ancient World," 21.

113. Gossen, *Chamulas in the World of the Sun*, 37.

114. Houston, Stuart, and Taube, *The Memory of Bones*, 156.

115. Stone and Zender, *Reading Maya Art*, 62.

116. Stuart and Stuart, *Palenque*, 175.

117. Tiesler, "Life and Death of the Ruler," 35.

118. González Cruz, *La reina roja*.

119. Fitzsimmons, *Death and the Classic Maya Kings*, 82.

120. Hall, "Realm of Death"; Coe, *Excavations in the Great Plaza, North Terrace, and North Acropolis of Tikal*; Folan et al., "Calakmul, Campeche"; Carrasco et al., "A Dynastic Tomb from Campeche, Mexico."

121. For such ambiguous examples, see Bell et al., "Tombs and Burials in the Early Classic Acropolis at Copan"; Harrison-Buck, McAnany, and Storey, "Empowered and Disempowered during the Late to Terminal Classic Transition."

122. Lori Wright, personal communication, 2012. Burials at Tikal with "PNT" in the enumeration refer to burials excavated by the Proyecto Nacional Tikal, directed by Juan Pedro Laporte. The PNT used a separate enumeration from the earlier Penn Museum Project. Burials without the PNT prefix from Tikal were excavated by the Penn Museum Project.

123. Tiesler, "Life and Death of the Ruler," 35.

124. Pitarch, *The Jaguar and the Priest*, 61. For discussion of bodily gesture and social opacity among the Tzotzil, see Groark, "Social Opacity and the Dynamics of Empathic In-Sight among the Tzotzil Maya of Chiapas, Mexico."

125. Miller, "Pose and Gesture in Classic Maya Monumental Sculpture."

126. Pitarch, *The Jaguar and the Priest*, 93, 37, 93.

127. Houston, Stuart, and Taube, *The Memory of Bones*, 190.

128. Welsh, *An Analysis of Classic Lowland Maya Burials*, table 19.

129. Sempowski and Spence, *Mortuary Practices and Skeletal Remains at Teotihuacan*.

130. Ibid., 139.

131. Martin and Grube, *Chronicle of the Maya Kings and Queens*, 151.

132. Tozzer, *Landa's Relación de las Cosas de Yucatan*, 130.

133. Duday, *The Archaeology of the Dead*.

134. Tiesler, "Life and Death of the Ruler," 34.

135. Coe, *Excavations in the Great Plaza, North Terrace, and North Acropolis of Tikal*, 217–220.

136. Hall, "Realm of Death," 62.

137. For a review of corpse wrapping and bundling, see Reese-Taylor, Zender, and Geller, "Fit to Be Tied." For the wrapping of bodies at Copan, see Bell et al., "Tombs and Burials in the Early Classic Acropolis at Copan."

138. Pitarch, *The Jaguar and the Priest*, 169.

139. Carrasco et al., "A Dynastic Tomb from Campeche, Mexico," 55; Folan et al., "Calakmul, Campeche," 321.

140. Hall, "Realm of Death," 62.

141. Fitzsimmons, *Death and the Classic Maya Kings*, 61–63.

142. Quoted in Miles, "The Sixteenth-Century Pokom-Maya," 749.

143. Quoted and translated in ibid.

144. Stephen Houston, personal communication, 2012.

145. Newborn maggots target the soft tissue of the eyes, mouth, and nose, which are typically the first areas of a corpse to be consumed by insects.

146. Reese-Taylor, Zender, and Geller, "Fit to Be Tied."

147. Pitarch, *Ch'ulel*, 38.

148. For further exploration of the conceptual continuity between Maya burials and caches, see Becker, "Burials as Caches; Caches as Burials."

149. For further information on the El Kinel burials, see Scherer et al., "Danse Macabre."

150. Looper, *To Be Like Gods*, 88; Maler, *Explorations in the Department of Petén, Guatemala, Tikal*, 134–135; Miller, "Pose and Gesture in Classic Maya Monumental Sculpture."

151. See especially images in Justin Kerr's vase database and in Looper, *To Be Like Gods*.

152. Miller, "The Image of People and Nature in Classic Maya Art and Architecture," 159.

153. Taube, "The Maya Maize God and the Mythic Origins of Dance," 49.

154. Houston, Stuart, and Taube, *The Memory of Bones*, 72.

155. Laporte and Fialko, "Un reencuentro con mundo perdido, Tikal, Guatemala," 60; Folan et al., "Calakmul, Campeche," 322; Ruz Lhuillier, *El Templo de las Inscripciones*; idem, "Exploraciones arqueológicas en Palenque: 1954"; López Jiménez, "El descubrimiento de la Tumba I del Templo de la Calavera y su contexto arquitectónico en Palenque, Chiapas"; Ruz Lhuillier, "Exploraciones arqueológicas en Palenque: 1956"; Nieto Calleja and Schiavon Signoret, "El Templo Olvidado de Palenque, Chiapas."

156. Coe, *Excavations in the Great Plaza, North Terrace, and North Acropolis of Tikal*, 51.

157. Taube, "The Classic Maya Maize God," fig. 5.

158. Gates, *Yucatan before and after the Conquest*, 120.

159. Tedlock, *Rabinal Achi*, 27.

160. Carlsen, "Footpath of the Dawn, Footpath of the Sun," 303.

161. See table 1 in Johnson et al., "Ancient Soil Resources of the Usumacinta River Region, Guatemala." See also Fernández et al., "Soil Resources of the Ancient Maya at Piedras Negras, Guatemala." "Neutral" to "very strong acid" corresponds to the pH range from 7.1 to 4.9 and is based on the soil pH classifications of the United States Department of Agriculture.

162. A soil pH range between 7.6 and 8.5 is reported by Fernández, Parnell, and Terry, "Análisis de los suelos del área de Piedras Negras."

163. Note that the skeletons in Tikal burial plans, which suggest good preservation, are often reconstructions of what the artists believed were the arrangements of the skeletons based on their observation of fragmentary remains.

164. The existence of drains, cisterns, and wells in Classic period patio groups is under-recognized in the Mayanist literature, but see Scherer and Golden, "Water in the West"; Ashmore, "Classic Maya Wells at Quirigua, Guatemala." As a result some scholars have assumed that water management was the purview of the royal lineages: for example, Lucero, "The Political and Sacred Power of Water in Classic Maya Society"; idem, "Water Control and Maya Politics in the Southern Maya Lowlands."

165. Christenson, *Popol Vuh*, 129.

166. Christenson, "K'iche'-English Dictionary and Guide to Pronunciation of the K'iche'-Maya Alphabet."

167. Carlsen and Prechtel, "The Flowering of the Dead."

168. Astor-Aguilera, *The Maya World of Communicating Objects*, 175.

169. Stone and Zender, *Reading Maya Art*, 74.

170. Ibid., 55.

171. Maffi, "A Linguistic Analysis of Tzeltal Maya Ethnosymptomatology," 190.

172. Ibid., 129.

173. Tedlock, *Rabinal Achi*.

174. For example, McAnany, *Living with the Ancestors*; Carlsen and Prechtel, "The Flowering of the Dead"; Gillespie, "Body and Soul among the Maya."

175. Astor-Aguilera, "Unshrouding the Communicating Cross," 177; Cook, "Quichean Folk Theology and Southern Maya Supernaturalism."

176. Astor-Aguilera, *The Maya World of Communicating Objects*, 164.

177. Tozzer, *Landa's Relación de las Cosas de Yucatan*, 131.

178. Chase and Chase, "Maya Multiples."

179. For important reviews, see Fitzsimmons, *Death and the Classic Maya Kings*; Eberl, *Muerte, entierro y ascensión*.

180. Fitzsimmons, *Death and the Classic Maya Kings*, 165–166.

181. Jones and Satterthwaite, *The Monuments and Inscriptions of Tikal*, 37.

182. Fitzsimmons, "Classic Maya Mortuary Anniversaries at Piedras Negras," 277.

183. Martin and Grube, *Chronicle of the Maya Kings and Queens*, 150.

184. For conjuring with bones, see Fitzsimmons, "Perspectives on Death and Transformation in Ancient Maya Society."

185. The conjuring skull depicted on Yaxchilan Lintel 25 wears earflares with the *k'an*-cross marking it as mature or ripe, like a fruit ready to bear seed. For epigraphic readings of this monument, see Mathews, "The Sculpture of Yaxchilan"; Martin and Grube, *Chronicle of the Maya Kings and Queens*, 125; Stuart, "Report."

186. Two unprovenanced vessels (K7289 and K7523) from the region of Calakmul, however, suggest that men may also have engaged in bone conjuring.

187. Redfield and Villa Rojas, *Chan Kom*; Astor-Aguilera, *The Maya World of Communicating Objects*, 160.

188. Astor-Aguilera, *The Maya World of Communicating Objects*, 160.

189. For another example from Xuenkal, Yucatán, see Tiesler et al., "A Taphonomic Approach to Late Classic Maya Mortuary Practices at Xuenkal, Yucatán, Mexico."

190. Inomata, "War, Violence, and Society in the Maya Lowlands," figs. 2.1 and 2.14.

191. Scherer and Golden, "War in the West."

192. Martin and Grube, *Chronicle of the Maya Kings and Queens*, 123–134.

193. Tedlock, *Rabinal Achi*, 104–105.

194. Laporte, "Contexto y función de los artefactos de hueso en Tikal, Guatemala," 42.

195. Milbrath and Peraza Lope, "Revisiting Mayapan," 18.

196. Consider the time and efforts spent on the recovery and identification of human remains from the 9/11 attack on the United States. For further information, see Sledzik et al., "Disaster Victim Recovery and Identification."

197. Chase and Chase, "Maya Veneration of the Dead at Caracol, Belize"; Chase and Chase, "Maya Multiples."

198. Vogt, *Zinacantan*, 219.

CHAPTER 3: RITUAL, LIMINALITY, AND THE MORTUARY SPACE

1. For example, Welsh, *An Analysis of Classic Lowland Maya Burials*; Krejci and Culbert, "Preclassic and Classic Burials and Caches in the Maya Lowlands"; Wright, *Diet, Health, and Status among the Pasión Maya*; Saxe, "Social Dimensions of Mortuary Practices"; Binford, "Mortuary Practices."

2. For general discussion of liminality, see van Gennep, *The Rites of Passage*. For explorations of Maya liminality in mortuary spaces, see Fitzsimmons, *Death and the Classic Maya Kings*.

3. Hertz, *Death and the Right Hand*; van Gennep, *The Rites of Passage*.

4. Maya cosmology has been exhaustively addressed in the literature. A few examples include Mathews and Garber, "Models of Cosmic Order"; Ashmore, "Site-Planning Principles and Concepts of Directionality among the Ancient Maya"; Coggins, "The Shape of Time"; Taube, "Where Earth and Sky Meet"; idem, "Creation and Cosmology"; Thompson, *Sky Bearers, Colors, and Directions in Maya and Mexican Religion*.

5. DeBoer, "Colors for a North American Past"; MacLaury, *Color and Cognition in Mesoamerica*; Riley, "Color-Direction Symbolism"; Thompson, *Sky Bearers, Colors, and Directions in Maya and Mexican Religion*, 10.

6. The association between colors and directions is given in a variety of colonial period sources. For example, Roys, *The Book of Chilam Balam of Chumayel*, 22–23; Christenson, *Popol Vuh*, 122.

7. Gossen, *Chamulas in the World of the Sun*, 33.

8. As, for example, in Tedlock, *Rabinal Achi.*

9. Vogt, *Zinacantan*, 298. See also Gossen, *Chamulas in the World of the Sun*, fig. 2.

10. Becker, "Maya Heterarchy as Inferred from Classic Period Plaza Plans," 131; Solari, *Maya Ideologies of the Sacred*, 66, 107.

11. Magnetic north actually does change with time and based on location on earth, but the declination can be corrected to find true north.

12. Girard, *People of the Chan*, 159–160.

13. Vogt, *Tortillas for the Gods*, 11.

14. Taube, "Ancient and Contemporary Maya Concepts about Field and Forest."

15. Taube, "Creation and Cosmology."

16. Tozzer, *Landa's Relación de las Cosas de Yucatan*, 135–136; Taube, *The Major Gods of Ancient Yucatan*, 92–99.

17. Taube, "Where Earth and Sky Meet."

18. Taube, "A Prehispanic Maya Katun Wheel"; idem, "Where Earth and Sky Meet," 210.

19. Gossen, *Chamulas in the World of the Sun*, 18.

20. Redfield and Villa Rojas, *Chan Kom*, 114; Taube, "Ancient and Contemporary Maya Concepts about Field and Forest"; idem, "Creation and Cosmology."

21. Hall, "Realm of Death."

22. Stuart, *The Inscription from Temple XIX at Palenque*, 168.

23. Crosses in general are often painted green, linking them to the concept of centrality. See, for example, Redfield and Villa Rojas, *Chan Kom*, 110.

24. Carlsen, "Footpath of the Dawn, Footpath of the Sun," 303.

25. For a general discussion of tree and centering symbolism, see Taube, "The Symbolism of Jade in Classic Maya Religion."

26. The crocodile is an appropriate symbol for the axis mundi and is like the *ceiba* in many regards. The surface of the earth was thought to be a crocodile. Like the crocodile, the surface of the *ceiba* is covered by a rough, spiky bark, especially in its youth. *Ceiba* trees were known as *yaxte'* (green trees) in the Classic period, matching the coloration of crocodiles. For a further discussion of these concepts, see Stone and Zender, *Reading Maya Art*, 183.

27. Tozzer, *Comparative Study of the Mayas and Lacandones*, 154.

28. Christenson, *Popol Vuh*, 189.

29. Nakamura, "Culto funerario de Copán en el siglo VI," fig. 3.

30. For similar depictions, see Taube, "The Symbolism of Jade in Classic Maya Religion," fig. 5.

31. See the various illustrations in Zender, "The Raccoon Glyph in Classic Maya Writing."

32. Schele and Miller, *Blood of Kings*, 77.

33. Stuart, "The Ceiba Tree on K1226."

34. Stone and Zender, *Reading Maya Art*, 71.

35. Moholy-Nagy, *The Artifacts of Tikal*, 27–28; Chase and Chase, "The Archaeological Context of Caches, Burials, and Other Ritual Activities for the Classic Period"; Coe, *Piedras Negras Archaeology*, fig. 51.

36. The "Charlie Chaplin" nickname originated in Thompson, *Excavations at San Jose, British Honduras*, 91.

37. Coe, *Excavations in the Great Plaza, North Terrace, and North Acropolis of Tikal*, 485.

38. Taschek and Ball, "Las Ruinas de Arenal," 221.

39. Moholy-Nagy, *The Artifacts of Tikal*, fig. 108c.

40. Ibid., fig. 108b.

41. Coe, *Piedras Negras Archaeology*, fig. 47e; Rich, "Ritual, Royalty, and Classic Period Politics," 278.

42. Ruz Lhuillier, "Exploraciones arqueológicas en Palenque: 1954," plate 47; López Jiménez, "El descubrimiento de la Tumba I del Templo de la Calavera y su contexto arquitectónico en Palenque, Chiapas," photograph 3.

43. Nakamura, "Culto funerario de Copán en el siglo VI"; Taube, "Cache Vessel with Directional Shells and Jades, 500–600, Copan, Honduras."

44. Houston, "Living Waters and Wondrous Beasts," 77.

45. Taube, "Cache Vessel with Directional Shells and Jades, 500–600, Copan, Honduras."

46. Ruz Lhuillier, *El Templo de las Inscripciones*, 202–204.

47. Tozzer, *Landa's Relación de las Cosas de Yucatan*, 102.

48. Saturno, Taube, and Stuart, *The Murals of San Bartolo, El Petén, Guatemala, Part 1*, 34.

49. Quenon and Le Fort, "Rebirth and Resurrection in Maize God Iconography," 888.

50. Carrasco et al., "A Dynastic Tomb from Campeche, Mexico," 53.

51. González Cruz, *La reina roja*, 181.

52. Hellmuth, "Structure 5D-73, Burial 196, Tikal, Peten, Guatemala," 180–183.

53. Among the burials with bivalves at the head are Tikal Burials 10, 23, 24, 116, 195, 196; Piedras Negras Burials 5, 13, and 110; and Río Azul Tombs 19 and 23. See Coe, *Excavations in the Great Plaza, North Terrace, and North Acropolis of Tikal*; idem, *Piedras Negras Archaeology*; Escobedo, "Tales from the Crypt"; Hall, "Realm of Death."

54. Coe, *Excavations in the Great Plaza, North Terrace, and North Acropolis of Tikal*, 566.

55. The second example is Moholy-Nagy, *The Artifacts of Tikal*, fig. 165f.

56. García Moll, "Shield Jaguar and Structure 23 at Yaxchilan," 270.

57. Stuart, *The Inscription from Temple XIX at Palenque*, 70.

58. Houston, "Living Waters and Wondrous Beasts," 78.

59. Fitzsimmons, *Death and the Classic Maya Kings*, 18.

60. Houston, "Panel Describing a King's Pilgrimage to the Sea."

61. Houston, Stuart, and Taube, *The Memory of Bones*, 186.

62. Guernsey, *Ritual and Power in Stone*, 138.

63. Taube, "The Jade Hearth," 441. A reptilian creature with three stones on its back is also shown on Copan Altar D'. This time, however, the animal is a crocodile. Based on texts at Palenque, David Stuart calls this crocodile the Starry Deer Crocodile and notes that it represents a dark sky, yet has terrestrial connotations and is perhaps understood as the sky of the underworld. See Stuart, *The Inscription from Temple XIX at Palenque*, 73.

64. Houston, Stuart, and Taube, *The Memory of Bones*, 186–188.

65. See, for example, Lucero, "Materialized Cosmology among Ancient Maya Commoners," 142; Reents-Budet, "The Art of Classic Vase Painting," 250.

66. McAnany, *Living with the Ancestors*, 57; Headrick, "The Quadripartite Motif and the Centralization of Power."

67. Headrick, "The Quadripartite Motif and the Centralization of Power," 368–369.

68. Ibid., 374.

69. Coe, "The Ideology of the Maya Tomb," 223.

70. Miller, "The Maya Ballgame," 86.

71. For further discussion of the hearthstones as axes mundi, see Taube, "The Jade Hearth," 432–446.

72. Hanks, *Referential Practice*, 357.

73. Wright, *Diet, Health, and Status among the Pasión Maya*, 45.

74. Houston and Stuart, "The Ancient Maya Self."

75. Houston and Taube, "An Archaeology of the Senses"; Houston, Stuart, and Taube, *The Memory of Bones*.

76. Houston, Stuart, and Taube, *The Memory of Bones*, 143.

77. Carlsen, "Footpath of the Dawn, Footpath of the Sun," 309.

78. Scherer et al., "Danse Macabre."

79. Cook, *Renewing the Maya World*; Cook and Offit, *Indigenous Religion and Cultural Performance in the New Maya World*.

80. Cook and Offit, *Indigenous Religion and Cultural Performance in the New Maya World*, 78–80.

81. Cook, *Renewing the Maya World*, 117.

82. Beekman, "Agricultural Pole Rituals and Rulership in Late Formative Central Jalisco."

83. Taube, *The Major Gods of Ancient Yucatan*, 41.

84. See, for example, accounts in Durán, *Book of the Gods and Rites and The Ancient Calendar*; Sahagún, *General History of the Things of New Spain: Florentine Codex*, book 2. Both of these are summarized in Headrick, "The Great Goddess at Teotihuacan."

85. Miller, "The Iconography of the Painting in the Temple of the Diving God, Tulum, Quintana Roo, Mexico." See also Looper and Guernsey, "The Cosmic Umbilicus in Mesoamerica"; Guernsey, *Ritual and Power in Stone*, 130–131.

86. Taube, "The Birth Vase," 659.

87. Klein, "Snares and Entrails," 88.

88. Maffi, "A Linguistic Analysis of Tzeltal Maya Ethnosymptomatology," 199.

89. Villa Rojas, "La imagen del cuerpo humano según los mayas de Yucatán."

90. Gossen, *Chamulas in the World of the Sun*, 18.

91. Carlsen, "Footpath of the Dawn, Footpath of the Sun."

92. Christenson, *Art and Society in a Highland Maya Community*, 77.

93. Farris, *Maya Society under Colonial Rule*, 287.

94. Tedlock, *Rabinal Achi*, 38.

95. Wisdom, *Materials on the Chorti Language*; Robertson, Law, and Haertel, *Colonial Ch'olti'*.

96. Stone and Zender, *Reading Maya Art*, 40.

97. Girard, *People of the Chan*, 342.

98. Hull, "Journey from the Ancient Maya Tomb."

99. The scene establishes a clever balance between the respective accession of the new lord and the death of the former lord. Dwarves sit as courtiers for the living lord; the dead lord is attended by rodents and a bat. They seem to communicate with one another: the dead lord passes a chord to a dancing attendant; a rodent comes to the living lord.

100. Redfield and Villa Rojas, *Chan Kom*, 200.

101. Rich, "Ritual, Royalty, and Classic Period Politics," fig. 7.80; Navarro-Farr, Pérez, and Menéndez, "Operación WK-1."

102. Coe, *Excavations in the Great Plaza, North Terrace, and North Acropolis of Tikal*, fig. 537; Vogt, *Zinacantan*, 218.

103. See Gutiérrez, Román, and Carteret, "Excavaciones en el grupo El Diablo (Operación 19)," fig. 1.20.

104. Taube, "The Jade Hearth," 441.

105. Houston, Stuart, and Taube, *The Memory of Bones*, 85.

106. Houston, "Impersonation, Dance, and the Problem of Spectacle among the Classic Maya," 148; Stuart, "'The Fire Enters His House,'" 408.

107. Taube, "The Turquoise Hearth," 292–293.

108. Vogt, *Zinacantan*, 44.

109. Stuart, "'The Fire Enters His House.'"

110. Maffi, "A Linguistic Analysis of Tzeltal Maya Ethnosymptomatology," 205.

111. Guiteras-Holmes, *Perils of the Soul*, 307.

112. Stuart, "'The Fire Enters His House,'" 398.

113. And also one year after the previous king was buried: see Martin and Grube, *Chronicle of the Maya Kings and Queens*, 152.

114. Escobedo, "Tales from the Crypt."

115. Histological analysis by Vera Tiesler confirmed that the skeleton was exposed to heat as dry bone: Tiesler Blos, "Reporte de los estudios de análisis de transición e histomorfológico de los restos humanos de los sitios Piedras Negras, Dos Pilas y Tamarindito."

116. Stephen Houston, personal communication, 2013.

117. Cuevas García, "The Cult of Patron and Ancestor Gods in Censers at Palenque," 253.

118. Taube, "The Jade Hearth," 450.

119. Taube, "The Iconography of Mirrors at Teotihuacan."

120. Kidder, Jennings, and Shook, *Excavations at Kaminaljuyu, Guatemala*.

121. Coe, *Piedras Negras Archaeology*; Moholy-Nagy, *The Artifacts of Tikal*, 141; Tovalín Ahumada, "La cripta funeraria asociada al edificio 4 de Bonampak, Chiapas."

122. Taube, "The Iconography of Mirrors at Teotihuacan."

123. For a summary of ancestor cartouches at Yaxchilan and other Maya sites, see Tate, *Yaxchilan*, 59–62. See also Taube, "Flower Mountain," 79.

124. See Stuart, "A Sun God Image from Dos Pilas, Guatemala."

125. Taube, "Maws of Heaven and Hell."

126. Stuart, *The Inscription from Temple XIX at Palenque*, 76; Taube, "The Temple of Quetzalcoatl and the Cult of Sacred War at Teotihuacan."

127. The upturned nose could refer to the blunt noses of the fer-de-lance and other Mesoamerican vipers.

128. Coggins, "New Fire at Chichen Itza," fig. 8.

129. Heizer and Gullberg, "Concave Mirrors from the Site of La Venta, Tabasco," 115.

130. Healy and Blainey, "Ancient Maya Mosaic Mirrors."

131. Taube, "The Iconography of Mirrors at Teotihuacan."

132. Miller and Martin, *Courtly Art of the Ancient Maya*, 44.

133. Bonampak burials are typically oriented with the head to the east-southeast, so a mirror at the foot end of the burial would be in the west-northwest.

134. Fitzsimmons et al., "Guardian of the Acropolis."

135. Coe, *Piedras Negras Archaeology*, fig. 64 description. Note that the Piedras Negras mortuary vessels probably were not made specifically for burial but were selected for the mortuary contexts because of their specific function in life or the symbolism of their decoration.

136. Coe, *Piedras Negras Archaeology*; Escobedo, "Tales from the Crypt."

137. Tozzer, *Landa's Relación de las Cosas de Yucatan*, 133.

138. Ibid., 114; Schele and Miller, *Blood of Kings*, 68.

139. Maroukis, *Peyote and the Yankton Sioux*, 20.

140. Germain, Smith, and Skelton, "The Cutaneous Cellular Infiltrate to Stingray Envenomization Contains Increased TIA+ Cells," 1074.

141. Tozzer, *Landa's Relación de las Cosas de Yucatan*, 191.

142. Benson, "A Knife in the Water," 183.

143. Pendergast, *Excavations at Altun Ha, Belize, 1964–1970*, 76; Kidder, Jennings, and Shook, *Excavations at Kaminaljuyu, Guatemala*, 156; Kidder, *The Artifacts of Uaxactun*, 159; Fitzsimmons et al., "Guardian of the Acropolis"; García Moll, "Shield Jaguar and Structure 23 at Yaxchilan"; Coe, *Excavations in the Great Plaza, North Terrace, and North Acropolis of Tikal.*

144. Coe, *Piedras Negras Archaeology*, fig. 64.

145. Escobedo, "Tales from the Crypt," 279.

146. Anaya Hernández, Guenter, and Mathews, "An Inscribed Wooden Box from Tabasco, Mexico."

147. Joralemon, "Ritual Blood Sacrifice among the Ancient Maya."

148. On the water serpent as *witz'*, see Stuart, "Reading the Water Serpent as WITZ'." For depictions of the water serpent gushing blood, see Hellmuth, *Monster und Menschen in der Maya-Kunst*, 322 and 23.

149. Occasionally the crossed band in the floral motif is replaced by the "percent" sign.

150. Fitzsimmons et al., "Guardian of the Acropolis"; Houston, "Engraved Stingray Spines and Bone Handles from Royal Burials."

151. On Piedras Negras Stelae 1 and 14 royal women hold a plumed object that is sheathed in paper or cloth. These could be bloodletters but are more likely sacrificial knives, considering the depiction of human sacrifice on Stela 14.

152. Chahk is sometimes shown with a barbel, but it is typically rendered like the whisker of a catfish rather than as a fin, which is more typical of G1 (perhaps a freshwater versus saltwater distinction for these two supernaturals).

153. Stuart, *The Inscription from Temple XIX at Palenque*, 164.

154. Scherer and Golden, "War in the West."

155. Taube, "The Birth Vase," 672.

156. For a complete report on the El Zotz royal tomb, see Houston, Newman, Román, and Garrison, *Temple of the Night Sun.*

157. Scherer and Garrett, "Osteología del complejo El Diablo."

158. Houston and Scherer, "La ofrenda máxima"; Houston, Stuart, and Taube, *The Memory of Bones*, 122–127.

159. Taube, "The Maize Tamale in Classic Maya Diet, Epigraphy, and Art."

160. Christenson, *Popol Vuh*, 133.

161. Love, "Yucatec Sacred Breads through Time."

162. Bunzel, *Chichicastenango*, 360; Vogt, *Tortillas for the Gods*, 91.

163. Taube, "The Birth Vase," 669–674.

164. Stuart, "Suicide by Proxy," 426, 434.

165. Taube, "The Birth Vase"; Martin, "The Baby Jaguar."

166. In the Kerr archives: K521, K1003, K1152 (a variant with the water lily jaguar), K1199, K1370, K1644, K1768, K1815, K2207, K2208, K2213, K3201, K4011, K4013, K4486, K8680. Elements of the myth appear on other vessels not listed here and may represent variations of the myth or different points in the story.

167. A Late Classic period funerary urn from the Guatemalan highlands in the Museo Popol Vuh shows the face of the Jaguar God of the Underworld on its side. The lid is topped by a Baby Jaguar (Museo Popol Vuh, Catalog No. 0015). Apparently myths pertaining to these two beings were at one time widely spread across the Maya area.

168. Martin, "The Baby Jaguar," 57–68.

169. Schele and Miller, *Blood of Kings*, 50–51.

170. Martin, "The Baby Jaguar," 51.

171. See also K521, K1003, K4385, and K4011 in the Maya Vase Database.

172. Martin, "The Baby Jaguar," 57.

173. Some of these vessels are retouched, so the interpretation that I offer here must be treated as preliminary and in need of better supporting evidence.

174. The lords on Altar 5 are engaged in the ritual exhumation of a royal woman. Their costumes also include paraphernalia associated with the Jaguar God of the Underworld, including sticks for fire drilling. Their headgear recalls the hat and backrack of the Death God, with blood-spattered cloth spilling from the top of their hats and bundled in front of their faces. The similarities in these costumes are explored in Clancy, *The Monuments of Piedras Negras, an Ancient Maya City*, 96.

175. Taube, "The Birth Vase," 672.

176. Guiteras-Holmes, *Perils of the Soul*, 298.

177. Pitarch, *The Jaguar and the Priest*, 33–35.

178. Guiteras-Holmes, *Perils of the Soul*, 174.

179. As Simon Martin shows, the myth of the Baby Jaguar can be traced into Late Preclassic times, as far back as the so-called were-jaguar of the Olmec. A fragment of Late Preclassic murals from the site of San Bartolo shows a person holding an infant with its arm cast in the sign of woe, presumably indicating imminent sacrifice, as in the Classic period depictions of the sacrifice of the Baby Jaguar. Unfortunately, both the head of the figure holding the infant and the baby's lower body are missing, so it is difficult to confirm that this is indeed the Baby Jaguar. The infant's head is the same as the adult Maize God. Both are depicted with a prominent incisor fang and catlike eyes that recall the Olmec were-jaguar babies. Indeed the Baby Jaguar and the Maize God are occasionally merged in Classic period imagery, perhaps suggesting that the Baby Jaguar, sacrificed in the west like the setting sun, is the replacement for the Maize God, a being who rises in the east like the dawning sun. To the right of the infant on the San Bartolo murals the adult Maize God dances within the earth-turtle. The pounding of the drum simulates the beating of rain as two rain deities look on (see fig. 1.21b). It seems that the infant will be offered as *k'ex* to allow the Maize God's resurrection. In the San Bartolo scene the figure holding the infant stands knee-deep in a rush of water, similar to the Late Classic period depictions of the sacrifice of the Baby Jaguar (see fig. 3.44). Framing the scene to the left, separated by a missing fragment of mural, is a scaffold upon which sits a lord dressed in jaguar skins, atop a jaguar pelt. He is faced by another being dressed in jaguar pelts. Taube and colleagues suggest that these are both maize deities. Structurally the scene foreshadows Classic period imagery that juxtaposes the death of one king with the accession of another, a transition marked by human sacrifice, just as we see on the scaffold monuments of Piedras Negras (compare with figs. 3.23, 3.50). As the reign of one king sets like the sun, the other rises like a new day. For some reason, the Maya saw *k'ex* offerings as necessary to mediate the process. For more information, see Martin, "The Baby Jaguar"; Saturno, Taube, and Stuart, *The Murals of San Bartolo, El Petén, Guatemala, Part 1.*

180. The scene from the ceramic vessel of the burning child also parallels the scattering scene on Yaxchilan Stela 7, where a *yax* and *k'an*-marked substance cascades from the hands of the king onto a subordinate. What unites all of these scenes is the movement or exchange of sacred essences.

181. Taube, "The Birth Vase," 672; Coe, *Piedras Negras Archaeology*, 95, 129.

182. Taube, "The Birth Vase," p. 668.

183. The figure at right wears a headdress associated with the Death God impersonators, recalling the paper-wrapped topknot of the Death God. Note that he wields a sacrificial knife, whereas the other lord, who is shown with a jaguar ear, wields a stone implement that is likely meant to suggest a jaguar paw. The wrap around the Death God's topknot can be identified as paper: in most depictions it is marked by a Ux Yop Huun diadem (an emblem of paper, as discussed earlier). For examples see K1003, K1152, K1200 K1370, 1380, K1646, K1650, K1652, and K2208. For Death God impersonators, see figures 3.46 and 3.47.

184. Not all Aztec knives impersonate the Death God, however. Two knives were recently excavated from in front of the Temple Mayor of Tenochtitlan that were "costumed" as Ehécatl-Quetzalcoatl: Chávez Balderas, Miramontes, and Robles, "Los cuchillos ataviados de la ofrenda 125."

185. Durán, *Book of the Gods and Rites and The Ancient Calendar*; Chávez Balderas, "Sacrifice at the Templo Mayor of Tenochtitlan and Its Role in Regard to Warfare," fig. 7.6.

186. Cucina and Tiesler, "The Companions of Janaab' Pakal and the 'Red Queen' from Palenque, Chiapas"; Tiesler and Cucina, "Procedures in Human Heart Extraction and Ritual Meaning."

187. Verano, "The Physical Evidence of Human Sacrifice in Ancient Peru."

188. Bray et al., "A Compositional Analysis of Pottery Vessels Associated with the Inca Ritual of Capacocha," 83.

189. Eeckhout and Owens, "Human Sacrifice at Pachacamac."

190. Coe, *Excavations in the Great Plaza, North Terrace, and North Acropolis of Tikal.*

191. Wright, "In Search of Yax Nuun Ayiin I."

192. Lori Wright, personal communication, 2013.

193. It is notable that the king's skull and bones of the hand were not identified by the excavator. It may be that the skull was retained or removed for use in conjuring or some other event. Yet the absence of the hands suggests that they may have been taken from the body prior to inhumation. Coe suggests that the absence of the first and second cervical vertebrae supports decapitation, though in general the skeleton is incomplete and was not articulated, judging from the burial plan: Coe, *Excavations in the Great Plaza, North Terrace, and North Acropolis of Tikal*, 118–123; Martin and Grube, *Chronicle of the Maya Kings and Queens*, 36.

194. All but one of the references to the Baby Jaguar (whether as supernatural being or historic person) date to the reign of Sihyaj Chan K'awiil II (AD 411–456). For more information, see Martin, "The Baby Jaguar."

195. Rich, "Ritual, Royalty, and Classic Period Politics," 276.

196. Cheetham, "The Role of 'Terminus Groups' in Lowland Maya Site Planning," 137; Chase and Chase, "The Archaeological Context of Caches, Burials, and Other Ritual Activities for the Classic Period"; Taschek and Ball, "Las Ruinas de Arenal," 228.

197. For example, Caches 14A, 14B, 14C, and 14D. See Coe, *Excavations in the Great Plaza, North Terrace, and North Acropolis of Tikal*, 490–494.

198. Chase and Chase, "The Archaeological Context of Caches, Burials, and Other Ritual Activities for the Classic Period," 319; Chase and Chase, *Heterogeneity in Residential Group Composition*, 10.

199. Chase and Chase, "Archaeological Perspectives on Classic Maya Social Organization from Caracol, Belize," 141; idem, "The Archaeological Context of Caches, Burials, and Other Ritual Activities for the Classic Period," 319.

200. Cheetham, "The Role of 'Terminus Groups' in Lowland Maya Site Planning," 137.

201. Ibid.

202. John Verano, personal communication, 2013.

203. One toe, consisting of an intermediate and distal pedal phalanx from the same digit, was found in one of the caches at El Zotz. The proximal phalanx has cut marks on its diaphysis.

204. Coe, *Excavations in the Great Plaza, North Terrace, and North Acropolis of Tikal*, 493.

205. Houston and Scherer, "La ofrenda máxima."

206. Leonard, *Narrative of the Adventures of Zenas Leonard*, 248–249; Curtis, *The North American Indian*; MacLeod, "Self-Sacrifice in Mortuary and Non-Mortuary Ritual in North America," 355–356.

207. Bonner, *The Life and Adventures of James P. Beckwourth*, 264–267.

208. MacLeod, "Self-Sacrifice in Mortuary and Non-Mortuary Ritual in North America," 356, 369.

209. Love, *Maya Shamanism Today*, 67–81. Recall that bone, teeth, and seeds are conceptually linked in contemporary Maya metaphor.

210. Stone and Zender, *Reading Maya Art*, 5.

211. Ibid., 43. Akan's name may also be meant as an allusion to the intense yet fleeting sting of the wasp.

212. Pitarch, *The Jaguar and the Priest*, 91–92.

213. Houston, Stuart, and Taube, *The Memory of Bones*, figs. 3.12–3.15, 5.14.

214. Stephen Houston, in a 2012 personal communication, first pointed out the possibility that the so-called Maya percentage sign may actually represent some sort of cutting on the body of Akan.

215. The other vessel is K2286 in the Maya Vase Database. See also Houston, Stuart, and Taube, *The Memory of Bones*, 195–196.

216. Kidder, Jennings, and Shook, *Excavations at Kaminaljuyu, Guatemala*.

217. Christenson, *Popol Vuh*, 130; Carlsen and Prechtel, "The Flowering of the Dead."

218. Christenson, *Popol Vuh*, 177.

219. Houston and Scherer, "La ofrenda máxima," fig. 6.

220. Christenson, *Popol Vuh*, 179.

221. Redfield and Villa Rojas, *Chan Kom*, 203.

222. For a complete discussion of Maya sweatbaths in epigraphic, ethnographic, and ethnohistoric perspectives, see Houston, "Symbolic Sweatbaths of the Maya."

223. Redfield and Villa Rojas, *Chan Kom*, 203.

224. Vogt, *Tortillas for the Gods*, 91.

225. Pitarch, *The Jaguar and the Priest*, 99.

226. Ibid., 23. Taxonomic identification was done by Sarah Newman, personal communication, 2014.

227. Coe, *Piedras Negras Archaeology*, 64.

228. Coggins, "Painting and Drawing Styles at Tikal," 216; Coe, *Excavations in the Great Plaza, North Terrace, and North Acropolis of Tikal*, 122.

229. López Bravo, "La veneración de los ancestros en Palenque," 41.

230. For further discussion of the bird avatars, see Zender, "The Raccoon Glyph in Classic Maya Writing"; Taube, *A Representation of the Principal Bird Deity in the Paris Codex*.

231. Taube, "Flower Mountain."

232. Culbert, *The Ceramics of Tikal*, fig. 31.

233. Vogt, *Tortillas for the Gods*, 91–93, 94.

234. Vogt, *Zinacantan*, 222.

235. Kerr, "The Last Journey."

236. Kidder, Jennings, and Shook, *Excavations at Kaminaljuyu, Guatemala*.

237. A rodent carrying a similar device is visible on K3061 from the Kerr archives.

238. I determined the initial minimum number of individuals based on 81 left rat mandibles. Ashley Sharpe confirmed the mandible count and provided the taxonomic designation (personal communication, 2012). Obviously, "Toltec" refers merely to the common contemporary name of the rat and has nothing to do with the ancient Toltec culture.

239. For rodent consumption of human corpses, see, for example, Tsokos et al., "Skin and Soft Tissue Artifacts Due to Postmortem Damage Caused by Rodents"; Haglund, Reay, and Swindler, "Tooth Mark Artifacts and Survival of Bones in Animal Scavenged Human Skeletons"; Klippel and Synstelien, "Rodents as Taphonomic Agents."

240. Taube, "The Maya Maize God and the Mythic Origins of Dance," 48.

241. Kidder, Jennings, and Shook, *Excavations at Kaminaljuyu, Guatemala*.

242. Moholy-Nagy, "Vertebrates in Tikal Burials and Caches," 202.

243. Navarro-Farr, Pérez, and Menéndez, "Operación WK-1."

244. Vogt, *Zinacantan*, 218; Nash, *In the Eyes of the Ancestors*, 132.

245. Houston, "A Splendid Predicament," 173.

246. Rands and Bishop, "The Dish-Plate Tradition at Palenque," 124.

247. Krejci and Culbert, "Preclassic and Classic Burials and Caches in the Maya Lowlands," 105.

248. Moholy-Nagy, *The Artifacts of Tikal*, figs. 223–227.

249. Coggins, "Painting and Drawing Styles at Tikal," 192, 215; García Moll, "Shield Jaguar and Structure 23 at Yaxchilan," 270.

250. Weiner, "Inalienable Wealth"; idem, *Inalienable Possessions*.

251. Vogt, *Zinacantan*, 220.

252. Miles, "The Sixteenth-Century Pokom-Maya," 750.

253. Houston, "Te-Mu and Te-Ma as 'Throne.'"

254. Coe and Van Stone, *Reading the Maya Glyphs*, 61.

255. Miles, "The Sixteenth-Century Pokom-Maya," 749–750.

256. Guillemin, "The Ancient Cakchiquel Capital of Iximche," 33; Kidder, Jennings, and Shook, *Excavations at Kaminaljuyu, Guatemala*.

257. The excavators of the Calakmul tomb surmised that the body was placed on a mat resting on a series of dishes. More likely the body was on a wooden bench with the vessel tucked underneath, as was the case for the El Zotz tomb. Coe, *Excavations in the Great Plaza, North Terrace, and North Acropolis of Tikal*, 480–487; Moholy-Nagy, *The Artifacts of Tikal*, 65; Folan et al., "Calakmul, Campeche."

258. Coe, *Excavations in the Great Plaza, North Terrace, and North Acropolis of Tikal*, 485.

259. Moholy-Nagy, *The Artifacts of Tikal*, 65.

260. Ibid.

261. Redfield and Villa Rojas, *Chan Kom*, 200.

262. Girard, *People of the Chan*, 342.

263. Barbara Arroyo and Lucia Henderson, personal communication, 2014; Fitzsimmons, *Death and the Classic Maya Kings*, 34.

264. Coe, *Excavations in the Great Plaza, North Terrace, and North Acropolis of Tikal*, 605.

265. Ibid.; Rich, "Ritual, Royalty, and Classic Period Politics."

266. Bell et al., "Tombs and Burials in the Early Classic Acropolis at Copan," fig. 8.3; Bell, Canuto, and Sharer, *Understanding Early Classic Copan*, plates 3, 6, 8.

267. López Jiménez, "El descubrimiento de la Tumba I del Templo de la Calavera y su contexto arquitectónico en Palenque, Chiapas," fig. 5.

268. López Bravo, "La veneración de los ancestros en Palenque."

269. Blom and La Farge, *Tribes and Temples*, 183.

270. Ibid., 116.

271. Coe, *Excavations in the Great Plaza, North Terrace, and North Acropolis of Tikal*, 644.

272. Carrasco et al., "A Dynastic Tomb from Campeche, Mexico," 53.

273. García Moll, "Shield Jaguar and Structure 23 at Yaxchilan," 269–270.

274. Guernsey, "A Consideration of the Quatrefoil Motif in Preclassic Mesoamerica"; Taube, "Gateways to Another World."

275. Fitzsimmons, *Death and the Classic Maya Kings*, 71.

CHAPTER 4: THE MORTUARY LANDSCAPE

1. The context of death and burial at Tikal has been widely discussed. See, for example, Ashmore, "Site-Planning Principles and Concepts of Directionality among the Ancient Maya"; idem, "The Idea of a Maya Town"; Ashmore and Sabloff, "Spatial Orders in Maya Civic Plans"; Weiss-Krejci, "Dépositos rituales en los complejos de pirámides gemelas de Tikal"; idem, "The Role of Dead Bodies in Late Classic Maya Politics."

2. Robin and Hammond, "Burial Practices."

3. Storey, "The Ancestors," 114–115.

4. See, for example, Gillespie, "Personhood, Agency, and Mortuary Ritual"; idem, "Body and Soul among the Maya"; McAnany, "Ancestor Worship and Sanctification of Place"; idem, *Living with the Ancestors*.

5. Fink, "Shadow and Substance," 410.

6. Redfield and Villa Rojas, *Chan Kom*, 198.

7. Pitarch, *The Jaguar and the Priest*, 38.

8. Nash, *In the Eyes of the Ancestors*, 136.

9. Redfield and Villa Rojas, *Chan Kom*, 204.

10. Pitarch, *The Jaguar and the Priest*, 38.

11. Nash, *In the Eyes of the Ancestors*, 134.

12. Tedlock, *Rabinal Achi*, 105.

13. Gillespie, "Body and Soul among the Maya."

14. Vogt, "Human Souls and Animal Spirits in Zinacantan," 1158.

15. Houston and McAnany, "Bodies and Blood," 32.

16. McAnany, *Living with the Ancestors*; idem, "Ancestors and the Classic Maya Built Environment"; McAnany, Storey, and Lockard, "Mortuary Ritual and Family Politics at Formative and Early Classic K'axob, Belize"; Gillespie, "Personhood, Agency, and Mortuary Ritual"; idem, "Body and Soul among the Maya."

17. Saxe, "Social Dimensions of Mortuary Practices." For ethnographic examples, see Glazier, "Mbeere Ancestors and the Domestication of Death"; Bloch and Parry, "Introduction," 33.

18. McAnany, *Living with the Ancestors*, 8, 161, 110.

19. Ibid., 14.

20. Ximénez, *Historia de la provincia de San Vicente Chiapa y Guatemala*, 1:100.

21. McAnany, *Living with the Ancestors*, 100. It is generally recognized that the relatively dispersed arrangement of Maya communities was to accommodate groves and in-fields adjacent to the domestic complexes.

22. Martin, "Cacao in Ancient Maya Religion," 161; Christenson, *Popol Vuh*, 126–127.

23. Morris, "The Archaeology of Ancestors," 152.

24. McAnany, *Living with the Ancestors*, 162.

25. Pitarch, *The Jaguar and the Priest*, 18.

26. Ibid., 38.

27. de Witte, "Money and Death," 535.

28. Haviland, "Musical Hammocks at Tikal."

29. Becker, "Plaza Plans at Tikal"; idem, "Maya Heterarchy as Inferred from Classic Period Plaza Plans"; idem, *Excavations in Residential Areas of Tikal*.

30. Becker, "Maya Heterarchy as Inferred from Classic Period Plaza Plans," 129.

31. Chase and Chase, "Maya Veneration of the Dead at Caracol, Belize," 55–56.

32. Becker, "Ancient Maya Houses and Their Identification."

33. Thompson, *Ethnology of the Mayas of Southern and Central British Honduras*, 82; Graham, *Maya Christians and Their Churches in Sixteenth-Century Belize*.

34. Douglass, *Death in Murelaga*.

35. Pitarch, *The Jaguar and the Priest*, 148.

36. McAnany, *Ancestral Maya Economies in Archaeological Perspective*, 247.

37. McAnany, *Living with the Ancestors*, 125.

38. Esposito, *Funerals, Festivals, and Cultural Politics in Porfirian Mexico.*

39. Verdery, *The Political Lives of Bodies.*

40. Ashmore, "Site-Planning Principles and Concepts of Directionality among the Ancient Maya"; idem, "The Idea of a Maya Town"; Ashmore and Sabloff, "Spatial Orders in Maya Civic Plans."

41. Roach, "Pyramid Tomb Found."

42. Inomata, "Plazas, Performers, and Spectators"; Houston et al., "The Moral Community."

43. Każdan and Constable, *People and Power in Byzantium*, 146.

44. Stuart, "The Arrival of Strangers"; Martin and Grube, *Chronicle of the Maya Kings and Queens*, 31.

45. Martin and Grube, *Chronicle of the Maya Kings and Queens*, 56.

46. Proskouriakoff, "Historical Data in the Inscriptions of Yaxchilan, Part 2"; Houston et al., "In the Land of the Turtle Lords"; Martin and Grube, *Chronicle of the Maya Kings and Queens*, 127, 51.

47. Escobedo, "Tales from the Crypt," 277.

48. Hruby, "The Organization of Chipped-Stone Economies at Piedras Negras, Guatemala," 71.

49. Martin and Grube, *Chronicle of the Maya Kings and Queens*, 150.

50. Metcalf and Huntington, *Celebrations of Death*, 140–141.

51. Handley, *The King Never Smiles*, 442–443.

52. O'Neil, *Engaging Ancient Maya Sculpture at Piedras Negras, Guatemala*, 148.

53. Fitzsimmons, "Classic Maya Mortuary Anniversaries at Piedras Negras."

54. Coe, *Piedras Negras Archaeology*, 126–127.

55. Martin and Grube, *Chronicle of the Maya Kings and Queens*, 145.

56. Metcalf and Huntington, *Celebrations of Death*, 149–150.

57. Golden and Scherer, "Territories and Trust, Growth and Collapse"; Stone and Zender, *Reading Maya Art*, 105.

58. Esposito, *Funerals, Festivals, and Cultural Politics in Porfirian Mexico*, 5.

59. Abrams, *How the Maya Built Their World.*

60. McAnany, *Living with the Ancestors*, 4.

61. Hertz, *Death and the Right Hand*, 71–72.

62. Chase and Chase, "Ghosts amid the Ruins."

63. Metcalf and Huntington, *Celebrations of Death*, 206.

64. García Moll, "Shield Jaguar and Structure 23 at Yaxchilan."

65. García Moll and Juárez Cosío, *Yaxchilán.*

66. For settlement history in the western Maya lowlands, see Golden et al., "Piedras Negras and Yaxchilan"; Golden and Scherer, "Territories and Trust, Growth and Collapse"; Liendo Stuardo, *La organización de la producción agrícola en un centro maya clásico.*

67. Aveni, *Skywatchers*, 245.

68. Carlsen, "Footpath of the Dawn, Footpath of the Sun."

69. Christenson, *Art and Society in a Highland Maya Community*, 102–106; Carlsen and Prechtel, "The Flowering of the Dead," 52.

70. Christenson, *Art and Society in a Highland Maya Community*, 49.

71. Vogt, *Zinacantan*, 350.

72. For Tzotzil ritual circuits, see Gossen, *Chamulas in the World of the Sun*, 32–35. For counterclockwise circuits at the time of the conquest, see Tedlock, *Rabinal Achi.*

73. For an overview of the logic of counterclockwise circuits, see O'Neil, *Engaging Ancient Maya Sculpture at Piedras Negras, Guatemala*, 97–102.

74. Scherer and Golden, "Tecolote, Guatemala."

75. Tate, *Yaxchilan*, 114.

76. These calculations were done in Google Earth and the NOAA (National Oceanic and Atmospheric Administration) Solar Calculator. The precise dates may be off by one day on either side of the date given here.

77. Aveni and Hartung, "Some Suggestions about the Arrangement of Buildings at Palenque"; Méndez et al., "Astronomical Observations at the Temple of the Sun."

78. Some of this cosmology may be evident on the Palenque sarcophagus lid. The top of the tree was oriented roughly north (specifically 17° E of N) and the sky-bands that run on the long sides of the lid may represent the eastern and western horizons. The eastward sky-band is marked in the middle by the *k'in* sign, emblematic of the sun. The opposing side shows the *uh* sign, a representation of the moon. Along the short axis of the upper edge of the lid, the northeast corner is marked by a second *k'in*

sign and the *ak'ab* sign for darkness in the northwest. That short axis of the lid is oriented 107° E of N and the eastward *k'in* sign corresponds to sunrise at Palenque around February 1 (near the start of the agricultural cycle) and November 8 (around the end of the agricultural cycle and roughly equivalent to modern celebrations of ancestors in the Maya area).

79. Girard, *People of the Chan*, 410–413.

80. Although modern Western observers make little observation of the zenith, when the sun passes overhead and at noon casts no shadow, contemporary Maya are quite interested in this phenomenon. Zenith dates vary by latitude and in the southern lowlands occur between April 30 and May 8 (earliest in communities to the south) and again between August 5 and August 13 (earliest in the north), corresponding to the same time of the year as the dates when the sun sets on the principal axes at Tikal, Palenque, and many other Maya sites. The contemporary Cho'rti' monitor the zenith with the use of a rod: when it casts no shadow, "this is the position that allows the god to descend vertically from the center of heaven to the center of the earth to fructify it" (ibid., 183). Celestial descent is well attested for the Precolumbian Maya as the so-called diving Maize God, as discussed. For zenith dates, see Aveni and Hartung, "Maya City Planning and the Calendar," table 3.

81. Vogt, *Zinacantan*, 51.

82. Pitarch, *The Jaguar and the Priest*, 10.

83. Stuart, *The Order of Days*, 71; Boot, *The Nominal Yaxha'(al) Chak on Classic Maya Ceramics and a Possible Cephalomorphic Variant for Yax*.

84. Anthony Aveni reached similar conclusions by looking at the spatial design of Copan and drawing on ethnographic work with the Cho'rti': Aveni, *Skywatchers*, 253.

85. Ashmore, "The Idea of a Maya Town," 42.

86. Gossen, *Chamulas in the World of the Sun*, 33.

87. Taube, "Flower Mountain."

88. Ashmore and Geller, "Social Dimensions of Mortuary Space," 88.

89. Understanding the meaning of north and south is complicated by the fact that many contemporary Maya languages have no words for these directions: Bricker, "Directional Glyphs in Maya Inscriptions and Codices," 352.

90. Freidel, Schele, and Parker, *Maya Cosmos*, 76, 116.

91. Girard, *People of the Chan*, 302.

92. Holland, "Contemporary Tzotzil Cosmological Concepts as a Basis for Interpreting Prehistoric Maya Civilization." This article has been curiously overlooked, perhaps because it was published in the same year as Holland's untimely death at the age of thirty-six: Spicer, "William R. Holland, 1928–1964."

93. Stone and Zender, *Reading Maya Art*, 38.

94. Stephen Houston, personal communication, 2012.

95. Vogt and Stuart, "Some Notes on Ritual Caves among the Ancient and Modern Maya," 166–168.

96. Coe, "The Ideology of the Maya Tomb"; Houston, "Classic Maya Depictions of the Built Environment," 351–352.

97. As, for example, the tombs at Piedras Negras named in the censing ritual, *el naah*: Stuart, "'The Fire Enters His House.'"

98. Bell, Canuto, and Sharer, *Understanding Early Classic Copan*, plate 2b.

99. Stone and Zender, *Reading Maya Art*, 55.

100. Taube, "Flower Mountain," 84–85.

101. See, for example, Stela 12 of Piedras Negras.

102. Gossen, *Chamulas in the World of the Sun*, 36.

103. Holland, "Contemporary Tzotzil Cosmological Concepts as a Basis for Interpreting Prehistoric Maya Civilization," 303.

104. Ibid., 304.

105. Christenson, *Art and Society in a Highland Maya Community*, 102.

106. Tokovinine, *Place and Identity in Classic Maya Narratives*, 117–118.

107. Hanks, *Referential Practice*, 336, 337.

108. For an extensive discussion of conjuring ancestral spirits with bone, see Fitzsimmons, "Perspectives on Death and Transformation in Ancient Maya Society."

109. Blom and La Farge, *Tribes and Temples*.

110. Ruz Lhuillier, "Exploraciones arqueológicas en Palenque: 1954," plate 43; López Jiménez, "Entierros humanos en el Templo de la Cruz y la Cruz Foliada de Palenque, Chiapas," 86.

111. For early descriptions of these tomb chambers, see Blom and La Farge, *Tribes and Temples*; Thompson, "Ancient Tombs of Palenque."

112. Ruz Lhuillier, *El Templo de las Inscripciones*, 95, 109, 270–273.

113. Palenque Cross Group Project, *The 2001 Season of the Cross Group Project*.

114. Stephen Houston reports two psychoducts at El Zotz, personal communication, 2014. For the Calakmul psychoduct, see Folan et al., "Calakmul, Campeche," 321.

115. Redfield and Villa Rojas, *Chan Kom*, 199.

116. Hull and Carrasco, "MAK-'Portal' Rituals Uncovered," 132.

117. Fitzsimmons, *Death and the Classic Maya Kings*, 276.

118. Compare with Tikal, for example, with its ritual-mortuary precinct (the North Acropolis) and the quotidian courtly precinct (the Central Acropolis). The removal of the royal dead from the palaces is an important distinction between elites and nonelites at most other Maya sites. Often royal palaces and similar structures used by royalty in their day-to-day affairs contain few (if any) burials.

119. Zender, "'Handspan' and 'Strike' in Classic Maya Ballgame Texts"; idem, "Sport, Spectacle and Political Theater."

120. Miller, "The Maya Ballgame"; Coe, "Another Look at the Maya Ballgame."

121. Taube and Zender, "American Gladiators."

122. I identified a similar bone handle, also crafted from the femur of a child, in Late Preclassic deposits at Rancho Búfalo, Mexico, suggesting selective treatment of children's femora for bone handles. The significance of this is not clear.

123. For images of ballcourts, see Stone and Zender, *Reading Maya Art*, 100.

124. Christenson, *Popol Vuh*.

125. Stephen Houston, "Deathly Sport," 2013.

126. There is no obvious explanation as to why the ballcourts of Yaxchilan are perpendicular to one another.

127. Satterthwaite, "Structure K-6," 237.

128. Houston and Escobedo, "Nuevas intervenciones en la estructura de Piedras Negras," 3.

129. For recent comment on the Bearded Dragon, see Taube, "Flower Mountain," 79.

130. Stuart, "'The Fire Enters His House,'" 397.

131. Becquelin and Baudez, *Tonina, une cité maya du Chiapas*, 1:79–87.

132. Guiteras-Holmes, *Perils of the Soul*, 287.

133. Scott and Brady, "Human Remains in Lowland Maya Caves."

134. Helmke and Wrobel, "Je'reftheel, Roaring Creek Works, Belize."

135. McGee, *Watching Lacandon Maya Lives*, 135–136.

136. Boremanse, *Hach Winik*, 16, 98.

137. The only other cave skeleton at Piedras Negras was found by the University of Pennsylvania project in a small cave behind Temple O-13. The body was found with two probable bone handles that Mary Butler describes as "decorated in each case on the front with a wide band form by two parallel lines enclosing a conventionalized snake head, shown in profile with simplicity and restraint. The backs were plain except for bands of rosettes encircling each end" (cited in Coe, *Piedras Negras Archaeology*, 61). No other such object has been found in a burial at Piedras Negras, though it does recall the bone handles found with the slain individual at Yaxha. The object may imply that the Piedras Negras person was also sacrificed, though the evidence is quite circumstantial.

138. Scott and Brady, "Human Remains in Lowland Maya Caves."

139. Pitarch, *The Jaguar and the Priest*, 39; Vogt, *Zinacantan*, 370. He is likely the contemporary equivalent of the Classic period God L.

140. Vogt and Stuart, "Some Notes on Ritual Caves among the Ancient and Modern Maya," 165.

141. McGee, *Watching Lacandon Maya Lives*, 135.

142. Tiesler, "What Can Bones Really Tell Us?" 356.

143. De Anda, Tiesler, and Zabala, "Cenotes, espacios sagrados y la práctica de sacrificio humano en Yucatán."

144. Taube, "Flower Mountain," 70.

145. Pitarch, *The Jaguar and the Priest*, 38.

146. Vogt, *Tortillas for the Gods*, 23.

147. Stuart and Stuart, *Palenque*, plate 35.

148. According to recent analyses of these beings in Stuart, *The Inscription from Temple XIX at Palenque*; Stuart and Stuart, *Palenque*.

149. Stuart and Stuart, *Palenque*, 193.

150. Ibid., 196.

151. Zender, "The Raccoon Glyph in Classic Maya Writing," 11.

152. Looper, *Lightning Warrior*, 159.

153. Stuart, "The Paintings of Tomb 12, Rio Azul," 167.

154. Zender, "The Raccoon Glyph in Classic Maya Writing," 13.

155. Thompson, "Ancient Tombs of Palenque"; López Jiménez, "Entierros humanos en el Templo de la Cruz y la Cruz Foliada de Palenque, Chiapas."

156. Cuevas García, "The Cult of Patron and Ancestor Gods in Censers at Palenque," 253. I suspect that the tomb within Temple XV belonged to either Kan Bahlam II or his brother Kan Joy Chitam II.

157. Clark and Hansen, "The Architecture of Early Kingship."

158. Folan, Gunn, and Domínguez Carrasco, "Triadic Temples, Central Plazas, and Dynastic Palaces."

159. The effect is lost on the modern visitor to Tikal because the final phases of Temple 5D-33 were largely dismantled and removed by the University of Pennsylvania Project.

160. To be more accurate, 117°.

161. Ashmore, "Site-Planning Principles and Concepts of Directionality among the Ancient Maya"; Weiss-Krejci, "Depósitos rituales en los complejos de pirámides gemelas de Tikal."

162. The name is derived from Group E at Uaxactun, the "type specimen" of Maya E-Groups.

163. Clark and Hansen, "The Architecture of Early Kingship"; Laporte and Fialko, "Un reencuentro con mundo perdido, Tikal, Guatemala"; Aimers and Rice, "Astronomy, Ritual and the Interpretation of the Maya 'E-Group' Architectural Assemblages"; Doyle, "Regroup on 'E-Groups.'"

164. The three-stone hearth is often rendered in two dimensions as a line of three stones. See illustrations in Taube, "The Jade Hearth."

165. Laporte and Fialko, "Un reencuentro con mundo perdido, Tikal, Guatemala."

166. Ruler 3's tomb was apparently not reentered, and there are no references to fire-events performed at his grave. See Fitzsimmons, "Classic Maya Mortuary Anniversaries at Piedras Negras"; Stuart, "'The Fire Enters His House.'"

167. Fitzsimmons, "Classic Maya Mortuary Anniversaries at Piedras Negras," 273–274.

168. Child, "The Symbolic Space of the Ancient Maya Sweatbath"; idem, "The Archaeology of Religious Movements."

169. Astor-Aguilera, *The Maya World of Communicating Objects*, 160.

170. Becker, "Burials as Caches; Caches as Burials."

171. Coe, *Piedras Negras Archaeology*; Hruby and Ware, "Painted Lithic Artifacts from Piedras Negras, Guatemala."

172. Stuart, "Blood Symbolism in Maya Iconography," 193–203; Taube, "The Womb of the World," 99–105.

173. Stuart, *The Inscription from Temple XIX at Palenque*, 164.

174. See, for example, Yaxchilan Lintels 14, 15, 24, and 25. As noted, a ceramic bowl with a painted decorative design was found at Tecolote that matched the step-fret design shown on the bowl used for bloodletting on Yaxchilan Lintel 24. This find suggests that Maya sculptors, at least at Yaxchilan, were quite specific in how they depicted pottery vessels.

175. Hruby and Ware, "Painted Lithic Artifacts from Piedras Negras, Guatemala," fig. 3.

176. Taube, "The Womb of the World," fig. 11.

177. Scherer, "The Classic Maya Sarcophagus."

178. Ruz Lhuillier, "Exploraciones arqueológicas en Palenque: 1952"; idem, *El Templo de las Inscripciones*; Schele and Mathews, *The Code of Kings*; Stuart and Stuart, *Palenque*.

179. Stuart and Stuart, *Palenque*, 176.

180. Mayakovsky, *Vladimir Ilyich Lenin*, 1.

181. Rakita and Buikstra, "Corrupting Flesh," 106.

182. González Cruz, "Trabajos recientes en Palenque"; idem, "El templo de la reina roja, Palenque, Chiapas"; idem, "The Red Queen"; idem, *La reina roja*.

183. Stuart and Stuart, *Palenque*, 198.

184. González Cruz, *La reina roja*, 81–82.

185. Thompson, "Ancient Tombs of Palenque"; Holmes, *Archeological Studies among the Ancient Cities of Mexico*.

186. Maudslay, *Biologia Centrali Americana*, 32.

187. Blom and La Farge, *Tribes and Temples*.

188. Becquelin and Baudez, *Tonina, une cité maya du Chiapas*, vols. 1–3.

189. Becquelin and Baudez, *Tonina, une cité maya du Chiapas*, 1:137–140.

190. Zender and Bassie, "The Wooden Offering Container of Aj K'ax B'ahlam of Tortugero."

BIBLIOGRAPHY

Abrams, Elliot M. *How the Maya Built Their World: Energetics and Ancient Architecture*. Austin: University of Texas Press, 1994.

Aimers, James J., and Prudence M. Rice. "Astronomy, Ritual and the Interpretation of the Maya 'E-Group' Architectural Assemblages." *Ancient Mesoamerica* 17, no. 1 (2006): 79–96.

Ambrose, Stanley H. "Isotopic Analysis of Paleodiets: Methodological and Interpretive Considerations." In *Investigations of Ancient Human Tissue: Chemical Analyses in Anthropology*, edited by Mary K. Sandford, 59–130. Langhorne, Pa.: Gordan and Breach, 1993.

Anaya Hernández, Armando, Stanley P. Guenter, and Peter Mathews. "An Inscribed Wooden Box from Tabasco, Mexico." *Mesoweb Reports and News* (2002), http://www.mesoweb.com/reports/box/index.html.

Ashmore, Wendy. "Classic Maya Wells at Quirigua, Guatemala: Household Facilities in a Water-Rich Setting." *American Antiquity* 49, no. 1 (1984): 147–153.

———. "The Idea of a Maya Town." In *Structure and Meaning in Human Settlements*, edited by Tony Atkin and Joseph Rykwert, 35–54. Philadelphia: University of Pennsylvania Museum of Archaeology and Anthropology, 2005.

———. "Site-Planning Principles and Concepts of Directionality among the Ancient Maya." *Latin American Antiquity* 2, no. 3 (1991): 199–226.

Ashmore, Wendy, and Pamela L. Geller. "Social Dimensions of Mortuary Space." In *Interacting with the Dead: Perspectives on Mortuary Archaeology for the New Millennium*, edited by Gordon F. M. Rakita, Jane E. Buikstra, Lane A. Beck, and Sloan R. Williams, 81–92. Gainesville: University Press of Florida, 2005.

Ashmore, Wendy, and Jeremy A. Sabloff. "Spatial Orders in Maya Civic Plans." *Latin American Antiquity* 13, no. 2 (2002): 201–215.

Astor-Aguilera, Miguel. "Unshrouding the Communicating Cross: The Iconology of a Maya Quadripartite Symbol." PhD diss., Department of Anthropology, State University of New York at Albany, 2004.

Astor-Aguilera, Miguel Angel. *The Maya World of Communicating Objects: Quadripartite Crosses, Trees, and Stones*. Albuquerque: University of New Mexico Press, 2011.

Aveni, Anthony F. *Skywatchers*. Austin: University of Texas Press, 2001.

Aveni, Anthony F., and Horst Hartung. "Maya City Planning and the Calendar." *Transactions of the American Philosophical Society* 76, no. 7 (1986): 1–87.

———. "Some Suggestions about the Arrangement of Buildings at Palenque." In *Proceedings of the Third Palenque Round Table*, edited by Merle Greene Robertson, 173–177. San Francisco: Pre-Columbian Art Research Institute, 1978.

Bahn, Paul, ed. *Written in Bone: How Human Remains Unlock the Secrets of the Dead*. Toronto: Firefly Books, 1998.

Bassie-Sweet, Karen. *Maya Sacred Geography and the Creator Deities*. Norman: University of Oklahoma Press, 2008.

Beaubien, Harriet F. "Ceramic Laminates: Introduction to the Craft Technology behind Aguateca's Mask." *Society for Archaeological Sciences Bulletin* 27, no. 4 (2004): 11–13.

Becker, Marshall J. "Ancient Maya Houses and Their Identification: An Evaluation of Architectural Groups at Tikal and Inferences Regarding Their Functions." *Revista Española de Antropología Americana* 12 (1982): 111–129.

———. "Burials as Caches; Caches as Burials: A New Interpretation of the Meaning of Ritual Deposits among the Classic Period Lowland Maya." In *New Theories on the Ancient Maya*, edited by E. C. Danien and Robert J. Sharer, 185–196. University Museum Monograph 77. Philadelphia: University of Pennsylvania, 1992.

———. *Excavations in Residential Areas of Tikal: Groups with Shrines*. Tikal Reports No. 21. Philadelphia: University Museum Publications, University of Pennsylvania, 1999.

———. "Maya Heterarchy as Inferred from Classic Period Plaza Plans." *Ancient Mesoamerica* 15, no. 1 (2004): 127–138.

———. "Plaza Plans at Tikal: A Research Strategy for Inferring Social Organization and Processes of Culture Change at Lowland Maya Sites." In *Tikal: Dynasties, Foreigners, and Affairs of State: Advancing Maya Archaeology*, edited by Jeremy A. Sabloff, 253–280. Santa Fe: School of American Research Press, 2003.

Becquelin, Pierre, and Claude Baudez. *Tonina, une cité maya du Chiapas*. Vol. 1. Paris: CEMCA, 1979.

———. *Tonina, une cité maya du Chiapas*. Vol. 2. Paris: CEMCA, 1982.

———. *Tonina, une cité maya du Chiapas*. Vol. 3. Paris: CEMCA, 1982.

Beekman, Christopher S. "Agricultural Pole Rituals and Rulership in Late Formative Central Jalisco." *Ancient Mesoamerica* 14, no. 2 (2003): 299–318.

———. "Recent Research in Western Mexican Archaeology." *Journal of Archaeological Research* 18, no. 1 (2010): 41–109.

Bell, Catherine. "Response: Defining the Need for a Definition." In *The Archaeology of Ritual*, edited by Evangelos Kyriakidis, 277–288. Cotsen Advanced Seminar 3. Los Angeles: Cotsen Institute of Archaeology, University of California, 2007.

———. *Ritual Theory, Ritual Practice*. Oxford: Oxford University Press, 1992.

Bell, Ellen E. "Engendering a Dynasty: A Royal Woman in the Margarita Tomb, Copan." In *Ancient Maya Women*, edited by Traci Arden, 89–104. Lanham, Md.: AltaMira, 2002.

Bell, Ellen E., Marcello A. Canuto, and Robert J. Sharer. *Understanding Early Classic Copan*. Philadelphia: University of Pennsylvania Museum of Archaeology and Anthropology, 2004.

Bell, Ellen E., Robert J. Sharer, Loa P. Traxler, David W. Sedat, Christine W. Carrelli, and Lynn A. Grant. "Tombs and Burials in the Early Classic Acropolis at Copan." In *Understanding Early Classic Copan*, edited by Ellen E. Bell, Marcello A. Canuto, and Robert J. Sharer, 131–158. Philadelphia: University of Pennsylvania Museum of Archaeology and Anthropology, 2004.

Benson, Elizabeth P. "A Knife in the Water: The Stingray in Mesoamerican Ritual Life." *Journal of Latin American Lore* 14, no. 2 (1988): 173–191.

Berlin, Elois Ann, and Brent Berlin. *Medical Ethnobiology of the Highland Maya of Chiapas, Mexico: The Gastrointestinal Diseases*. Princeton, N.J.: Princeton University Press, 1996.

Berlo, Janet C. "Artistic Specialization at Teotihuacan: The Ceramic Incense Burner." In *Pre-Columbian Art History: Selected Readings*, edited by A. Cordy-Collins, 83–100. Palo Alto, Calif.: Peek Publications, 1982.

Binford, Lewis R. "Mortuary Practices: Their Study and Potential." In *Approaches to the Social Dimensions of Mortuary Practices*, edited by James A. Brown. Memoirs of the Society for American Archaeology no. 25. Washington, D.C.: Society for American Archaeology, 1971.

Bloch, Maurice, and Jonathan Parry. "Introduction: Death and the Regeneration of Life." In *Death and the Regeneration of Life*, edited by Maurice Bloch and Jonathan Parry, 1–44. Cambridge: Cambridge University Press, 1982.

Blom, Frans. *A Maya Skull from the Uloa Valley, Republic of Honduras*. Middle American Research Institute, publication 5. New Orleans: Middle American Research Institute, Tulane University, 1933.

Blom, Frans, and Oliver La Farge. *Tribes and Temples: A Record of the Expedition to Middle America Conducted by the Tulane University of Louisiana in 1925*. Vol. 1 of 2. New Orleans: Middle American Research Institute, Tulane University, 1926.

Bogin, B., P. Smith, A. B. Orden, M. I. Varela Silva, and J. Loucky. "Rapid Change in Height and Body Proportions of Maya American Children." *American Journal of Human Biology* 14, no. 6 (2002): 753–761.

Bonner, T. D. *The Life and Adventures of James P. Beckwourth, Mountaineer, Scout, Pioneer and Chief of the Crow Nation of Indians*. London: Harper and Brothers, 1856.

Boot, Erik. *The Nominal Yaxha'(al) Chak on Classic Maya Ceramics and a Possible Cephalomorphic Variant for Yax*. Available online at Mayavase.com, http://www.mayavase.com/YaxhaalChak.pdf.

———. "What Happened on the Date 7 Manik' 5 Woh?: An Analysis of Text and Image on Kerr Nos. 0717, 7447, and 8457." *Wayeb Notes* 20 (2006): 1–21.

Boremanse, Didier. *Hach Winik: The Lacandon Maya of Chiapas, Southern Mexico*. Institute for Mesoamerican Studies, monograph 11. Albany: Institute for Mesoamerican Studies, State University of New York at Albany, 1998.

Bray, Tamara L., Leah D. Minc, María Constanza Ceruti, José Antonio Chávez, Ruddy Perea, and Johan Reinhard. "A Compositional Analysis of Pottery Vessels Associated with the Inca Ritual of Capacocha." *Journal of Anthropological Archaeology* 24, no. 1 (2005): 82–100.

Bricker, Victoria R. "Directional Glyphs in Maya Inscriptions and Codices." *American Antiquity* 48, no. 2 (1983): 347–353.

Brooks, William E. "Industrial Use of Mercury in the Ancient World." In *Mercury in the Environment: Pattern and Process*, edited by Michael S. Bank, 19–26. Berkeley: University of California Press, 2010.

Buikstra, Jane E., and Lane A. Beck, eds. *Bioarchaeology: The Contextual Analysis of Human Remains*. Burlington, Mass.: Academic Press, 2006.

Buikstra, Jane E., T. Douglas Price, Lori E. Wright, and James H Burton. "Tombs and Burials in the Early Classic Acropolis at Copan." In *Understanding Early Classic Copan*, edited by Ellen E. Bell, Marcello A. Canuto, and Robert J. Sharer, 191–212. Philadelphia: University of Pennsylvania Museum of Archaeology and Anthropology, 2004.

Bunzel, Ruth. *Chichicastenango: A Guatemalan Village*. Seattle: University of Washington Press, 1952.

Butler, Judith. *Bodies That Matter: On the Discursive Limits of "Sex."* New York: Routledge, 1993.

———. *Gender Trouble: Feminism and the Subversion of Identity*. New York: Routledge, 1990.

Calvin, Inga. "Where the Wayob Live: A Further Examination of Classic Maya Supernaturals." In *The Maya Vase Book: A Corpus of Rollout Photographs of Maya Vases*, vol. 5, edited by Justin Kerr, 868–883. New York: Kerr Associates, 1997.

Carlsen, Robert S. "Footpath of the Dawn, Footpath of the Sun: Worldviews at Lake Atitlán." In *Maya Worldviews at Conquest*, edited by Leslie G. Cecil and Timothy W. Pugh, 299–316. Boulder: University Press of Colorado, 2009.

Carlsen, Robert S., and Martín Prechtel. "The Flowering of the Dead." In *The War for the Heart and Soul of a Highland Maya Town*, edited by Robert S. Carlsen, 47–67. Austin: University of Texas Press, 1997.

Carr, Robert F., and James E. Hazard. *Tikal Report No. 11: Map of the Ruins of Tikal, El Peten, Guatemala*. Philadelphia: University of Pennsylvania Museum of Archaeology and Anthropology, 1961.

Carrasco, Ramón, Sylviane Boucher, Paula Alvarez, Vera Tiesler, Valeria García, Renata García, and Javier Vásquez. "A Dynastic Tomb from Campeche, Mexico: New Evidence on Jaguar Paw, a Ruler from Calakmul." *Latin American Antiquity* 10 (1999): 47–58.

Carrasco Vargas, Ramón, Verónica Vázquez López, and Simon Martin. "Daily Life of the Ancient Maya Recorded on Murals at Calakmul, Mexico." *Proceedings of the National Academy of Science* 106, no. 46 (2009): 19245–19249.

Casagrande, David Gregory. "Ecology, Cognition, and Cultural Transmission of Tzeltal Maya Medicinal Plant Knowledge." PhD diss., Department of Anthropology, University of Georgia, 2002.

Chase, Arlen F., and Diane Z. Chase. *Heterogeneity in Residential Group Composition: Continued Investigation in and near Caracol's Epicenter: Caracol Archaeological Project Investigations for 2011*. Belize City: Belize Institute of Archaeology, 2011.

———. "Maya Veneration of the Dead at Caracol, Belize." In *Seventh Palenque Round Table, 1989*, edited by Merle Greene Robertson and Virginia M. Fields, 55–62. San Francisco: Pre-Columbian Art Research Institute, 1994.

Chase, Diane Z., and Arlen F. Chase. "The Archaeological Context of Caches, Burials, and Other Ritual Activities for the Classic Period (as Reflected at Caracol, Belize)." In *Function and Meaning in Classic Maya Architecture*, edited by Stephen D. Houston, 299–332. Washington, D.C.: Dumbarton Oaks, 1998.

———. "Archaeological Perspectives on Classic Maya Social Organization from Caracol, Belize." *Ancient Mesoamerica* 15, no. 1 (2004): 139–147.

———. "Ghosts amid the Ruins." In *Living with the Dead: Mortuary Ritual in Mesoamerica*, edited by James L. Fitzsimmons and Izumi Shimada, 78–101. Tucson: University of Arizona Press, 2011.

———. "Maya Multiples: Individuals, Entries, and Tombs in Structure A34 of Caracol, Belize." *Latin American Antiquity* 7, no. 1 (1996): 61–71.

———. "Secular, sagrado, y 'revisitado': La profanación, alteración y reconsagración de los antiguas entierros mayas." In *Antropología de la eternidad: La muerte en la cultura maya*, edited by Andrés Ciudad Ruiz, Mario Humberto Ruz Sosa, and María Josefa Iglesias Ponce de León, 255–278. Publicaciones de la Sociedad Española de Estudios Mayas, number 7. Madrid: Universidad Compultense, 2004.

Chávez Balderas, Ximena. "Sacrifice at the Templo Mayor of Tenochtitlan and Its Role in Regard to Warfare." In *Embattled Bodies, Embattled Places: War in Pre-Columbian Mesoamerica and the Andes*, edited by Andrew K. Scherer and John W. Verano, 173–199. Washington, D.C.: Dumbarton Oaks Research Library and Collection, 2014.

Chávez Balderas, Ximena, Ana Miramontes, and Erika Robles. "Los cuchillos ataviados de la ofrenda 125." *Arqueología Mexicana* 17, no. 103 (2010): 70–75.

Cheetham, David T. "The Role of 'Terminus Groups' in Lowland Maya Site Planning: An Example from Cahal Pech." In *The Ancient Maya of the Belize Valley: Half a Century of Archaeological Research*, edited by James F. Garber, 125–148. Gainesville: University Press of Florida, 2004.

Chevalier, Jacques M., and Andrés Sánchez Bain. *The Hot and the Cold: Ills of Humans and Maize in Native Mexico*. Toronto: University of Toronto Press, 2003.

Child, Mark B. "The Archaeology of Religious Movements: The Maya Sweatbath Cult of Piedras Negras." PhD diss., Department of Anthropology, Yale University, 2006.

———. "The Symbolic Space of the Ancient Maya Sweatbath." In *Space and Spatial Analysis in Archaeology*, edited by Elizabeth C. Robertson, Jeffrey D. Seibert, Deepika C. Fernandez, and Mark U. Zender, 157–168. Albuquerque: University of New Mexico Press, 2006.

Christenson, Allen J. *Art and Society in a Highland Maya Community: The Altarpiece of Santiago Atitlán*. Austin: University of Texas Press, 2001.

———. "K'iche'-English Dictionary and Guide to Pronunciation of the K'iche'-Maya Alphabet." Unpublished document available online at the Foundation for the Advancement of Mesoamerican Studies, Inc., 2003: www.famsi.org/mayawriting/dictionary/Christenson.

———. "Maize Was Their Flesh: Ritual Feasting in the Maya Highlands." In *Pre-Columbian Foodways: Interdisciplinary Approaches to Food, Culture, and Markets in Ancient Mesoamerica*, edited by John E. Stall and Michael Carrasco, 577–600. New York: Springer, 2010.

———. *Popol Vuh: The Sacred Book of the Maya—The Great Classic of Central American Spirituality, Translated from the Original Maya Text*. Norman: University of Oklahoma Press, 2007.

Clancy, Flora Simmons. *The Monuments of Piedras Negras, an Ancient Maya City*. Albuquerque: University of New Mexico Press, 2009.

Clark, John E., and Richard D. Hansen. "The Architecture of Early Kingship: Comparative Perspectives on the Origins of the Maya Royal Court." In *Royal Courts of the Ancient Maya*, vol. 2, *Data and Case Studies*, edited by Takeshi Inomata, Héctor Escobedo, and Stephen D. Houston, 1–45. Boulder, Colo.: Westview, 2001.

Coe, Michael D. "Another Look at the Maya Ballgame." In *Il sacro e il paesaggio nell'America indigena*, edited by Davide Domenici, Carolina Orsini, and Sofia Venturoli, 197–204. Bologna: CLUEB, 2003.

———. *The Art of the Maya Scribe*. New York: Harry N. Abrams, 1998.

———. "The Ideology of the Maya Tomb." In *Maya Iconography*, edited by Elizabeth Benson and Gillett Griffen, 222–235. Princeton, N.J.: Princeton University Press, 1988.

———. *Lords of the Underworld: Masterpieces of Classic Maya Ceramics*. Princeton, N.J.: Princeton University Press, 1978.

———. *Old Gods and Young Heroes: The Pearlman Collection of Maya Ceramics*. Jerusalem: Israel Museum, 1982.

Coe, Michael D., and Mark Van Stone. *Reading the Maya Glyphs*. 2nd ed. New York: Thames and Hudson, 2005.

Coe, William R. *Excavations in the Great Plaza, North Terrace, and North Acropolis of Tikal*. Tikal Report No. 14. Philadelphia: University Museum, University of Pennsylvania, 1990.

———. *Piedras Negras Archaeology: Artifacts, Caches, and Burials*. Museum Monographs. Philadelphia: University Museum, University of Pennsylvania, 1959.

Coggins, Clemency C. "New Fire at Chichen Itza." In *Memorias del Primer Coloquio Internacional de Mayistas*, 427–484. Mexico City: Centro de Estudios Mayas, Universidad Nacional Autónoma de México, 1987.

———. "Painting and Drawing Styles at Tikal: An Historical and Iconographic Reconstruction." PhD diss., Department of Fine Arts, Harvard University, 1975.

———. "The Shape of Time: Some Political Implications of a Fourt-Part Figure." *American Antiquity* 45 (1980): 727–739.

Cook, Garrett W. "Quichean Folk Theology and Southern Maya Supernaturalism." In *Symbol and Meaning beyond the Closed Corporate Community*, edited by Gary H. Gossen, 139–153. Albany: Institute for Mesoamerican Studies, 1986.

———. *Renewing the Maya World: Expressive Culture in a Highland Town*. Austin: University of Texas Press, 2000.

Cook, Garrett W., and Thomas A. Offit (with a contribution from Rhonda Taube). *Indigenous Religion and Cultural Performance in the New Maya World*. Albuquerque: University of New Mexico Press, 2013.

Cucina, Andrea, Cristina Perera Cantillo, Thelma Sierra Sosa, and Vera Tiesler. "Carious Lesions and Maize Consumption among the Prehispanic Maya: An Analysis of a Coastal Community in Northern Yucatan." *American Journal of Physical Anthropology* 145, no. 4 (2011): 560–567.

Cucina, Andrea, and Vera Tiesler. "The Companions of Janaab' Pakal and the 'Red Queen' from Palenque, Chiapas: Meanings of Human Companion Sacrifice in Classic Maya Society." In *Janaab' Pakal of Palenque: Reconstructing the Life and Death of a Maya Ruler*, edited by Vera Tiesler and Andrea Cucina, 102–125. Tucson: University of Arizona Press, 2006.

———. "Dental Caries and Antemortem Tooth Loss in the Northern Peten Area, Mexico: A Biocultural Perspective on Social Status Differences among the Classic Maya." *American Journal of Physical Anthropology* 122 (2003): 1–10.

Cuevas García, Martha. "The Cult of Patron and Ancestor Gods in Censers at Palenque." In *Courtly Art of the Ancient Maya*, edited by Mary Miller and Simon Martin, 253–255. New York: Thames and Hudson, 2004.

Culbert, T. Patrick. *The Ceramics of Tikal: Vessels from the Burials, Caches, and Special Deposits*. Tikal Reports 25. Philadelphia: University Museum, 1993.

Cummings, Byron. "Apache Puberty Ceremony for Girls." *Kiva* 5, no. 1 (1939): 1–4.

Curtis, Edward S. *The North American Indian*. Vol. 4. Cambridge, Mass.: University Press, 1909.

Danforth, Marie Elaine. "Coming Up Short: Stature and Nutrition among the Ancient Maya of the Southern Lowlands." In *Reconstructing Ancient Maya Diet*, edited by C. D. White, 103–118. Salt Lake City: University of Utah Press, 1999.

Danziger, Eve. "On Trying and Lying: Cultural Configurations of the Gricean Maxim of Quality." *Intercultural Pragmatics* 7, no. 2 (2010): 199–219.

de Anda, Guillermo, Vera Tiesler, and Pilar Zabala. "Cenotes, espacios sagrados y la práctica de sacrificio humano en Yucatán." *Los Investigadores de la Cultura Maya* 12, no. 2 (2004): 376–386.

DeBoer, Warren R. "Colors for a North American Past." *World Archaeology* 37, no. 1 (2005): 66–91.

de León, Juan. *Diccionario Quiche-Español*. Guatemala City: Editorial Landivar, 1954.

de Witte, Marleen. "Money and Death: Funeral Business in Asante, Ghana." *Africa: Journal of the International African Institute* 73, no. 4 (2003): 531–559.

Díaz del Castillo, Bernal. *The History of the Conquest of New Spain*. Albuquerque: University of New Mexico Press, 2008 [1632].

Douglass, William A. *Death in Murelaga: Funerary Rituals in a Spanish Basque Village*. Seattle: University of Washington Press, 1969.

Doyle, James A. "Regroup on 'E-Groups': Monumentality and Early Centers in the Middle Preclassic Maya Lowlands." *Latin American Antiquity* 23, no. 4 (2012): 355–379.

Duday, Henri. *The Archaeology of the Dead: Lectures in Archaeothanatology*. Oxford: Oxbow Books, 2009.

Duncan, William N., and Charles Andrew Hofling. "Why the Head? Cranial Modification as Protection and Ensoulment among the Maya." *Ancient Mesoamerica* 22, no. 1 (2011): 199–210.

Durán, Diego. *Book of the Gods and Rites and The Ancient Calendar*. Translated by Fernando Horcasitas and Doris Heyden. Norman: University of Oklahoma Press, 1971 [1574–1576].

Eberl, Marcus. *Muerte, entierro y ascensión: Ritos funerarios entre los antiguos mayas*. Libros Scientíficos, tratados 21. Mérida: Universidad Autónoma de Yucatán, 2005.

Edmonson, Munro S. *The Ancient Future of the Itza: The Book of Chilam Bilam of Tizimin*. Austin: University of Texas Press, 1982.

Edwards, Carolyn P. "The Funeral Ceremony in Zinacantan." Honors thesis, Department of Anthropology, Radcliffe College, 1969.

Eeckhout, Peter, and Lawrence Stewart Owens. "Human Sacrifice at Pachacamac." *Latin American Antiquity* 19, no. 4 (2008): 375–398.

Escobedo, Héctor L. "Tales from the Crypt: The Burial Place of Ruler 4, Piedras Negras." In *Courtly Art of the Ancient Maya*, edited by Mary Miller and Simon Martin, 277–280. New York: Thames and Hudson, 2004.

Esposito, Matthew D. *Funerals, Festivals, and Cultural Politics in Porfirian Mexico*. Albuquerque: University of New Mexico Press, 2010.

Estrada-Belli, Francisco. *The First Maya Civilization: Ritual and Power before the Classic Period*. New York: Routledge, 2011.

———. "Lightning Sky, Rain, and the Maize God: The Ideology of Preclassic Maya Rulers at Cival, Peten, Guatemala." *Ancient Mesoamerica* 17 (2006): 57–78.

Farris, Nancy. *Maya Society under Colonial Rule: The Collective Enterprise of Survival*. Princeton, N.J.: Princeton University Press, 1984.

Fernández, Fabián G., Kristofer D. Johnson, Richard E. Terry, Sheldon D. Nelson, and David L. Webster. "Soil Resources of the Ancient Maya at Piedras Negras, Guatemala." *Soil Science Society of America Journal* 69 (2005): 2020–2032.

Fernández, Fabián, John Jacob Parnell, and Richard E. Terry. "Análisis de los suelos del área de Piedras Negras." In *Proyecto arqueológico Piedras Negras: Informe preliminar no. 4, cuarta temporada, 2000*, edited by Héctor Escobedo and Stephen D. Houston, 567–586. Guatemala City: Instituto de Antropología e Historia de Guatemala, 2000.

Fink, Ann E. "Shadow and Substance: A Mopan View of Human Existence." *Canadian Journal of Native Studies* 7, no. 2 (1987): 399–414.

Fischer, Clare B., and Luh Estiti Andarawati. "Tooth-Filing in Bali: One Woman's Experience." *Journal of Ritual Studies* 12, no. 1 (1998): 39–46.

Fitzsimmons, James L. "Classic Maya Mortuary Anniversaries at Piedras Negras." *Ancient Mesoamerica* 9, no. 2 (1998): 271–278.

———. *Death and the Classic Maya Kings*. Austin: University of Texas Press, 2009.

———. "Perspectives on Death and Transformation in Ancient Maya Society." In *Living with the Dead: Mortuary Ritual in Mesoamerica*, edited by James L. Fitzsimmons and Izumi Shimada, 51–77. Tucson: University of Arizona Press, 2011.

Fitzsimmons, James, L., Andrew K. Scherer, Stephen D. Houston, and Héctor L. Escobedo. "Guardian of the Acropolis: The Sacred Space of a Royal Burial at Piedras Negras, Guatemala." *Latin American Antiquity* 14, no. 4 (2003): 449–468.

Folan, William J., Joel D. Gunn, and María del Rosario Domínguez Carrasco. "Triadic Temples, Central Plazas, and Dynastic Palaces: A Diachronic Analysis of the Royal Court Complex, Calakmul, Campeche, Mexico." In *Royal Courts of the Ancient Maya*, vol. 2: *Data and Case Studies*, edited by Takeshi Inomata, Héctor Escobedo, and Stephen D. Houston, 223–265. Boulder, Colo.: Westview, 2001.

Folan, William J., Joyce Marcus, Sophia Pincemin, María del Rosario Domínguez Carrasco, Laraine A. Fletcher, and Abel Morales Lopéz. "Calakmul, Campeche: New Data from an Ancient Maya Capital in Campeche, Mexico." *Latin American Antiquity* 6 (1995): 310–334.

Foster, G. M. *Hippocrates' Latin American Legacy: Humoral Medicine in the New World*. Langhorne, Pa.: Gordon and Breach, 1994.

Foucault, Michel. *Discipline and Punish: The Birth of the Prison*. London: Penguin, 1977.

———. *The History of Sexuality*, vol. 1: *An Introduction*. London: Penguin, 1978.

Freidel, David, Linda Schele, and Joy Parker. *Maya Cosmos: Three Thousand Years on the Shaman's Path*. New York: William Morrow, 1993.

García Barrios, Ana. "Chaahk, el dios de la lluvia, en el periodo clásico maya: Aspectos religiosos y políticos." PhD diss., Facultad de Geografía e Historia, Departamento de Historia del América II (Antropología de América), Madrid, 2008.

García Moll, Roberto. "Shield Jaguar and Structure 23 at Yaxchilan." In *Courtly Art of the Ancient Maya*, edited by Mary Ellen Miller and Simon Martin, 268–270. San Francisco: Fine Arts Museums of San Francisco, 2004.

García Moll, Roberto, and Daniel Juárez Cosío. *Yaxchilán: Antología de su descubrimiento y estudios*. Colección Científica, Serie Arqueología, no. 152. Mexico City: Instituto Nacional de Antropología e Historia, 1986.

Gates, William. *Yucatan before and after the Conquest*. New York: Dover, 1978.

Geller, Pamela L. "Altering Identities: Body Modification and the Pre-Columbian Maya." In *Social Archaeology of Funerary Remains*, edited by Rebecca Gowland and Christopher J. Knüsel, 279–291. Oxford: Oxbow Books, 2006.

———. "Skeletal Analysis and Theoretical Complications." *World Archaeology* 37, no. 4 (2005): 597–609.

Germain, M., K. J. Smith, and H. Skelton. "The Cutaneous Cellular Infiltrate to Stingray Envenomization Contains Increased TIA+ Cells." *British Journal of Dermatology* 143 (2000): 1074–1077.

Gerry, John P., and Harold W. Krueger. "Regional Diversity in Classic Maya Diets." In *Bones of the Maya: Studies of Ancient Skeletons*, edited by S. L. Whittington and D. M. Reed, 196–207. Washington, D.C.: Smithsonian Institution Press, 1997.

Gillespie, Susan D. "Body and Soul among the Maya: Keeping the Spirits in Place." *Archeological Papers of the American Anthropological Association* 11 (2002): 67–78.

———. "Personhood, Agency, and Mortuary Ritual: A Case Study from the Ancient Maya." *Journal of Anthropological Archaeology* 20 (2001): 73–112.

Girard, Raphael. *People of the Chan*. Chino Valley, Ariz.: Continuum Foundation, 1995.

Glazier, Jack. "Mbeere Ancestors and the Domestication of Death." *Man* 19, no. 1 (1984): 133–147.

Golden, Charles, and Andrew K. Scherer. "Territories and Trust, Growth and Collapse: Political Dynamics in Classic Period Maya Kingdoms." *Current Anthropology* (2013): 397–417.

Golden, Charles, Andrew K. Scherer, A. René Muñoz, and Zachary Hruby. "Polities, Boundaries, and Trade in the Classic Period Usumacinta River Basin." *Mexicon* 34, no. 1 (2012): 11–19.

Golden, Charles, Andrew K. Scherer, A. René Muñoz, and Rosaura Vásquez. "Piedras Negras and Yaxchilan: Divergent Political Trajectories in Adjacent Maya Polities." *Latin American Antiquity* 19, no. 2 (2008): 249–274.

Goldstein, Lynne G. "Mortuary Analysis and Bioarchaeology." In *Bioarchaeology: The Contextual Analysis of Human Remains*, edited by Jane E. Buikstra and Lane A. Beck, 375–387. Burlington, Mass.: Academic Press, 2006.

González Cruz, Arnoldo. "El templo de la reina roja, Palenque, Chiapas." *Arqueología Mexicana* 5, no. 30 (1998): 61.

———. *La reina roja: Una tumba real de Palenque*. Mexico City: Insituto Nacional de Antropología e Historia, 2011.

———. "The Red Queen." *Mesoweb* (2000), www.mesoweb.com/palenque/features/red_queen/01.html.

———. "Trabajos recientes en Palenque." *Arqueología Mexicana* 2 (1994): 39–45.

González Lauck, Rebecca. "La Venta: An Olmec Capital." In *Olmec Art of Ancient Mexico*, edited by E. P. Benson and B. de la Fuente, 73–81. Washington, D.C.: National Gallery of Art, 1996.

Gossen, Gary H. "Animal Souls and Human Destiny in Chamula." *Man* 10, no. 3 (1975): 448–461.

———. *Chamulas in the World of the Sun: Time and Space in Maya Oral Tradition*. Prospect Heights, Ill.: Waveland, 1974.

———. *Four Creations: An Epic Story of the Chiapas Mayas*. Norman: University of Oklahoma Press, 2002.

———. "From Olmecs to Zapatistas: A Once and Future History of Souls." *American Anthropologist* 96, no. 3 (1994): 553–570.

Graham, Elizabeth. *Maya Christians and Their Churches in Sixteenth-Century Belize*. Gainesville: University Press of Florida, 2011.

Graham, Ian, Lucia Henderson, Peter Mathews, and David Stuart. *Corpus of Maya Hieroglyphic Inscriptions*, vol. 9, part 2, *Tonina*. Cambridge, Mass.: Peabody Museum of Archaeology and Ethnology, Harvard University, 2006.

Graham, Ian, and Peter Mathews. *Corpus of Maya Hieroglyphic Inscriptions*, vol. 6, part 2, *Tonina*. Cambridge, Mass.: Peabody Museum of Archaeology and Ethnology, Harvard University, 1996.

Graham, Ian, and Eric von Euw. *Corpus of Maya Hieroglyphic Inscriptions*, vol. 3, part 1, *Yaxchilan*. Cambridge, Mass.: Peabody Museum of Archaeology and Ethnology, Harvard University, 1977.

Groark, Kevin P. "Discourses of the Soul: The Negotiation of Personal Agency in Tzotzil Maya Dream Narrative." *American Ethnologist* 36, no. 4 (2009): 705–721.

———. "Pathogenic Emotions: Sentiment, Sociality, and Sickness among the Tzotzil Maya of San Juan Chamula, Chiapas, Mexico." PhD diss., Department of Anthropology, University of California, Los Angeles, 2005.

———. "Social Opacity and the Dynamics of Empathic In-Sight among the Tzotzil Maya of Chiapas, Mexico." *Ethos* 36, no. 4 (2008): 427–448.

———. "Toward a Cultural Phenomenology of Intersubjectivity: The Extended Relational Field of the Tzotzil Maya of Highland Chiapas, Mexico." *Language and Communication* (2012): 278–291.

Grube, Nikolai, and Werner Nahm. "A Census of Xibalba: A Complete Inventory of *Way* Characters on Maya Ceramics." In *The Maya Vase Book: A Corpus of Rollout Photographs of Maya Vases*, vol. 4, edited by Barbara Kerr, photographs by Justin Kerr, 688–715. New York: Kerr Associates, 1994.

Guernsey, Julia. "A Consideration of the Quatrefoil Motif in Preclassic Mesoamerica." *RES: Anthropology and Aesthetics* 57–58 (Spring–Autumn 2010): 75–96.

———. *Ritual and Power in Stone: The Performance of Rulership in Mesoamerican Izapan Style Art*. Austin: University of Texas Press, 2006.

Guernsey, Julia, and F. Kent Reilly, eds. *Sacred Bundles: Ritual Acts of Wrapping and Binding in Mesoamerica*. Barnardsville, N.C.: Boundary End Resarch Center, 2006.

Guillemin, George F. "The Ancient Cakchiquel Capital of Iximche." *Expedition* (Winter 1967): 22–35.

Guiteras-Holmes, C. *Perils of the Soul: The World View of a Tzotzil Indian*. New York: Free Press of Glencoe, 1961.

Gutiérrez, Yeny, Edwin Román, and Alyce de Carteret. "Excavaciones en el grupo El Diablo (Operación 19)." In *Proyecto arqueológico El Zotz: Informe no. 7, temporada 2012*, edited by José Luis Garrido López, Thomas G. Garrison, Edwin Román, and Stephen Houston, 17–46. Guatemala City: Instituto de Antropología e History, 2012.

Haglund, W. D., D. T Reay, and D. R. Swindler. "Tooth Mark Artifacts and Survival of Bones in Animal Scavenged Human Skeletons." *Journal of Forensic Sciences* 33, no. 4 (1988): 985–997.

Hall, Grant D. "Realm of Death: Royal Mortuary Customs and Polity Interaction in the Classic Maya Lowlands." PhD diss., Department of Anthropology, Harvard University, 1989.

Handley, Paul M. *The King Never Smiles: A Biography of Thailand's Bhumibol Adulyadej*. New Haven, Conn.: Yale University Press, 2006.

Hanks, William F. *Referential Practice: Language and Lived Space among the Maya*. Chicago: University of Chicago Press, 1990.

Harris, Grace Gredys. "Concepts of Individual, Self, and Person in Description and Analysis." *American Anthropologist* 91, no. 3 (1989): 599–612.

Harrison-Buck, Eleanor, Patricia A. McAnany, and Rebecca Storey. "Empowered and Disempowered during the Late to Terminal Classic Transition: Maya Burial and Termination Rituals in the Sibun Valley." In *New Perspectives on Human Sacrifice and Ritual Body Treatments in Ancient Maya Society*, edited by Vera Tiesler and Andrea Cucina, 74–101. New York: Springer, 2007.

Haviland, Leslie K., and John B. Haviland. "Privacy in a Mexican Indian Village." In *Public and Private in Social Life*, edited by Stanley I. Benn and Gerald F. Gaus, 341–361. New York: St. Martin's Press/Croom Helm, 1983.

Haviland, William A. "Musical Hammocks at Tikal: Problems with Reconstructing Household Composition." In *Household and Community in the Mesoamerican Past*, edited by Richard R. Wilk and Wendy Ashmore, 121–134. Albuquerque: University of New Mexico Press, 1988.

Headland, Thomas N. "Teeth Mutilation among the Casiguran Dumagat." *Philippine Quarterly of Culture and Society* 5, nos. 1–2 (1977): 54–64.

Headrick, Annabeth. "The Great Goddess at Teotihuacan." In *Heart of Creation: The Mesoamerican World and the Legacy of Linda Schele*, edited by A. Stone, 83–100. Tuscaloosa: University of Alabama Press, 2002.

———. "The Quadripartite Motif and the Centralization of Power." In *K'axob: Ritual, Work, and Family in an Ancient Maya Village*, edited by Patricia A. McAnany, 367–378. Monumenta Archaeologica 22. Los Angeles: Cotsen Instititute of Archaeology, University of California, 2004.

Healy, Paul F., and Marc G. Blainey. "Ancient Maya Mosaic Mirrors: Function, Symbolism, and Meaning." *Ancient Mesoamerica* 22, no. 2 (2011): 229–244.

Heizer, Robert F., and Jonas E. Gullberg. "Concave Mirrors from the Site of La Venta, Tabasco: Their Occurrence, Mineralogy, Optical Description, and Function." In *The Olmec and Their Neighbors: Essays in Memory of Matthew W. Sterling*, edited by Elizabeth P. Benson, 109–116. Washington, D.C.: Dumbarton Oaks Rsearch Library and Collection, 1981.

Hellmuth, Nicholas. *Monster und Menschen in der Maya-Kunst*. Graz, Austria: Adeva Akademische Druck- u. Verlaganstalt, 1987.

———. "Structure 5D-73, Burial 196, Tikal, Peten, Guatemala: A Preliminary Report." Honors thesis, Department of Anthropology, Harvard College, 1967.

Helmke, Christophe. "Mesoamerican Lexical Calques in Ancient Maya Writing and Imagery." *PARI Journal* 14, no. 2 (2013): 1–15.

Helmke, Christophe, and Axel E. Nielsen. "Hidden Identity and Power in Ancient Mesoamerica: Supernatural Alter Egos as Personified Diseases." *Acta Americana* 17, no. 2 (2009): 49–98.

Helmke, Christophe G. B., and Gabriel D. Wrobel. "Je'reftheel, Roaring Creek Works, Belize." In *Heart of Earth: Studies in Maya Ritual Cave Use*, edited by James E. Brady, 69–81. Association for Mexican Cave Studies, bulletin 23. Austin: Association for Mexican Cave Studies, 2012.

Hendrickson, Carol. *Weaving Identities: Construction of Dress and Self in a Highland Maya Town*. Austin: University of Texas Press, 1995.

Hertz, Robert. *Death and the Right Hand*. London: Cohen and West, 1960 [1928].

Holland, William R. "Contemporary Tzotzil Cosmological Concepts as a Basis for Interpreting Prehistoric Maya Civilization." *American Antiquity* 29, no. 3 (1964): 301–336.

Holmes, William Henry. *Archeological Studies among the Ancient Cities of Mexico*, part II: *Monuments of Chiapas, Oaxaca, and the Valley of Mexico*. Field Columbian Museum publication 16; Anthropological Series, vol. 1, no. 1. Chicago: Field Museum, 1897.

Houston, Stephen D. "Classic Maya Depictions of the Built Environment." In *Function and Meaning in Classic Maya Architecture*, edited by Stephen D. Houston, 333–372. Washington, D.C.: Dumbarton Oaks, 1998.

———. "Deathly Sport." *Maya Decipherment: Ideas in Ancient Maya Writing* (2014), decipherment.wordpress.com/2014/07/29/deathly-sport/.

———. "Engraved Stingray Spines and Bone Handles from Royal Burials." In *The Fiery Pool: The Maya and the Mythic Sea*, edited by Daniel Finamore and Stephen D. Houston, 138–139. New Haven, Conn.: Yale University Press; Salem, Mass.: Peabody Essex Museum, 2010.

———. "Impersonation, Dance, and the Problem of Spectacle among the Classic Maya." In *Archaeology of Performance: Theaters of Power, Community, and Politics*, edited by Takeshi Inomata and Lawrence S. Coben, 135–155. Lanham, Md.: AltaMira, 2006.

———. *The Life Within: Classic Maya and the Matter of Impermanence*. New Haven, Conn.: Yale University Press, 2014.

———. "Living Waters and Wondrous Beasts." In *The Fiery Pool: The Maya and the Mythic Sea*, edited by Daniel Finamore and Stephen D. Houston, 66–82. New Haven, Conn.: Yale University Press; Salem, Mass.: Peabody Essex Museum, 2010.

———. "Panel Describing a King's Pilgrimage to the Sea." In *Fiery Pool: The Maya and the Mythic Sea*, edited by Daniel Finamore and Stephen D. Houston, 201. New Haven, Conn.: Yale University Press; Salem, Mass.: Peabody Essex Museum, 2010.

———. "A Splendid Predicament: Young Men in Classic Maya Society." *Cambridge Archaeological Journal* 19, no. 2 (2009): 149–178.

———. "Symbolic Sweatbaths of the Maya: Architectural Meaning in the Cross Group at Palenque, Mexico." *Latin American Antiquity* 72, no. 2 (1996): 132–151.

———. "Te-Mu and Te-Ma as 'Throne.'" *Maya Decipherment: Ideas on Ancient Maya Writing and Iconography* (2008), http://decipherment.wordpress.com/2008/04/24/te-mu-and-te-ma-as-throne/.

Houston, Stephen D., and Héctor L. Escobedo. "Nuevas intervenciones en la estructura de Piedras Negras." In *Informe de las intervenciones en la estructura K-5 de Piedras Negras*, edited by Héctor L. Escobedo and Stephen D. Houston, 1–3. Guatemala City: Dirección General del Patrimonio Cultural y Natural de Guatemala, 2005.

Houston, Stephen D., Héctor Escobedo, Mark Child, Charles Golden, and René Muñoz. "The Moral Community: Maya Settlement Transformation at Piedras Negras, Guatemala." In *The Social Construction of Ancient Cities*, edited by Monica L. Smith, 212–253. Washington, D.C.: Smithsonian Institution Press, 2003.

Houston, Stephen D., Héctor Escobedo, Mark Child, Charles Golden, Richard Terry, and David L. Webster. "In the Land of the Turtle Lords: Archaeological Investigations at Piedras Negras, Guatemala, 2000." *Mexicon* 22 (2000): 97–110.

Houston, Stephen D., Héctor Escobedo, Andrew K. Scherer, James L. Fitzsimmons, and Mark Child. "Classic Maya Death at Piedras Negras, Guatemala." In *Antropología de la eternidad: La muerte en la cultura maya*, edited by Andrés Ciudad Ruiz, Mario Humberto Ruz Sosa, and María Josefa Iglesias Ponce de León, 113–144. Madrid: Sociedad Española de Estudios Mayas, 2003.

Houston, Stephen D., and Takeshi Inomata. *The Classic Maya*. Cambridge: Cambridge University Press, 2009.

Houston, Stephen D., and Patricia A. McAnany. "Bodies and Blood: Critiquing Social Construction in Maya Archaeology." *Journal of Anthropological Archaeology* 22 (2003): 26–41.

Houston, Stephen, Sarah Newman, Edwin Román, and Thomas Garrison (with contributions by Nicholas Carter, Alyce de Carteret, Andrew Scherer, and Karl Taube). *Temple of the Night Sun: A Royal Tomb at El Diablo, Guatemala*. San Francisco: Precolumbia Mesoweb Press, in press.

Houston, Stephen D., and Andrew K. Scherer. "La ofrenda máxima: El sacrificio humano en la parte central del área maya." In *Nuevas perspectivas sobre el sacrificio humano entre los mexicas*, edited by Leonardo López Luján and Guilhem Olivier, 167–191. Mexico City: Instituto Nacional de Antropología e Historia, 2010.

Houston, Stephen D., and David Stuart. "The Ancient Maya Self: Personhood and Portraiture in the Classic Period." *RES: Anthropology and Aesthetics* 33 (1998): 73–101.

———. "Of Gods, Glyphs, and Kings: Divinity and Rulership among the Classic Maya." *Antiquity* 70 (1996): 289–312.

———. "The *Way* Glyph: Evidence for 'Co-Essences' among the Classic Maya." *Research Reports on Ancient Maya Writing* 30 (1989): 1–16.

Houston, Stephen, David Stuart, and Karl Taube. *The Memory of Bones: Body, Being, and Experience among the Classic Maya*. Austin: University of Texas Press, 2006.

Houston, Stephen D., and Karl A. Taube. "An Archaeology of the Senses: Perceptions and Cultural Expression in Ancient Mesoamerica." *Cambridge Archaeological Journal* 10, no. 2 (2000): 261–294.

Hruby, Zachary X. "The Organization of Chipped-Stone Economies at Piedras Negras, Guatemala." PhD diss., University of California at Riverside, 2006.

Hruby, Zachary X., and Gene Ware. "Painted Lithic Artifacts from Piedras Negras, Guatemala." In *Maya Archaeology I*, edited by Charles Golden, Stephen Houston, and Joel Skidmore, 77–85. San Francisco: Precolumbia Mesoweb Press, 2009.

Hull, Kerry. "Journey from the Ancient Maya Tomb: Ropes, Roads, and the Point of Departure." In *Jaws of the Underworld: Life, Death, and Rebirth among the Ancient Maya*, edited by Pierre R. Colas, Geneviève Le Fort, and Bodil Liljefors Persson, 43–52. Acta Mesoamericana, vol. 16. Markt Schwaben, Germany: Verlag Anton Saurwein, 2006.

Hull, Kerry, and Michael David Carrasco. "*Mak*-'Portal' Rituals Uncovered: An Approach to Interpreting Symbolic Architecture and the Creation of Sacred Space among the Maya." In *Continuity and Change: Maya Religious Practices in Temporal Perspective*, edited by Daniel Graña Behrens, Nikolai Grube, Christian M. Prager, Frauke Sachse, Stefanie Teufel, and Elisabeth Wagner, 131–142. Acta Mesoamericana, vol. 14. Markt Schwaben, Germany: Verlag Anton Saurwein, 2000.

Ingold, Tim. "Materials against Materiality." *Archaeological Dialogues* 14, no. 1 (2007): 1–16.

———. "Technology, Language, and Intelligence: A Reconsideration of Basic Concepts." In *Tools, Language, and Cognition in Human Evolution*, edited by Kathleen R. Gibson and Tim Ingold, 449–472. Cambridge: Cambridge University Press, 1993.

Inomata, Takeshi. "Plazas, Performers, and Spectators: Political Theaters of the Classic Maya." *Current Anthropology* 47, no. 5 (2006): 805–842.

———. "War, Violence, and Society in the Maya Lowlands." In *Embattled Places, Embattled Bodies: War in Pre-Columbian America*, edited by Andrew K. Scherer and John W. Verano, 25–56. Washington, D.C.: Dumbarton Oaks Research Library and Collection, 2014.

Johnson, Kristofer D., Richard E. Terry, Mark W. Jackson, and Charles Golden. "Ancient Soil Resources of the Usumacinta River Region, Guatemala." *Journal of Archaeological Science* 34 (2007): 1117–1129.

Jones, Christopher, and Linton Satterthwaite. *The Monuments and Inscriptions of Tikal*. Tikal Report no. 33, part A. Philadelphia: University Museum, University of Pennsylvania, 1982.

Joralemon, David. "Ritual Blood Sacrifice among the Ancient Maya: Part I." In *Primera mesa redonda de Palenque, Part II*, edited by Merle Greene Robertson, 59–66. Pebble Beach, Calif.: Robert Louis Stevenson School, 1974.

Joyce, Rosemary A. *Ancient Bodies, Ancient Lives: Sex, Gender, and Archaeology*. New York: Thames and Hudson, 2008.

———. "Archaeology of the Body." *Annual Review of Anthropology* 34 (2005): 139–158.

———. "The Construction of Gender in Classic Maya Monuments." In *Gender and Archaeology: Essays in Research and Practice*, edited by R. P. Wright, 167–195. Philadelphia: University of Pennsylvania Press, 1996.

———. *Gender and Power in Prehispanic Mesoamerica*. Austin: University of Texas Press, 2001.

———. "Girling the Girl and Boying the Boy: The Production of Adulthood in Ancient Mesoamerica." *World Archaeology* 31, no. 3 (2000): 473–483.

Kaplan, Jonathan. "El trono incienso y otros tronos de Kaminaljuyu, Guatemala." In *VIII Simposio de Investigaciones Arqueológicas en Guatemala, 1994*, edited by Juan Pedro Laporte and Héctor Escobedo, 338–361. Guatemala City: Museo Nacional de Arqueología y Etnología, 1995.

Každan, Aleksandr Petrovič, and Giles Constable. *People and Power in Byzantium: An Introduction to Modern Byzantine Studies*. Washington, D.C.: Dumbarton Oaks, Trustees for Harvard University, 1982.

Kerr, Justin. "The Last Journey: Reflections on the Ratinlinxul Vase and Others of the Same Theme." Maya Vase Database (2001), http://www.mayavase.com/jour/journey.html.

Kertzer, David. *Ritual, Politics, and Power*. New Haven, Conn.: Yale University Press, 1988.

Kidder, Alfred V. *The Artifacts of Uaxactun*. Carnegie Institution of Washington, publication 576. Washington, D.C.: Carnegie Institution of Washington, 1947.

Kidder, Alfred V., Jesse D. Jennings, and Edwin M. Shook. *Excavations at Kaminaljuyu, Guatemala*. Washington, D.C.: Carnegie Institution of Washington, 1946.

Kintz, Ellen R. *Life under the Tropical Canopy*. Fort Worth, Tex.: Holt, Rinehart and Winston, 1990.

Klein, Cecelia F. "Snares and Entrails: Mesoamerican Symbols of Sin and Punishment." *RES: Anthropology and Aesthetics*, nos. 19–20 (1990): 81–103.

Klippel, Walter E., and Jennifer A. Synstelien. "Rodents as Taphonomic Agents: Bone Gnawing by Brown Rats and Gray Squirrels." *Journal of Forensic Sciences* 52, no. 4 (2007): 765–773.

Kovacevich, Brigitte. "Ritual, Crafting, and Agency at the Classic Maya Kingdom of Cancuen." In *Mesoamerican Ritual Economy: Archaeological and Ethnological Perspectives*, edited by E. Christian Wells and Karla L. Davis-Salazar, 67–114. Boulder: University Press of Colorado, 2007.

Krejci, Estella, and T. Patrick Culbert. "Preclassic and Classic Burials and Caches in the Maya Lowlands." In *The Emergence of Lowland Maya Civilization*, edited by Nikolai Grube, 103–116. Acta Mesoamericana, vol. 8. Möchmühl, Germany: Verlag Anton Saurwein, 1995.

Laporte, Juan Pedro. "Contexto y función de los artefactos de hueso en Tikal, Guatemala." *Revista Española de Antropología Americana* 29 (1999): 31–64.

Laporte, Juan Pedro, and Vilma Fialko. "Un reencuentro con mundo perdido, Tikal, Guatemala." *Ancient Mesoamerica* 6 (1995): 41–94.

Larsen, Clark S. *Bioarchaeology: Interpreting Behavior from the Human Skeleton*. Cambridge: Cambridge University Press, 1997.

Lau, George F. "Feasting and Ancestor Veneration at Chinchawas North Highlands of Ancash, Peru." *Latin American Antiquity* 13, no. 3 (2002): 279–304.

Leonard, Zenas. *Narrative of the Adventures of Zenas Leonard*. Clearfield, Pa.: D. W. Moore, 1839.

Liendo Stuardo, Rodrigo. *La organización de la producción agrícola en un centro maya clásico: Patrón de asentamiento en la región de Palenque, Chiapas, México*. Mexico City: Instituto de Antropología e Historia, 2002.

Little, Michael A., and Kenneth A. R. Kennedy. *Histories of American Physical Anthropology in the Twentieth Century*. Lanham, Md.: Lexington Books, 2010.

Looper, Matthew G. *Lightning Warrior: Maya Art and Kingship at Quirigua*. Austin: University of Texas Press, 2003.

———. *To Be Like Gods: Dance in Ancient Maya Civilization*. Austin: University of Texas Press, 2009.

———. "Women-Men (and Men-Women): Classic Maya Rulers and the Third Gender." In *Ancient Maya Women*, edited by Traci Ardren, 171-202. Walnut Creek, Calif.: AltaMira Press, 2002.

Looper, Matthew, and Julia Kappelman Guernsey. "The Cosmic Umbilicus in Mesoamerica: A Floral Metaphor for the Source of Life." *Journal of Latin American Lore* 21, no. 1 (2001): 3–54.

López Bravo, Roberto. "La veneración de los ancestros en Palenque." *Arqueología Mexicana* 8, no. 45 (2000): 38–43.

López Jiménez, Fanny. "El descubrimiento de la Tumba I del Templo de la Calavera y su contexto arquitectónico en Palenque, Chiapas." *Pueblos y Fronteras* 1 (2001): 115–129.

———. "Entierros humanos en el Templo de la Cruz y la Cruz Foliada de Palenque, Chiapas." In *Cuarto foro de arqueología de Chiapas: Serie memorias; Instituto Chiapaneco de Cultura*, 83–97. Tuxtla Gutiérrez, Mexico: Instituto Chiapaneco de Cultura, 1994.

Love, Bruce. *Maya Shamanism Today*. 2nd rev. ed. San Francisco: Precolumbia Mesoweb Press, 2012.

———. "Yucatec Sacred Breads through Time." In *Word and Image in Maya Culture: Explorations in Language, Writing, and Representations*, edited by William F. Hanks and Don S. Rice, 336–350. Salt Lake City: University of Utah Press, 1989.

Luber, George E. "The Biocultural Epidemiology of 'Second Hair' Illness in Two Mesoamerican Societies." PhD diss., Department of Anthropology, University of Georgia, 2002.

Lucero, Lisa J. "Materialized Cosmology among Ancient Maya Commoners." *Journal of Social Archaeology* 10, no. 1 (2010): 138–167.

———. "The Political and Sacred Power of Water in Classic Maya Society." In *Precolumbian Water Management: Ideology, Ritual, and Power*, edited by Lisa J. Lucero and Barbara W. Fash, 116–128. Tucson: University of Arizona Press, 2006.

———. "Water Control and Maya Politics in the Southern Maya Lowlands." In *Complex Polities in the Ancient Tropical World*, edited by E. A. Bacus and Lisa J. Lucero, 35–49. Archaeological Papers of the American Anthropological Association, no. 9. Arlington, Va.: American Anthropological Association, 1999.

Lukacs, J. R. "Dental Trauma and Antemortem Tooth Loss in Prehistoric Canary Islanders: Prevalence and Contributing Factors." *International Journal of Osteoarchaeology* 17, no. 2 (2007): 157–173.

MacLaury, R. E. *Color and Cognition in Mesoamerica: Constructing Categories as Vantages*. Austin: University of Texas Press, 1997.

MacLeod, William Christie. "Self-Sacrifice in Mortuary and Non-Mortuary Ritual in North America." *Anthropos* 33, nos. 3–4 (1938): 349–400.

Maffi, Luisa. "A Linguistic Analysis of Tzeltal Maya Ethnosymptomatology." PhD diss., Department of Anthropology, University of California, Berkeley, 1994.

Maler, Teobert. *Explorations in the Department of Petén, Guatemala, Tikal: Report of Explorations for the Museum.* Memoirs of the Peabody Museum of American Archaeology and Ethnology, Harvard University, vol. 5, no. 1. Cambridge, Mass.: Peabody Museum, Harvard University, 1911.

Maples, William R. *Dead Men Do Tell Tales: The Strange and Fascinating Cases of a Forensic Anthropologist.* New York: Broadway Books, 1995.

Marcus, Joyce, and Kent M. Flannery. "Ancient Zapotec Ritual and Religion: An Application of the Direct Historical Approach." In *The Ancient Mind: Elements of Cognitive Archaeology*, edited by Colin Renfrew and Ezra B. W. Zubrow, 55–74. Cambridge: Cambridge University Press, 1994.

Maroukis, Thomas C. *Peyote and the Yankton Sioux: The Life and Times of Sam Necklace.* Norman: University of Oklahoma Press, 2004.

Martin, Debra L., Ryan P. Harrod, and Ventura R. Pérez. *Bioarchaeology: An Integrated Approach to Working with Human Remains.* New York: Springer, 2013.

Martin, Simon. "The Baby Jaguar: An Exploration of Its Identity and Origins in Maya Art and Writing." In *La organización social entre los mayas: Memoria de la tercera mesa redonda de Palenque*, vol. 1, edited by Vera Tiesler Blos, Rafael Cobos Palma, and Merle Greene Robertson, 50–78. Mexico City: Consejo Nacional para la Cultura y las Artes, Instituto Nacional de Antropología e Historia, Universidad Autónoma de Yucatán, 2002.

———. "Cacao in Ancient Maya Religion: First Fruit from the Maize Tree and Other Tales from the Underworld." In *Chocolate in Mesoamerica: A Cultural History of Cacao*, edited by Cameron McNeil, 154–183. Gainesville: University Press of Florida, 2006.

———. "Hieroglyphs from the Painted Pyramid: The Epigraphy of Chiik Nahb Structure Sub 1-4, Calakmul, Mexico." In *Maya Archaeology* 2, edited by Charles Golden, Stephen Houston, and Joel Skidmore, 60–81. San Francisco: Precolumbia Mesoweb Press, 2012.

Martin, Simon, and Nikolai Grube. *Chronicle of the Maya Kings and Queens.* 2nd ed. New York: Thames and Hudson, 2008.

Martínez del Campo, Sofía. *Rostros de la divinidad: Los mosaicos mayas de piedra verde.* Mexico City: Instituto Nacional de Antropología e Historia, 2010.

Mascia-Lees, Frances E., ed. *A Companion to the Anthropology of the Body and Embodiment.* New York: Wiley-Blackwell, 2011.

Mathews, Jennifer P., and James F. Garber. "Models of Cosmic Order: Physical Expression of Sacred Space among the Ancient Maya." *Ancient Mesoamerica* 15 (2004): 49–59.

Mathews, Peter. "The Sculpture of Yaxchilan." PhD diss., Yale University, 1988.

Maudslay, Alfred P. *Biologia Centrali Americana: Archaeology.* London: R. H. Porter and Dulau, 1889–1902.

Mayakovsky, Vladimir. *Vladimir Ilyich Lenin: A Poem.* Honolulu: University of Hawaii Press, 2003 [1924].

McAnany, Patricia A. "Ancestors and the Classic Maya Built Environment." In *Function and Meaning in Classic Maya Architecture*, edited by Stephen D. Houston, 271–298. Washington, D.C.: Dumbarton Oaks Research Library and Collection, 1998.

———. "Ancestor Worship and Sanctification of Place: Excavations at K'axob, Belize." *Context* 9 (1991): 12–16.

———. *Ancestral Maya Economies in Archaeological Perspective.* Cambridge: Cambridge University Press, 2010.

———. *Living with the Ancestors: Kinship and Kingship in Ancient Maya Society.* Rev. ed. Cambridge: Cambridge University Press, 2013.

McAnany, Patricia A., Rebecca Storey, and Angela K. Lockard. "Mortuary Ritual and Family Politics at Formative and Early Classic K'axob, Belize." *Ancient Mesoamerica* 10 (1999): 129–146.

McGee, R. Jon. *Watching Lacandon Maya Lives.* Boston, Mass.: Allyn and Bacon, 2002.

Mead, George Herbert. *Mind, Self, and Society.* Chicago: University of Chicago Press, 1934.

Méndez, Alonso, Edwin L. Barnhart, Christopher Powell, and Carol Karasik. "Astronomical Observations at the Temple of the Sun." *Maya Exploration Center* online publication (2005), www.mayaexploration.org/pdf/observations_temple_sun.pdf.

Meskell, Lynn, and Rosemary A. Joyce. *Embodied Lives: Figuring Ancient Maya and Egyptian Experience*. London: Routledge, 2003.

Metcalf, Peter, and Richard Huntington. *Celebrations of Death: The Anthropology of Mortuary Ritual*. Cambridge: Cambridge University Press, 1991.

Mickenautsch, Steffen, Soraya Coelho Leal, Veerasamy Yengopal, Ana Cristina Bezerra, and Vanessa Cruvinel. "Sugar-Free Chewing Gum and Dental Caries: A Systematic Review." *Journal of Applied Oral Science* 15 (2007): 83–88.

Milbrath, Susan, and Carlos Peraza Lope. "Revisiting Mayapan: Mexico's Last Maya Capital." *Ancient Mesoamerica* 14 (2003): 1–46.

Miles, S. W. "The Sixteenth-Century Pokom-Maya: A Documentary Analysis of Social Structure and Archaeological Setting." *Transactions of the American Philosophical Society* 47, no. 4 (1957): 733–781.

Miller, Arthur. "The Iconography of the Painting in the Temple of the Diving God, Tulum, Quintana Roo, Mexico: The Twisted Cords." In *Mesoamerican Archaeology: New Approaches*, edited by Norman Hammond, 167–186. Austin: University of Texas Press, 1974.

Miller, Daniel. "Materiality: An Introduction." In *Materiality*, edited by Daniel Miller, 1–50. Durham, N.C.: Duke University Press, 2005.

Miller, Mary. "Extreme Makeover." *Archaeology* 62, no. 1 (2009): 36–42.

———. "The Maya Ballgame: Rebirth in the Court of Life and Death." In *The Sport of Life and Death: The Mesoamerican Ballgame*, edited by E. Michael Whittington, 79–87. New York: Thames and Hudson, 2001.

Miller, Mary, and Simon Martin. *Courtly Art of the Ancient Maya*. London: Thames and Hudson, 2004.

Miller, Mary, and Karl Taube. *The Gods and Symbols of Ancient Mexico and the Maya: An Illustrated Dictionary of Mesoamerican Religion*. New York: Thames and Hudson, 1993.

Miller, Mary Ellen. "The Image of People and Nature in Classic Maya Art and Architecture." In *The Ancient Americas: Art from Sacred Landscapes*, edited by Richard F. Townsend, 158–169. Chicago: Art Institute of Chicago, 1992.

Miller, Virginia E. "Pose and Gesture in Classic Maya Monumental Sculpture." PhD diss., University of Texas at Austin, 1981.

Moholy-Nagy, Hattula. *The Artifacts of Tikal: Ornamental and Ceremonial Artifacts and Unworked Material*. Tikal Report no. 27, part A. Philadelphia: University of Pennsylvania Museum of Archaeology and Anthropology, 2008.

———. "Vertebrates in Tikal Burials and Caches." In *Maya Zooarchaeology: New Direction in Method and Theory*, edited by Kitty F. Emery, 193–208. Los Angeles: Cotsen Institute of Archaeology, UCLA, 2004.

Montero López, Coral, Luis Fernando Núñez, Pedro Morales, Edith Cienfuegos, and Francisco Otero. "Diet and Health at Chinikihá, Chiapas, Mexico: Some Preliminary Results." *Environmental Archaeology* 16, no. 2 (2011): 82–96.

Morris, Ian. "The Archaeology of Ancestors: The Saxe/Goldstein Hypothesis Revisited." *Cambridge Archaeological Journal* 1, no. 2 (1991): 147–169.

Nakamura, Seiichi. "Culto funerario de Copán en el siglo VI: Un estudio de caso en el conjunto 10j-45." In *Culto funerario en la sociedad maya*, edited by Rafael Cabos, 245–253. Mexico City: Instituto Nacional de Antropología e Historia, 2004.

Nash, June. *In the Eyes of the Ancestors: Belief and Behavior in a Mayan Community*. Prospect Heights, Ill.: Waveland, 1970.

———. "The Reassertion of Indigenous Identity: Mayan Responses to State Intervention in Chiapas." *Latin American Research Review* 30, no. 3 (1995): 7–41.

Navarro-Farr, Olivia, Griselda Pérez, and Damaris Menéndez. "Operación WK-1: Excavaciones en la estructura M13-1." In *Proyecto arqueológico el Perú-Waka': Informe no. 10, temporada 2012*, edited by Juan Carlos Perez, 3–91. Guatemala City: Dirección General del Patrimonio Cultural y Natural de Guatemala, 2013.

Neiburger, E. J. "Jaded Smiles: Ancient Mayan Dental Inlays and Today's Central American Patients." *Journal of the Massachusetts Dental Society* (Winter 2012), mydigimag.rrd.com/article/Jaded_Smiles%3A_Ancient_Mayan_Dental_Inlays_And_Today's_Central_American_Patients/987427/101959/article.html.

Nieto Calleja, Rosalba, and Humberto Schiavon Signoret. "El Templo Olvidado de Palenque, Chiapas." *Arqueología* 5 (1989): 191–210.

O'Neil, Megan E. *Engaging Ancient Maya Sculpture at Piedras Negras, Guatemala*. Norman: University of Oklahoma Press, 2012.

Palenque Cross Group Project. *The 2001 Season of the Cross Group Project: Archaeological Excavation and Preliminary Results of Operations at the Temples XIX and XX*. Project report published online at http://www.mesoweb.com/palenque/dig/report/report_01_text.html.

Paul, Benjamin D. "The Maya Bonesetter as Sacred Specialist." *Ethnology* 15, no. 1 (1976): 77–81.

Pendergast, David M., ed. *Excavations at Altun Ha, Belize, 1964–1970*. Vol. 1. Toronto: Royal Ontario Museum, 1979.

Pitarch, Pedro. "Almas y cuerpo en una tradición indígena tzeltal." *Archives de Sciences Sociales des Religions* 112 (2000): 31–48.

———. *Ch'ulel: Una etnografía de las almas tzeltales*. Mexico City: Fondo de Cultura Económica, 1996.

———. *The Jaguar and the Priest: An Ethnography of Tzeltal Souls*. Austin: University of Texas Press, 2010.

Pohl, Mary E. D., Kevin O. Pope, and Christopher von Nagy. "Olmec Origins of Mesoamerican Writing." *Science* 298, no. 5600 (December 6, 2002): 1984–1987.

Proskouriakoff, Tatiana. "Historical Data in the Inscriptions of Yaxchilan, Part 2." *Estudios de Cultura Maya* 4 (1964): 177–201.

Prowse, Tracy, Henry P. Schwarcz, Shelley Saunders, Roberto Macchiarelli, and Luca Bondioli. "Isotopic Paleodiet Studies of Skeletons from the Imperial Roman-Age Cemetery of Isola Sacra, Rome, Italy." *Journal of Archaeological Science* 31, no. 3 (2004): 259–272.

Quenon, Michel, and Geneviève Le Fort. "Rebirth and Resurrection in Maize God Iconography." In *The Maya Vase Book: A Corpus of Rollout Photographs of Maya Vases*, vol. 5, edited by Barbara Kerr, photographs by Justin Kerr, 884–902. New York: Kerr Associates, 1997.

Rakita, Gordon F. M., and Jane E. Buikstra. "Corrupting Flesh: Reexaming Hertz's Perspective on Mummification and Cremation." In *Interacting with the Dead: Perspective on Mortuary Archaeology for the New Millennium*, edited by Gordon F. M. Rakita, Jane Buikstra, Lane A. Beck, and Sloan R. Williams, 97–106. Gainesville: University Press of Florida, 2005.

Rands, Robert L., and Ronald L. Bishop. "The Dish-Plate Tradition at Palenque: Continuity and Change." In *Patterns and Process: A Festschrift in Honor of Dr. Edward V. Sayre*, edited by Lambertus van Zelst, 109–132. Suitland, Md.: Smithsonian Center for Materials Research and Education, 2003.

Redfield, Robert, and Alfonso Villa Rojas. *Chan Kom: A Maya Village*. Publication 448. Chicago: University of Chicago Press, 1962.

Reents-Budet, Dorie. "The Art of Classic Vase Painting." In *Maya: Divine Kings of the Rainforest*, edited by Nikolai Grube, 247–259. Cologne: Könemann, 2001.

Reese-Taylor, Kathryn, Marc Zender, and Pamela L. Geller. "Fit to Be Tied: Funerary Practices among the Prehispanic Maya." In *Sacred Bundles: Ritual Acts of Wrapping and Binding in Mesoamerica*, edited by Julia Guernsey and F. Kent Reilly, 40–58. Barnardsville, N.C.: Boundary End Archaeology Research Center, 2006.

Reilly, F. Kent. "Middle Formative Origins of the Mesoamerican Ritual Act of Bundling." In *Sacred Bundles: Ritual Acts of Wrapping and Binding in Mesoamerica*, edited by Julia Guernsey and F. Kent Reilly, 1–21. Barnardsville, N.C.: Boundary End Resarch Center, 2006.

Restall, Matthew. *The Maya World: Yucatec Culture and Society, 1550–1850*. Stanford, Calif.: Stanford University Press, 1999.

Rich, Michelle E. "Ritual, Royalty, and Classic Period Politics: The Archaeology of the Mirador Group at El Perú-Waka', Petén, Guatemala." PhD diss., Department of Anthropology, Southern Methodist University, 2011.

Riley, Carroll L. "Color-Direction Symbolism: An Example of Mexican Southwestern Contacts." *América Indígena* 23, no. 1 (1963): 49–60.

Roach, John. "Pyramid Tomb Found: Sign of a Civilization's Birth?" *National Geographic News* (2010), http://news.nationalgeographic.com/news/2010/05/100518-oldest-pyramid-tomb-zoque-mexico-science/.

Robb, John E. "Time and Biography." In *Thinking through the Body: Archaeologies of Corporeality*, edited by Y. Hamilakis, M. Pluciennik, and S. Tarlow, 145–163. London: Kluwer/Academic, 2002.

Robertson, John, Danny Law, and Robbie A. Haertel. *Colonial Ch'olti': The Seventeenth-Century Morán Manuscript.* Norman: University of Oklahoma Press, 2010.

Robertson, Merle Greene. *The Sculpture of Palenque*, vol. 1: *The Temple of the Inscriptions.* Princeton, N.J.: Princeton University Press, 1983.

Robicsek, Frances, and Donald Hales. *The Maya Book of the Dead, the Ceramic Codex: The Corpus of Codex Style Ceramics of the Late Classic Period.* Norman: University of Oklahoma Press, 1981.

Robin, Cynthia, and Norman Hammond. "Burial Practices." In *Cuello: An Early Maya Community in Belize*, edited by Norman Hammond, 204–225. Cambridge: Cambridge University Press, 1991.

Rodolfo Cid Beziez, José, and Liliana Torres Sanders. "Patrones de mutilación dental en el sector oeste de Teotihuacan." *Estudios de Antropología Biológica* 9 (2012): 281–294.

Romero Molina, Javier. *Catálogo de la colección de dientes mutilados prehispánicos: IV parte.* Mexico City: Instituto Nacional de Antropología e Historia, 1986.

——. "Dental Mutilation, Trephination, and Cranial Deformation." In *Physical Anthropology*, edited by T. Dale Stewart, 50–67. *Handbook of Middle American Indians*, vol. 9. Austin: University of Texas Press, 1970.

Roys, Ralph L. *The Book of Chilam Balam of Chumayel.* Washington, D.C.: Carnegie Institution, 1933.

Ruz Lhuillier, Alberto. *El Templo de las Inscripciones.* Mexico City: Instituto Nacional de Antropología e Historia, 1973.

——. "Exploraciones arqueológicas en Palenque: 1952." *Anales del Instituto Nacional de Antropología e Historia*, series 6, vol. 6, no. 34 (1954): 79–110.

——. "Exploraciones arqueológicas en Palenque: 1953." *Anales del Instituto Nacional de Antropología e Historia*, series 6, vol. 10, no. 39 (1958): 69–116.

——. "Exploraciones arqueológicas en Palenque: 1954." *Anales del Instituto Nacional de Antropología e Historia*, series 6, vol. 10, no. 39 (1958): 117–185.

——. "Exploraciones arqueológicas en Palenque: 1956." *Anales del Instituto Nacional de Antropología e Historia*, series 6, vol. 10, no. 39 (1958): 241–299.

——. "Exploraciones Arqueológicas en Palenque: 1957." *Anales del Instituto Nacional de Antropología e Historia*, series 6, vol. 14, no. 43 (1962): 35–90.

Sahagún, Bernadino de. *General History of the Things of New Spain: Florentine Codex*, book 2: *The Ceremonies.* Translated by Arthur J. O. Anderson and Charles E. Dibble. 2nd ed. Salt Lake City: University of Utah Press, 1981.

——. *General History of the Things of New Spain: Florentine Codex*, book 10: *The People.* Translated by Arthur J. O. Anderson and Charles E. Dibble. Salt Lake City: University of Utah Press, 1961.

——. *General History of the Things of New Spain: Florentine Codex*, book 11: *Earthly Things.* Translated by Arthur J. O. Anderson and Charles E. Dibble. Salt Lake City: University of Utah Press, 1963.

Satterthwaite, Linton. "Structure K-6: The West Group Ballcourt." In *Piedras Negras Archaeology: 1931–1939*, edited by John M. Weeks, Jane A. Hill, and Charles W. Golden, 228–240. Philadelphia: University of Pennsylvania Museum Press, 2005.

Saturno, William A., Karl A. Taube, and David Stuart. *The Murals of San Bartolo, El Petén, Guatemala, Part 1: The North Wall.* Ancient America No. 7. Barnardsville, N.C.: Center for Ancient American Studies, 2005.

Saul, Frank P. "Osteobiography: Life History Recorded in Bone." In *The Measures of Man: Methodologies in Biological Anthropology*, edited by Eugene Giles and Jonathan S. Friedlander, 372–382. Cambridge, Mass.: Peabody Museum Press, 1976.

Saxe, Arthur A. "Social Dimensions of Mortuary Practices." PhD diss., University of Michigan, 1971.

Schele, Linda. *Notebook for the XVIth Maya Hieroglyphic Workshop at Texas.* Austin: Department of Art and Art History and the Institute of the Latin American Studies, University of Texas, 1992.

Schele, Linda, and Peter Mathews. *The Code of Kings: The Language of Seven Sacred Maya Temples and Tombs.* New York: Scribner, 1998.

Schele, Linda, and Mary Ellen Miller. *Blood of Kings: Dynasty and Ritual in Maya Art.* New York: George Braziller, 1986.

Schellhas, Paul. *Representations of Deities of the Maya Manuscripts*. Papers of the Peabody Museum of American Archaeology and Ethnology, vol. 4, no. 1. Cambridge, Mass.: Peabody Museum of American Archaeology and Ethnology, Harvard University, 1904.

Scherer, Andrew K. "The Classic Maya Sarcophagus: Rebirth and Renewal at Palenque and Tonina." *RES: Anthropology and Aesthetics* 61–62 (2012): 242–261.

Scherer, Andrew K., and Chelsea Garrett. "Osteología del complejo El Diablo: Escondites y entierro 9, El Zotz." In *Proyecto arqueológico "El Zotz": Informe no. 5, temporada 2010*, edited by Jose Luis Garrido López, Stephen Houston, and Edwin Román, 425–458. Guatemala City, Guatemala: Instituto de Antropología e Historia de Guatemala, 2011.

Scherer, Andrew K., and Charles Golden. "Tecolote, Guatemala: Archaeological Evidence for a Fortified Late Classic Maya Political Border." *Journal of Field Archaeology* 34, no. 3 (2009): 285–304.

———. "War in the West: History, Landscape, and Classic Maya Conflict." In *Embattled Places, Embattled Bodies: War in Pre-Columbian America*, edited by Andrew K. Scherer and John W. Verano, 57–92. Washington, D.C.: Dumbarton Oaks Research Library and Collection, 2014.

———. "Water in the West: Chronology and Collapse of the Classic Maya River Kingdoms." In *The Great Maya Droughts in Cultural Context*, edited by Gyles Iannone. Boulder: University Press of Colorado, 2014.

Scherer, Andrew K., Charles Golden, Ana Lucía Arroyave Prera, and Griselda Pérez Robles. "Danse Macabre: Death, Community, and Kingdom at El Kinel, Guatemala." In *The Bioarchaeology of Space and Place: Ideology, Power and Meaning in Maya Mortuary Contexts*, edited by Gabriel D. Wrobel, 193–224. New York: Springer, 2014.

Scherer, Andrew K., Lori E. Wright, and Cassady J. Yoder. "Bioarchaeological Evidence for Social and Temporal Differences in Subsistence at Piedras Negras, Guatemala." *Latin American Antiquity* 18, no. 1 (2007): 85–104.

Schieber de Lavarreda, Christa, and Miguel Orrego Corzo. "El descubrimiento del Altar 48 de Tak'alik' Ab'aj." In *XXII Simposio de Investigaciones Arqueológicas en Guatemala, 2008*, edited by Juan Pedro Laporte, Barbara Arroyo, and Héctor Mejía, 456–470. Guatemala City: Museo Nacional de Arqueología y Etnología, 2009.

Schneider, Renata. *Fardo funerario de la Tumba 1 de Calakmul: Conservación, restauración y montaje*. Mexico City: Instituto Nacional de Antropología e Historia, 2008.

Scott, Ann M., and James E. Brady. "Human Remains in Lowland Maya Caves: Problems of Interpretation." In *Stone Houses and Earth Lords: Maya Religion in the Cave Context*, edited by Keith M. Prufer and James E. Brady, 263–284. Boulder: University Press of Colorado, 2005.

Sempowski, Martha L., and Michael W. Spence. *Mortuary Practices and Skeletal Remains at Teotihuacan*. Salt Lake City: University of Utah Press, 1994.

Serrano S., Carlos, Martha Pimienta Merlin, and Alfonso Gallardo Velázquez. "Mutilación dentaria y filiación étnica en los entierros del templo de Quetzalcoatl, Teotihuacan." In *II Coloquio Pedro Bosch-Gimpera*, edited by Maria Teresa Cabrero G., 263–276. Mexico City: Universidad Nacional Autónoma de México, 1993.

Silver, Daniel B., and Horacio Fábrega. *Illness and Shamanistic Curing in Zinacantan: An Ethnomedical Analysis*. Stanford, Calif.: Stanford University Press, 1973.

Sledzik, Paul S., Dennis Dirkmaat, Robert W. Mann, Thomas D. Holland, Amy Zelson Mundorff, Bradley J. Adams, Christian Crowder, and Frank DePaolo. "Disaster Victim Recovery and Identification: Forensic Anthropology in the Aftermath of September 11." In *Hard Evidence: Case Studies in Forensic Anthropology*, edited by Dawnie Wolfe Steadman, 289–302. Upper Saddle River, N.J.: Prentice Hall, 2009.

Smith, A. Ledyard. *Excavations at Altar de Sacrificios: Architecture, Settlement, Burials, and Caches*. Papers of the Peabody Museum, vol. 62. Cambridge, Mass.: Harvard University, 1972.

Smith, B. H. "Dental Attrition in Hunter-Gatherers and Agriculturalists." PhD diss., University of Michigan, 1983.

———. "Patterns of Molar Wear in Hunter-Gatherers and Agriculturalists." *American Journal of Physical Anthropology* 63 (1984): 39–56.

Sofaer, Joanna R. *The Body as Material Culture: A Theoretical Osteoarchaeology*. Cambridge: Cambridge University Press, 2006.

Solari, Amara. *Maya Ideologies of the Sacred: The Transfiguration of Space in Colonial Yucatan*. Austin: University of Texas Press, 2013.

Spence, Michael W., and Grégory Pereira. "The Human Skeletal Remains of the Moon Pyramid, Teotihuacan." *Ancient Mesoamerica* 18, no. 1 (2007): 147–157.

Spicer, Edward H. "William R. Holland, 1928–1964." *American Anthropologist* 67, no. 1 (1965): 80–82.

Spinden, Herbert J. *Ancient Civilizations of Mexico and Central America*. New York: American Museum of Natural History, 1917.

Stone, Andrea. "Disconnection, Foreign Insignia, and Political Expansion: Teotihuacan and the Warrior Stelae of Piedras Negras." In *Mesoamerica after the Decline of Teotihuacan*, edited by Richard A. Diehl and Janet C. Berlo, 153–172. Washington, D.C.: Dumbarton Oaks, Trustees for Harvard University, 1989.

Stone, Andrea, and Marc Zender. *Reading Maya Art: A Hieroglyphic Guide to Ancient Maya Painting and Sculpture*. New York: Thames and Hudson, 2011.

Storey, Rebecca. "The Ancestors: The Bioarchaeology of the Human Remains of K'axob. In *K'axob: Ritual, Work and Family in an Ancient Maya Village*, edited by Patricia A. McAnany, 109–138. Monograph 51. Los Angeles: Cotsen Institute of Archaeology, University of California, 2004.

Strong, William Duncan. "From History to Prehistory in the Northern Great Plains." *Smithsonian Miscellaneous Collections* 100 (1940): 353–394.

Stuart, David. "The Arrival of Strangers: Teotihuacan and Tollan in Classic Maya History." In *Mesoamerica's Classic Heritage: From Teotihuacan to the Aztecs*, edited by D. Carrasco, L. Jones, and J. S. Sessions, 465–513. Boulder: University Press of Colorado, 2000.

———. "The Beginnings of the Copan Dynasty: A Review of the Hieroglyphic and Historical Evidence." In *Understanding Early Classic Copan*, edited by Ellen E. Bell, Marcello A. Canuto, and Robert J. Sharer, 215–248. Philadelphia: University of Pennsylvania Museum of Archaeology and Anthropology, 2004.

———. "Blood Symbolism in Maya Iconography." In *Maya Iconography*, edited by Elizabeth P. Benson and Gillett G. Griffen, 175–221. Princeton, N.J.: Princeton University Press, 1988.

———. "The Ceiba Tree on K1226." *Maya Decipherment: Ideas on Ancient Maya Writing and Iconography* (2007), decipherment.wordpress.com/2007/04/14/the-ceiba-tree-on-K1226/.

———. "'The Fire Enters His House': Architecture and Ritual in Classic Maya Texts." In *Form and Meaning in Classic Maya Architecture*, edited by Stephen D. Houston, 373–425. Washington, D.C.: Dumbarton Oaks, 1998.

———. *The Inscription from Temple XIX at Palenque: A Commentary*. San Francisco: Precolumbian Art Research Institute, 2005.

———. "Kings of Stone: A Consideration of Stelae in Ancient Maya Ritual Representation." *RES: Anthropology and Aesthetics* 29–30 (1996): 148–171.

———. "Maya Spooks." *Maya Decipherment: Ideas on Ancient Maya Writing and Iconography* (2012), decipherment.wordpress.com/2012/10/26/maya-spooks/.

———. "The Name of Paper: The Mythology of Crowning and Royal Nomenclature on Palenque's Palace Tablet." In *Maya Archaeology* 2, edited by Charles Golden, Stephen Houston, and Joel Skidmore, 116–142. San Francisco: Pre-Columbia Mesoweb Press, 2012.

———. *The Order of Days: The Maya World and the Truth about 2012*. New York: Harmony Books, 2011.

———. "The Paintings of Tomb 12, Rio Azul." In *Rio Azul Reports*, no. 3, *The 1985 Season*, edited by Richard E. W. Adams, 161–167. San Antonio: Center for Archaeological Research, University of Texas at San Antonio, 1987.

———. "Reading the Water Serpent as WITZ'." Online at *Maya Decipherment: Ideas on Ancient Maya Writing and Iconography* (2007), http://decipherment.wordpress.com/2007/08/18/the-captives-on-piedras-negras-panel-12/.

———. "Report: Two Inscribed Bones from Yaxchilan." Online at *Maya Decipherment: Ideas on Ancient Maya Writing and Iconography* (2013), http://decipherment.wordpress.com/2007/08/18/the-captives-on-piedras-negras-panel-12/.

———. "A Sun God Image from Dos Pilas, Guatemala." Online at *Maya Decipherment: Ideas on Ancient Maya Writing and Iconography* (2009), http://decipherment.wordpress.com/2007/08/18/the-captives-on-piedras-negras-panel-12/.

Stuart, David, and Ian Graham. *Corpus of Maya Hieroglyphic Inscriptions*, vol. 9, part 1, *Piedras Negras*. Cambridge, Mass.: Peabody Museum of Archaeology and Ethnology, Harvard University, 2003.

Stuart, David, and George Stuart. *Palenque: Eternal City of the Maya*. London: Thames and Hudson, 2008.

Stuart, Kathy. "Suicide by Proxy: The Unintended Consequences of Public Executions in Eighteenth-Century Germany." *Central European History* 41, no. 3 (2008): 413–445.

Sugiyama, Saburo. *Human Sacrifice, Militarism, and Rulership: Materialization of State Ideology at the Feathered Serpent Pyramid, Teotihuacan*. Cambridge: Cambridge University Press, 2005.

Taschek, Jennifer T., and Joseph W. Ball. "Las Ruinas de Arenal: Preliminary Report on a Subregional Major Center in the Western Belize Valley (1991–1992 Excavations)." *Ancient Mesoamerica* 10, no. 2 (1999): 215–235.

———. "Lord Smoke-Squirrel's Cacao Cup: The Archaeological Context and Socio-Historical Significance of the Buenavista 'Jauncy-Vase.'" In *The Maya Vase Book: A Corpus of Rollout Photographs of Maya Vases*, vol. 3, edited by Barbara Kerr, photographs by Justin Kerr, 490–497. New York: Justin Kerr Associates, 1992.

Tate, Carolyn E. *Yaxchilan: The Design of a Maya Ceremonial City*. Austin: University of Texas Press, 1992.

Taube, Karl A. "Ancient and Contemporary Maya Concepts about Field and Forest." In *The Lowland Maya Area: Three Millennia at the Human-Wildland Interface*, edited by Arturo Gómez-Pompa, Michael F. Allen, Scott L. Fedick, and Juan J. Jiménez-Osornio, 461–494. New York: Food Products Press, 2003.

———. *Aztec and Maya Myths*. London: British Museum Press, 1993.

———. "The Birth Vase: Natal Imagery in Ancient Maya Myth and Ritual." In *The Maya Vase Book: A Corpus of Rollout Photographs of Maya Vases*, vol. 4, edited by Barbara Kerr, photographs by Justin Kerr, 650–685. New York: Kerr Associates, 1994.

———. "Cache Vessel with Directional Shells and Jades, 500–600, Copan, Honduras." In *Fiery Pool: The Maya and the Mythic Sea*, edited by Daniel Finamore and Stephen D. Houston, 266–267. New Haven, Conn.: Yale University Press; Salem, Mass.: Peabody Essex Museum, 2010.

———. "The Classic Maya Maize God: A Reappraisal." In *Fifth Palenque Round Table, 1983*, edited by Virginia M. Fields, 171–181. San Francisco: Pre-Columbian Art Research Institute, 1985.

———. "Creation and Cosmology: Gods and Mythic Origins in Ancient Mesoamerica." In *The Oxford Handbook of Mesoamerican Archaeology*, edited by Deborah Nichols and Christopher Pool, 741–751. Oxford: Oxford University Press, 2012.

———. "Flower Mountain: Concepts of Life, Beauty, and Paradise among the Classic Maya." *RES: Anthropology and Aesthetics* 45 (Spring 2004): 69–98.

———. "Gateways to Another World: The Symbolism of Flowers in Mesoamerica and the American Southwest." In *The Land Brightened with Flowers: The Hopi Iconography Project*, edited by Kelley Hays-Gilpin and Polly Schaafsma, 73–120. Flagstaff: Museum of Northern Arizona, 2010.

———. "The Iconography of Mirrors at Teotihuacan." In *Art, Ideology, and the City of Teotihuacan*, edited by Janet C. Berlo, 169–204. Washington, D.C.: Dumbarton Oaks, 1992.

———. "The Jade Hearth: Centrality, Rulership, and the Classic Maya Temple." In *Function and Meaning in Classic Maya Architecture*, edited by Stephen D. Houston, 427–478. Washington, D.C.: Dumbarton Oaks Research Library and Collection, 1998.

———. "The Maize Tamale in Classic Maya Diet, Epigraphy, and Art." *American Antiquity* 54, no. 1 (1989): 31–51.

———. *The Major Gods of Ancient Yucatan*. Washington, D.C.: Dumbarton Oaks Research Library and Collection, 1992.

———. "Maws of Heaven and Hell: The Symbolism of the Centipede and Serpent in Classic Maya Religion." In *Antropología de la eternidad: La muerte en la cultura maya*, edited by Andrés Ciudad Ruiz, Mario Humberto Ruz Sosa, and María Josefa Iglesias Ponce de León, 404–442. Madrid: Sociedad de Española de Estudios Mayas, 2003.

———. "The Maya Maize God and the Mythic Origins of Dance." In *The Maya and Their Sacred Narratives: Text and Context in Maya Mythologies*, edited by Geneviève Le Fort, Raphael Gardiol, Sebastian Matteo, and Christophe Helmke, 41–52. Acta Mesoamericana, vol. 20. Markt Schwaben, Germany: Verlag Anton Saurwein, 2009.

———. "The Olmec Maize God: The Face of Corn in Formative Mesoamerica." *RES: Anthropology and Aesthetics* 29-30 (Spring/Autumn 1996): 39–81.

———. "A Prehispanic Maya Katun Wheel." *Journal of Anthropological Research* 44, no. 2 (1988): 183–203.

———. "A Representation of the Principal Bird Deity in the Paris Codex." In *Research Reports on Ancient Maya Writing*, vol. 6 (1987), 1–10. Washington, D.C.: Center for Maya Research.

———. "The Symbolism of Jade in Classic Maya Religion." *Ancient Mesoamerica* 16 (2005): 23–50.

———. "The Temple of Quetzalcoatl and the Cult of Sacred War at Teotihuacan." *RES: Anthropology and Aesthetics* 21 (1992): 53–87.

———. "The Turquoise Hearth: Fire, Self-Sacrifice, and the Central Mexican Cult of War." In *Mesoamerica's Classic Heritage: From Teotihuacan to the Aztecs*, edited by Davíd Carrasco, Lindsay Jones, and Scott Sessions, 269–340. Boulder: University Press of Colorado, 2000.

———. "Where Earth and Sky Meet: The Sea in Ancient and Contemporary Maya Cosmology." In *Fiery Pool: The Maya and the Mythic Sea*, edited by Daniel Finamore and Stephen D. Houston, 202–213. New Haven, Conn.: Yale University Press; Salem, Mass.: Peabody Essex Museum, 2010.

———. "The Womb of the World: The Cuauhxicalli and Other Offering Bowls of Ancient and Contemporary Mesoamerica." In *Maya Archaeology 1*, edited by Charles Golden, Stephen Houston, and Joel Skidmore, 86–106. San Francisco: Precolumbia Mesoweb Press, 2009.

Taube, Karl A., William Saturno, David Stuart, and Heather Hurst. *The Murals of San Bartolo, El Petén, Guatemala, Part 2: The West Wall.* Barnardsville, N.C.: Boundary End Archaeology Research Center, 2010.

Taube, Karl A., and Marc Zender. "American Gladiators: Ritual Boxing in Ancient Mesoamerica." In *Blood and Beauty: Organized Violence in the Art and Archaeology of Mesoamerica and Central America*, edited by Heather Orr and Rex Koontz, 161–220. Los Angeles: Cotsen Institute of Archaeology Press, 2009.

Tedlock, Dennis. *Rabinal Achi: A Mayan Drama of War and Sacrifice*. Oxford: Oxford University Press, 2003.

Thompson, Edward H. "Ancient Tombs of Palenque." *Proceedings of the American Antiquarian Society* 10, no. 2 (1895): 418–421.

Thompson, J. Eric S. *Ethnology of the Mayas of Southern and Central British Honduras*. Anthropological Series, publication 274/Anthropological Series, vol. 17, no. 2. Chicago: Field Museum of Natural History, 1930.

———. *Excavations at San Jose, British Honduras*. Publication 506. Washington, D.C.: Carnegie Institution of Washington, 1939.

———. *Sky Bearers, Colors, and Directions in Maya and Mexican Religion*. Contributions to American Archaeology no. 10, Carnegie Institution of Washington publication no. 436. Washington, D.C.: Carnegie Institution of Washington, 1934.

Tiesler, Vera. *The Bioarchaeology of Artificial Cranial Modifications: New Approaches to Head Shaping and Its Meanings in Pre-Columbian Mesoamerica and Beyond*. New York: Springer, 2013.

———. *La costumbre de la deformación cefálica entre los antiguos mayas: Aspectos morfológicos y culturales*. Mexico City: Instituto Nacional de Antropología e Historia, 1998.

———. "Life and Death of the Ruler: Recent Bioarchaeological Findings." In *Janaab' Pakal of Palenque*, edited by Vera Tiesler and Andrea Cucina, 21–47. Tucson: University of Arizona Press, 2006.

———. "'Olmec' Head Shapes among the Preclassic Period Maya and Cultural Meanings." *Latin American Antiquity* 21, no. 3 (2010): 290–311.

———. "Studying Cranial Vault Modifications in Ancient Mesoamerica." *Journal of Anthropological Sciences* 90 (2012): 1–26.

———. *Tranformarse en Maya: El modelado cefálico entre los mayas prehispánicos y coloniales*. Mexico City: Universidad Nacional Autónoma de México, 2012.

———. "What Can Bones Really Tell Us?" In *Stone Houses and Earth Lords: Maya Religion in the Cave Context*, edited by Keith M. Prufer and James E. Brady, 341–363. Boulder: University Press of Colorado, 2005.

Tiesler, Vera, and Andrea Cucina. "La deformación craneana como emblema de identidad, etnicidad y reproducción cultural entre los mayas del clásico." In *Identidad y vida cotidiana entre los mayas*, edited by H. Hernández, 111–134. Mérida, Mexico: Universidad Autónoma de Yucatán, 2010.

———. "Procedures in Human Heart Extraction and Ritual Meaning: A Taphonomic Assessment of Anthropogenic Marks in Classic Maya Skeletons." *Latin American Antiquity* 17, no. 4 (2006): 494–510.

Tiesler, Vera, Andrea Cucina, T. Kam Manahan, T. Douglas Price, Traci Arden, and James H. Burton. "A Taphonomic Approach to Late Classic Maya Mortuary Practices at Xuenkal, Yucatán, Mexico." *Journal of Anthropological Archaeology* 35, no. 4 (2010): 365–379.

Tiesler Blos, Vera. *Decoraciones dentales entre los antiguos mayas*. Mexico City: Instituto Nacional de Antropología e Historia, 2001.

———. "Reporte de los estudios de análisis de transición e histomorfológico de los restos humanos de los sitios Piedras Negras, Dos Pilas y Tamarindito." In *Proyecto regional arqueológico Sierra Lacandon: Informe preliminar número 7*, edited by Ana Lucía Arroyave Prera, Charles Golden, and Andrew K. Scherer, 99–116. Report Presented to the Instituto de Antropología e Historia de Guatemala (IDAEH), Guatemala City, 2009, online at usumacinta-archaeology.blogspot.com/p/research-reports.html.

Tiesler Blos, Vera, Marco Ramírez Salomón, and Iván Oliva Arias. "Técnicas de decoración dental en México: Un acercamiento experimental." *Actualidades Arqueológicas: Pasado en Presente* 2: 18–24.

Tokovinine, Alexandre. *Place and Identity in Classic Maya Narratives*. Washington, D.C.: Dumbarton Oaks Research Library and Collection, 2013.

Tovalín Ahumada, Alejandro. "La cripta funeraria asociada al edificio 4 de Bonampak, Chiapas." *Lakamha'* 1, no. 5 (2002): 4–10.

Tozzer, Alfred M. *Comparative Study of the Mayas and Lacandones*. London: Archaeological Institute of America/Macmillan, 1907.

———. *Landa's Relación de las Cosas de Yucatan: A Translation*. Papers of the Peabody Museum of American Archaeology and Ethnology, vol. 18. Cambridge, Mass.: Harvard University Press, 1941.

Tsokos, Michael, Jakob Matschke, Axel Gehl, Erwin Koops, and Klaus Püschel. "Skin and Soft Tissue Artifacts Due to Postmortem Damage Caused by Rodents." *Forensic Science International* 104, no. 1 (1999): 47–57.

Tykot, Robert H. "Contribution of Stable Isotope Analysis to Understanding Dietary Variation among the Maya." In *Archaeological Chemistry: Materials, Methods, and Meaning*, edited by K. Jakes, 214–230. ACS Symposium Series 831. Washington, D.C.: American Chemical Society, 2002.

van Gennep, Arnold. *The Rites of Passage*. Chicago: University of Chicago Press, 1960.

Verano, John W. "The Physical Evidence of Human Sacrifice in Ancient Peru." In *Ritual Sacrifice in Ancient Peru*, edited by Elizabeth Benson and Anita Cook, 165–184. Austin: University of Texas Press, 2001.

Verdery, Katherine. *The Political Lives of Bodies: Reburial and Postsocialist Change*. New York: Columbia University Press, 1999.

Villa Rojas, Alfonso. "La imagen del cuerpo humano según los mayas de Yucatán." *Anales de Antropología* 17 (1980): 31–46.

Vogt, Evon Z. "Human Souls and Animal Spirits in Zinacantan." In *Echanges et communications: Mélanges offerts à Claude Lévi-Strauss à l'occasion de son 60ème anniversaire*, edited by J. Pouillon and P. Maranda, 1148–1167. The Hague: Mouton, 1970.

———. *Tortillas for the Gods: A Symbolic Analysis of Zinacanteco Rituals*. Norman: University of Oklahoma Press, 1992.

———. *Zinacantan: A Maya Community in the Highlands of Chiapas*. Cambridge, Mass.: Belnap, 1969.

———. "Zinacanteco Astronomy." *Mexicon* 119 (1997): 110–117.

———. *The Zinacantecos of Mexico: A Modern Maya Way of Life*. New York: Holt, Rinehart and Winston, 1970.

Vogt, Evon Z., and David Stuart. "Some Notes on Ritual Caves among the Ancient and Modern Maya." In *In the Maw of the Earth Monster: Mesoamerican Ritual Cave Use*, edited by James E. Brady and Keith M. Prufer, 155–185. Austin: University of Texas Press, 2005.

von Winning, Haso. *La iconografía de Teotihuacan: Los dioses y los signos*. 2 vols. Mexico City: Universidad Nacional Autónoma de México.

Walker, Phillip L., and Della Collins Cook. "Brief Communication: Gender and Sex: Vive la Difference." *American Journal of Physical Anthropology* 106 (1998): 255–259.

Warren, Kay B. *Indigenous Movements and Their Critics: Pan-Maya Activism in Guatemala*. Princeton, N.J.: Princeton University Press, 1998.

Watanabe, John M. "From Saints to Shibboleths: Image, Structure, and Identity in Maya Religious Syncretism." *American Ethnologist* 17, no. 1 (1990): 131–150.

———. *Maya Saints and Souls in a Changing World*. Austin: University of Texas Press, 1992.

———. "Unimagining the Maya: Anthropologists, Others, and the Inescapable Hubris of Authorship." *Bulletin of Latin American Research* 14, no. 1 (1995): 25–45.

Webster, David. *The Fall of the Ancient Maya: Solving the Mystery of the Maya Collapse*. New York: Thames and Hudson, 2002.

Wedel, Waldo Rudolph. "The Direct-Historical Approach in Pawnee Archaeology." *Smithsonian Miscellaneous Collections* 97, no. 7 (1938): 1–21.

Weiner, Annette B. *Inalienable Possessions: The Paradox of Keeping-While-Giving*. Berkeley: University of California Press, 1992.

———. "Inalienable Wealth." *American Ethnologist* 12, no. 2 (1985): 210–227.

Weiss-Krejci, Estella. "Depósitos rituales en los complejos de pirámides gemelas de Tikal." In *El ritual en el mundo maya: De lo privado a lo público*, edited by Andrés Ciudad Ruiz and María Josefa Iglesias Ponce de León, 83–105. Publicaciones de la SEEM, vol. 9. Madrid: Grupo de Investigación Andalucía-América, CEPHCIS-UNAM, 2010.

———. "The Role of Dead Bodies in Late Classic Maya Politics: Cross-Cultural Reflections on the Meaning of Tikal Altar 5." In *Living with the Dead: Mortuary Ritual in Mesoamerica*, edited by James L. Fitzsimmons and Izumi Shimada, 17–52. Tucson: University of Arizona Press, 2011.

Welsh, W. Bruce M. *An Analysis of Classic Lowland Maya Burials*. International Series 409. Oxford: British Archaeological Reports, 1988.

White, Christine D., David M. Pendergast, Fred J. Longstaffe, and Kimberly R. Law. "Social Complexity and Food Systems at Altun Ha, Belize: The Isotopic Evidence." *Latin American Antiquity* 12, no. 4 (2001): 371–394.

Whittington, Stephen L. "Analysis of Human Skeletal Material Excavated by Guillemin." In *Archaeology and Ethnohistory of Iximché*, edited by C. Roger Nance, Stephen L. Whittington, and Barbara E. Borg, 241–304. Gainesville: University Press of Florida, 2003.

———. "Description of Human Remains and Burial Structures." In *Archaeology and Ethnohistory of Iximché*, edited by C. Roger Nance, Stephen L. Whittington, and Barbara E. Borg, 205–240. Gainesville: University Press of Florida, 2003.

Williams, Jocelyn S., Christine D. White, and Fred J. Longstaffe. "Maya Marine Subsistence: Isotopic Evidence from Marco González and San Pedro, Belize." *Latin American Antiquity* 20, no. 1 (2009): 37–56.

Wilson, Richard. *Maya Resurgence in Guatemala: Q'eqchi' Experiences*. Norman: University of Oklahoma Press, 1995.

Wisdom, Charles. *The Chorti Indians of Guatemala*. Chicago: University of Chicago Press, 1940.

———. *Materials on the Chorti Language*. Microfilm Collection of Manuscript Materials on Middle American Cultural Anthropology, no. 28. Chicago: University of Chicago Library, 1950.

Wright, Lori E. *Diet, Health, and Status among the Pasión Maya: A Reappraisal of the Collapse*. Nashville, Tenn.: Vanderbilt University Press, 2006.

———. "Identifying Immigrants to Tikal, Guatemala: Defining Local Variability in Strontium Isotope Ratios of Human Tooth Enamel." *Journal of Archaeological Science* 32, no. 4 (2005): 555–566.

———. "In Search of Yax Nuun Ayiin I: Revisiting the Tikal Project's Burial 10." *Ancient Mesoamerica* 16 (2005): 89–100.

———. "La muerte y el estatus económico: Investigando el simbolismo mortuorio y el acceso a los recursos alimenticios entre los mayas." In *Antropología de la eternidad: La muerte en la cultura maya*, edited by Andrés Ciudad Ruiz, Mario Humberto Ruz Soza, and María Josefa Iglesias Ponce de León, 175–193. Madrid: Sociedad Española de Estudios Mayas, 2003.

Wright, Lori E., and Christine D. White. "Human Biology in the Classic Maya Collapse: Evidence from Paleopathology and Paleodiet." *Journal of World Prehistory* 10 (1996): 147–198.

Ximénez, Francisco. *Historia de la provincia de San Vicente Chiapa y Guatemala*. Biblioteca "Goathemala" de la Sociedad de Geografía e Historia de Guatemala, publication 1. 3 vols. Guatemala City: Guatemala Tipografía Nacional, 1929–1931.

Zender, Marc. "'Handspan' and 'Strike' in Classic Maya Ballgame Texts." PARI *Journal* 4, no. 4 (2004): 1–9.

———. "The Raccoon Glyph in Classic Maya Writing." PARI *Journal* 5, no. 4 (2005): 6–16.

———. "Sport, Spectacle and Political Theater: New Views of the Classic Maya Ballgame." PARI *Journal* 4, no. 4 (2004): 10–12.

Zender, Marc, and Karen Bassie. "The Wooden Offering Container of Aj K'ax B'ahlam of Tortugero." N.d. Jay I. Kislak Foundation, online at http://www.kislakfoundation.org/tortuguero.html.

INDEX